Advances in Intelligent Systems and Computing

Volume 682

Series editor

Janusz Kacprzyk, Polish Academy of Sciences, Warsaw, Poland
e-mail: kacprzyk@ibspan.waw.pl

About this Series

The series "Advances in Intelligent Systems and Computing" contains publications on theory, applications, and design methods of Intelligent Systems and Intelligent Computing. Virtually all disciplines such as engineering, natural sciences, computer and information science, ICT, economics, business, e-commerce, environment, healthcare, life science are covered. The list of topics spans all the areas of modern intelligent systems and computing.

The publications within "Advances in Intelligent Systems and Computing" are primarily textbooks and proceedings of important conferences, symposia and congresses. They cover significant recent developments in the field, both of a foundational and applicable character. An important characteristic feature of the series is the short publication time and world-wide distribution. This permits a rapid and broad dissemination of research results.

More information about this series at http://www.springer.com/series/11156

Pavel Krömer · Enrique Alba
Jeng-Shyang Pan · Václav Snášel
Editors

Proceedings of the Fourth Euro-China Conference on Intelligent Data Analysis and Applications

Editors
Pavel Krömer
Department of Computer Science
VŠB-Technical University of Ostrava
Ostrava-Poruba
Czech Republic

Enrique Alba
Dept. Lenguajes y Ciencias de la Computación
University of Málaga
Málaga
Spain

Jeng-Shyang Pan
Fujian University of Technology
Fuzhou, Fujian
China

Václav Snášel
Department of Computer Science
VŠB-Technical University of Ostrava
Ostrava-Poruba
Czech Republic

ISSN 2194-5357 ISSN 2194-5365 (electronic)
Advances in Intelligent Systems and Computing
ISBN 978-3-319-68526-7 ISBN 978-3-319-68527-4 (eBook)
DOI 10.1007/978-3-319-68527-4

Library of Congress Control Number: 2017954900

Printed on acid-free paper

This Springer imprint is published by Springer Nature
The registered company is Springer International Publishing AG
The registered company address is: Gewerbestrasse 11, 6330 Cham, Switzerland

Preface

This volume of Advances in Intelligent Systems and Computing contains accepted papers presented at ECC 2017, the Fourth Euro-China Conference on Intelligent Data Analysis and Applications. The aim of ECC is to provide an internationally respected forum for scientific research in the broad areas of intelligent data analysis, computational intelligence, signal processing, and all associated applications of AIs.

ECC puts a special emphasis on promoting research and scientific collaboration between Europe and China, two major centers of development on the contemporary scientific landscape. It aims at strengthening research partnerships and providing an opportunity for joint efforts leading to higher quality fundamental and applied research.

The fourth edition of ECC was organized jointly by the University of Málaga, Spain, VŠB—Technical University of Ostrava, Czech Republic, and Fujian University of Technology, Fuzhou, China. The conference will take place on October 9–11, 2017, in the beautiful Mediterranean city of Málaga, Spain.

The organization of the ECC 2017 conference was entirely voluntary. The review process required an enormous effort from the members of the International Technical Program Committee, and we would therefore like to thank all its members for their contribution to the success of this conference. We would like to express our sincere thanks to the host of ECC 2017, University of Málaga, and to the publisher, Springer, for their hard work and support in organizing the conference. Finally, we would like to thank all the authors for their high-quality contributions. The friendly and welcoming attitude of conference supporters and contributors made this event a success!

July 2017

Pavel Krömer
Enrique Alba
Jeng-Shyang Pan
Václav Snášel

Organization

Honorary Chair

Ivo Vondrak	VŠB-TU Ostrava, Czech Republic

Advisory Committee Chair

Yihan Zhang	Computer Network Information Center, Chinese Academy of Sciences, Guangzhou, China

Conference Chairs

Enrique Alba	University of Málaga, Spain
Jeng-Shyang Pan	Fujian University of Technology, China
Václav Snášel	VŠB-TU Ostrava, Czech Republic

Program Committee Chairs

Ivo Hlavatý	VŠB-TU Ostrava, Czech Republic
Pavel Krömer	VŠB-TU Ostrava, Czech Republic

Invited Session Chairs

Karel Frydrýšek	VŠB-TU Ostrava, Czech Republic
Tien-Wen Sung	Fujian University of Technology, China

Local Organizing Chairs

Christian Cintrano	University of Málaga, Spain
Jan Platoš	VŠB-TU Ostrava, Czech Republic

Publicity and Electronic Media Chair

Christian Cintrano	University of Málaga, Spain

Publication Chairs

Pavel Krömer	VŠB-TU Ostrava, Czech Republic
Radek Martínek	VŠB-TU Ostrava, Czech Republic

International Program Committee

Aarti Singh	Maharishi Markandeshwar University, India
Abdel hamid Bouchachia	University of Klagenfurt, Austria
Abdelhameed Ibrahim	Mansoura University, Egypt
AbdElrahman Shabayek	Suez Canal University, Egypt
Abd. Samad Hasan Basari	Universiti Teknikal Malaysia Melaka, Malaysia
Abraham Duarte	Universidad Rey Juan Carlos, Spain
Ahmed Anter	Beni-Suef University, Egypt
Akira Asano	Kansai University, Japan
Alaa Tharwat	Suez Canal University, Egypt
Alberto Alvarez	European Centre for Soft Computing, Spain
Alberto Cano	University of Cordoba, Spain
Alberto Fernandez	Universidad de Jaen, Spain
Alberto Bugarin	University of Santiago de Compostela, Spain
Alex James	Indian Institute of Information Technology and Management - Kerala, India
Alexandru Floares	Romania
Alma Gomez	University of Vigo, Spain
Amelia Zafra Gomez	University of Cordoba, Spain
Amira S. Ashour	Tanta University, Egypt
Amparo Fuster-Sabater	Institute of Applied Physics (C.S.I.C.), Spain
Ana Lorena	Federal University of ABC, Brazil
Anazida Zainal	Universiti Teknologi Malaysia, Malaysia
Andre Carvalho	University of Sao Paulo, Brazil
Andreas Koenig	Technische Universitat Kaiserslautern, Germany
Anna Bartkowiak	University of Wroclaw, Poland

Anna Fanelli	Universita di Bari, Italy
Antonio Peregrin	University of Huelva, Spain
Antonio J. Tallon-Ballesteros	University of Seville, Spain
Anusuriya Devaraju	Forschungszentrum Julich GmbH, Germany
Aranzazu Jurio	Universidad Publica de Navarra, Spain
Ashish Umre	University of Sussex, UK
Ashraf Saad	Armstrong Atlantic State University, USA
Ayeley Tchangani	University Toulouse III, France
Aymeric Histace	Universite Cergy-Pontoise, France
Azah Kamilah Muda	Universiti Teknikal Malaysia Melaka, Malaysia
Bartosz Krawczyk	Virginia Commonwealth University, USA
Beatriz Pontes	University of Seville, Spain
Brijesh Verma	Central Queensland University, Australia
Bing-Huang Chen	Fujian University of Technology, China
Carlos Barranco	Pablo de Olavide University, Spain
Carlos Cano	University of Granada, Spain
Carlos Fernandes	GeNeura Team, Spain
Carlos Garcia-Martinez	University of Cordoba, Spain
Carlos Lopezmolina	Universidad Publica de Navarra, Spain
Carlos Morell	Universidad Central Marta Abreu de Las Villas, Cuba
Cesar Hervas-Martinez	University of Cordoba, Spain
Chang-Shing Lee	National University of Tainan, Taiwan
Chao-Chun Chen	Southern Taiwan University, Taiwan
Chia-Feng Juang	National Chung-Hsing University, Taiwan
Chin-Chen Chang	Feng Chia University, Taiwan
Chris Cornelis	Ghent University, Belgium
Chun-Wei Lin	Harbin Institute of Technology Shenzhen Graduate School, China
Chuan-Kang Ting	National Chung Cheng University, Taiwan
Chuan-Yu Chang	National Yunlin University of Science and Technology, Taiwan
Chu-Hsing Lin	Tunghai University, Taiwan
Coral del Val	University of Granada, Spain
Crina Grosan	Norwegian University of Science and Technology, Norway
Cristina Rubio-Escudero	University of Sevilla, Spain
Cristobal Romero	University of Cordoba, Spain
Cristobal J. Carmona	University of Jaen, Spain
Chia-Hung Wang	Fujian University of Technology, China
Chia-Jung Lee	Fujian University of Technology, China
Dalia Kriksciuniene	Vilnius University, Lithuania
David Becerra-Alonso	ETEA-INSA, Spain
Detlef Seese	Karlsruhe Institute of Technology (KIT), Germany

Edurne Barrenechea	Universidad Publica de Navarra, Spain
Eiji Uchino	Yamaguchi University, Japan
Eliska Ochodkova	VSB—Technical University of Ostrava, Czech Republic
Elizabeth Goldbarg	Federal University of Rio Grande do Norte, Brazil
Emaliana Kasmuri	Universiti Teknikal Malaysia Melaka, Malaysia
Enrique Herrera-Viedma	University of Granada, Spain
Enrique Yeguas	University of Cordoba, Spain
Eulalia Szmidt	Systems Research Institute Polish Academy of Sciences, Poland
Eva Gibaja	University of Cordoba, Spain
Federico Divina	Pablo de Olavide University, Spain
Fernando Bobillo	University of Zaragoza, Spain
Fernando Delaprieta	University of Salamanca, Spain
Fernando Gomide	University of Campinas, Brazil
Fernando Jimenez	University of Murcia, Spain
Francesc J. Ferri	Universitat de Valencia, Spain
Francesco Marcelloni	University of Pisa, Italy
Francisco Fernandez Navarro	University of Cordoba, Spain
Francisco Herrera	University of Granada, Spain
Francisco Martinez-Alvarez	Pablo de Olavide University, Spain
Francisco Martinez-Estudillo	University Loyola Andalucia, Spain
Frank Klawonn	University of Applied Sciences Braunschweig, Germany
Gabriel Luque	University of Malaga, Spain
Gede Pramudya	Universiti Teknikal Malaysia Melaka, Malaysia
Giacomo Fiumara	University of Messina, Italy
Giovanna Castellano	Universita di Bari, Italy
Giovanni Acampora	University of Salerno, Italy
Girijesh Prasad	University of Ulster, UK
Gladys Castillo	University of Aveiro, Portugal
Gloria Bordogna	CNR IDPA, Italy
Gregg Vesonder	AT&T Labs – Research, USA
Huiyu Zhou	Queen's University belfast, UK
Hai-Yan Yang	Fujian University of Technology, China
Ilkka Havukkala	Intellectual Property Office of New Zealand, New Zealand
Imre Lendak	University of Novi Sad, Serbia
Intan Ermahani A. Jalil	Universiti Teknikal Malaysia Melaka, Malaysia
Isabel Nunes	UNL/FCT, Portugal
Isabel S. Jesus	Instituto Superior de Engenharia do Porto, Portugal
Ivan Garcia-Magarino	Universidad a Distancia de Madrid, Spain
Jae Oh	Syracuse University, USA

Leon Wang	National University of Kaohsiung, Taiwan
Liang Zhao	University of Sao Paulo, Brazil
Liliana Ironi	IMATI-CNR, Italy
Lincoln faria	Universidade Federal Fluminense, Brazil
Luciano Stefanini	University of Urbino "Carlo Bo", Italy
Ludwig Simone	North Dakota State University, USA
Luigi Troiano	University of Sannio, Italy
Luka Eciolaza	European Centre for Soft Computing, Spain
Liang-Cheng Shiu	National Pingtung University, Taiwan
Lydia Ogiela	AGH University of Science and Technology, Poland
Macarena Espinilla Estevez	Universidad de Jaen, Spain
Manuel Grana	University of Basque Country, Spain
Manuel Lama	Universidade de Santiago de Compostela, Spain
Manuel Mucientes	University of Santiago de Compostela, Spain
Marco Cococcioni	University of Pisa, Italy
Marek R. Ogiela	AGH University of Science and Technology, Poland
Maria Nicoletti	Federal University of Sao Carlos, Brazil
Maria Torsello	Universita di Bari, Italy
Maria Jose Del Jesus	Universidad de Jaen, Spain
Mariantonietta Noemi La Polla	IIT-CNR, Italy
Maria Teresa Lamata	University of Granada, Spain
Mario Giovanni C.A. Cimino	University of Pisa, Italy
Mario Koeppen	Kyushu Institute of Technology, Japan
Martine De Cock	Ghent University, Belgium
Michael Blumenstein	Griffith University, Australia
Michal Kratky	VSB—Technical University of Ostrava, Czech Republic
Michal Musilek	University of Hradec Kralove, Czech Republic
Michal Wozniak	Wroclaw University of Technology, Poland
Michela Antonelli	University of Pisa, Italy
Mikel Galar	Universidad Publica de Navarra, Spain
Milos Kudelka	VSB—Technical University of Ostrava, Czech Republic
Min Wu	Oracle, USA
Mohamed Eltoukhy	Suez Canal University, Egypt
Mohamed Khairy	Suez Canal University, Egypt
Mohamed Tahoun	Suez Canal University, Egypt
Mona Solyman	Cairo university, Egypt
Nilanjan Dey	Techno India College of Technology, India
Noor Azilah Muda	Universiti Teknikal Malaysia Melaka, Malaysia
Norberto Diaz-Diaz	Pablo de Olavide University, Spain
Norton Gonzalez	University of Fortaleza, Brazil

Noura Semary	Menofia University, Egypt
Nurulakmar Emran	Universiti Teknikal Malaysia Melaka, Malaysia
Olgierd Unold	Wroclaw University of Technology, Poland
Oscar Castillo	Tijuana Institute of Technology, Mexico
Ovidio Salvetti	ISTI-CNR, Italy
Ozgur Koray Sahingoz	Turkish Air Force Academy, Turkey
Pablo Villacorta	University of Granada, Spain
Patrick Siarry	Universite de Paris, France
Paulo Carrasco	Universidade do Algarve, Portugal
Paulo Moura Oliveira	University of Tras-os-Montes and Alto Douro, Portugal
Pedro Gonzalez	University of Jaen, Spain
Petr Musilek	University of Alberta, Canada
Philip Samuel	Cochin University of Science and Technology, India
Pierre-Francois Marteau	Universite de Bretagne Sud, France
Pietro Ducange	University of Pisa, Italy
Punam Bedi	University of Delhi, India
Qieshi Zhang	Waseda University, Japan
Qinghan Xiao	Defence R&D Canada, Canada
Radu-Codrut David	Politehnica University of Timisoara, Romania
Rafael Bello	Universidad Central de Las Villas, Cuba
Ramin Halavati	Sharif University of Technology, Iran
Ramiro Barbosa	Instituto Superior de Engenharia do Porto, Portugal
Ramon Sagarna	University of Birmingham, UK
Richard Jensen	Aberystwyth University, UK
Robert Berwick	Massachusetts Institute of Technology, USA
Roberto Armenise	Poste Italiane, Italy
Robiah Yusof	Universiti Teknikal Malaysia Melaka, Malaysia
Rodrigo Martins	University of Alberta, Canada
Roman Neruda	Institute of Computer Science, Czech Republic
S. Ramakrishnan	Dr. Mahalingam College of Engineering and Technology, India
Sabrina Ahmad	Universiti Teknikal Malaysia Melaka, Malaysia
Sadaaki Miyamoto	University of Tsukuba, Japan
Santi Llobet	Universitat Oberta de Catalunya, Spain
Sarwar kamal	East West University, Bangladesh
Satrya Fajri Pratama	Universiti Teknikal Malaysia Melaka, Malaysia
Saurav Karmakar	Georgia State University, USA
Sazalinsyah Razali	Universiti Teknikal Malaysia Melaka, Malaysia
Sebastian Basterrech	Czech Technical University, Czech Republic
Sebastian Ventura	University of Cordoba, Spain
Selva Rivera	Institute of C.E.D.I.A.C, Argentina
Shang-Ming Zhou	University of Wales Swansea, UK

Siby Abraham	University of Mumbai, India
Silvia Poles	EnginSoft, Italy
Silvio Bortoleto	Federal University of Rio de Janeiro, Brazil
Siti Rahayu Selamat	Universiti Teknikal Malaysia Melaka, Malaysia
Steven Guan	Xi'an Jiaotong-Liverpool University, China
Sung-Bae Cho	Yonsei University, Korea
Swati V. Chande	International School of Informatics and Management, India
Sylvain Piechowiak	Universite de Valenciennes et du Hainaut-Cambresis, France
Subhas Mukhopadhyay	Massey University, New Zealand
Takashi Hasuike	Osaka University, Japan
Taras Kotyk	Ivano-Frankivsk National Medical University, Ukraine
Tarek Gaber	Suez Canal University, Egypt
Tay Kai Meng	Universiti Malaysia Sarawak, Malaysia
Teresa Ludermir	Federal University of Pernambuco, Brazil
Thomas Hanne	University of Applied Sciences Northwestern Switzerland, Switzerland
Tzung-Pei Hong	National University of Kaohsiung, Taiwan
Ting-Ting Wu	National Yunlin University of Science and Technology, Taiwan
Vaclav Snasel	VSB—Technical University of Ostrava, Czech Republic
Valentina Colla	Scuola Superiore Sant'Anna, Italy
Varun Ojha	ETH Zurich, Switzerland
Victor Hugo Menendez Dominguez	Universidad Autonoma de Yucatan, Mexico
Vincenzo Loia	University of Salerno, Italy
Vincenzo Piuri	University of Milan, Italy
Virgilijus Sakalauskas	Vilnius University, Lithuania
Vivek Deshpande	MIT College of Engineering, India
Vladimir Filipovic	University of Belgrade, Serbia
Wahiba Ben Abdessalem Karâa	Taif University, KSA
Wei Wei	Xi'an University of Technology, China
Wei-Chiang Hong	Oriental Institute of Technology, Taiwan
Wen-Yang Lin	National University of Kaohsiung, Taiwan
Wilfried Elmenreich	University of Klagenfurt, Austria
Yasuo Kudo	Muroran Institute of Technology, Japan
Ying-Ping Chen	National Chiao Tung University, Taiwan
Yun-Huoy Choo	Universiti Teknikal Malaysia Melaka, Malaysia
Yunyi Yan	Xidian University, China
Yusuke Nojima	Osaka Prefecture University, Japan
Feng-Cheng Chang	Tamkang University, Taiwan

Sponsoring Institutions

University of Málaga, Spain
VŠB—Technical University of Ostrava, Czech Republic

Contents

Image Processing and Applications

Research on Eye Detection and Fatigue Early Warning Technologies

S.H. Meng[1,2(✉)], S.B. Hu[3(✉)], A.C. Huang[4], T.J. Huang[5], Zhixuan Xie[1], and Chen Jian[1]

[1] School of Information Science and Engineering, Fujian University of Technology, Fuzhou 350118, Fujian, China
menghui@fjut.edu.cn

[2] Key Laboratory of Big Data Mining and Application, School of Information Science and Engineering, Fujian University of Technology, Minhou, Fuzhou City 350118, Fujian, China

[3] Institute of Electro-Optical Science and Engineering, National Cheng Kung University, Tainan 70101, Taiwan

[4] Department of Electrical Engineering, National Sun Yat-sen University, Kaohsiung, Taiwan

[5] Kaohsiung Municipal ChungShan Senior High School, Kaohsiung, Taiwan

Abstract. Utilizing non-contact eye-detection- and image-processing-based fatigue recognition and early warning technology, this study established a human-computer interaction system based on dynamic expression recognition. The degree of fatigue was determined based on the percentage of eyelid closure (PERCLOS). Facial and eye areas in the key frames for facial expression identification were adopted to determine the expression of fatigue. A skin-color technique was used for face detection. A skin model was later established using the hue, saturation, value (HSV) color model, which was then used to detect the skin color of any given face. The improved circle Hough transform algorithm was applied for use in eye detection. Blink rate (pupil region extraction) and PERCLOS were combined in the detection of fatigue.

Keywords: Fatigue assessment · Skin color detection · Face recognition · Eye localization

1 Introduction

With the rapid development of modern transportation, traffic accidents have become a critical concern of countries throughout the world. When drivers are tired, their sensitivity to the environment, their accuracy and speed of judgment, and their control of the vehicle tend to decrease, resulting in traffic accidents. The system presented in this study can alert and remind the driver, as well as transmit feedback and other information collected from human-computer interaction, to help prevent driving fatigue and thus reduce traffic accidents.

P. Krömer et al. (eds.), *Proceedings of the Fourth Euro-China Conference on Intelligent Data Analysis and Applications*, Advances in Intelligent Systems and Computing 682, DOI 10.1007/978-3-319-68527-4_1

Previous studies on the detection of the eye area and eye status usually adopted an image or infrared radiation (IR)-light-based method [1]. Furthermore, image-based eye detection can be divided into template-, feature-, and appearance-based methods shown in Fig. 1.

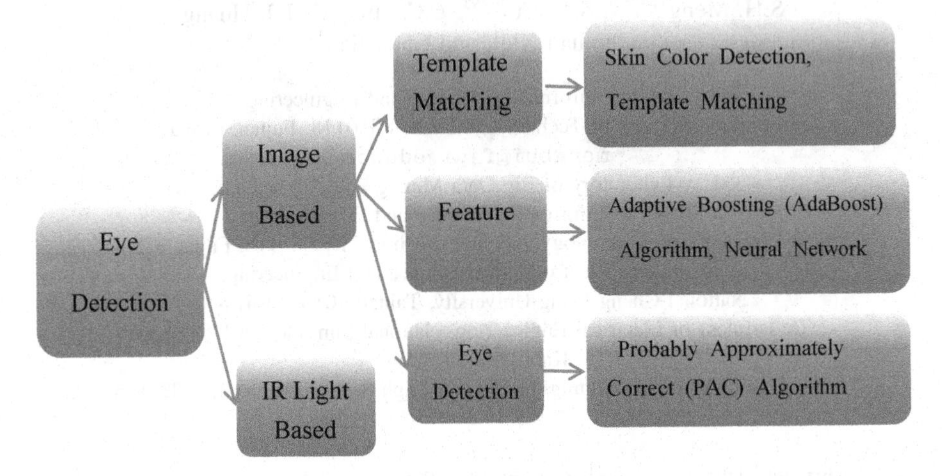

Fig. 1. Methods of eye detection.

(1) Template-matching method: In this method, a standard human eye template is initially constructed, which is then compared to the images that contain a human face in order to identify and determine whether the image has a similar area to that in the template. In addition, the template is used to determine the center of the image if a high degree of similarity is found. Fast template matching requires a simple background and images with explicitly distinctive facial features for testing, as factors such as a complicated background may lead to substantial computations and longer processing time. In addition, a human eye template must meet certain prerequisites for clarity. This is because successful matching of the template and image depends on the generality of the eye template and the accuracy and clarity of the images in the digital image libraries.
(2) Image-feature statistical learning method: In this method, a mathematical model is constructed in order to train computers to recognize human facial features using numerous facial object detection processes. A classifier is then established based on the model to enhance the system's ability to analyze and classify data related to human eyes. Finally, a well-trained classifier is used to identify the region in which an input image contains a human face. The AdaBoost algorithm is the most widely employed method among the various statistical methods. The method is used to detect each region and section of an image. Because the eigenvalues of each image section must be calculated, the computational complexity is high and real-time performance is rather poor. However, during the actual detection process, several trained classifiers are usually combined into a cascading classifier, which, in addition to having a high degree of accuracy, can accelerate the detection speed and reduce detection time.

(3) Geometrical-feature-based method: Based on the characteristics of the location and geometric features of the human eyes on the face, a mathematical model of the position of the facial organs can be established. In addition, the mathematical rules corresponding to the layout of the facial organs can be obtained through analysis of the mathematical model. Studies have shown that human eyes are geometrically symmetric. Thus, eye-specific geometric shapes can be used to locate human eyes. The advantage of such an eye detection method is in its ability to analyze clearly the geometric characteristics of the human face. However, the method requires that the image have a non-complex background and be in a well-lit environment. Visual interference in other areas of the image must also be reduced. The geometric-feature method is usually used with the template-matching method in order to create the mathematical model function.

2 System Overviewfirst

Using circle Hough transforms in image analysis requires major computations. Hence, real-time performance is often poor and cannot be applied in a timely manner [2]. Near-infrared imaging is limited to certain applications in which infrared lighting can be applied [3]. To achieve better analytical results of the feature model, this study adopted a color-information method. Specifically, we constructed a skin-color-model based on an HSV color model [4] for use in detecting facial images based on a skin-color detection algorithm. Data from image processing were then compared to standard physical reactions of facial organs that occur during fatigue. Statistical analysis was then conducted before determining fatigue. Next, selected frames were extracted from a video clip to act as images, and corresponding image processing and image detection techniques were then applied to identify the facial and eye areas. PERCLOS information was detected and analyzed. Finally, the level of fatigue assessed using a facial dynamic system.

2.1 Research and Analytical

(1) Review and analyze the various face detection algorithms, establish a simulation model using induction and abstraction, and conduct an analytical simulation based on the principles of existing algorithms.
(2) Apply the skin-color detection technique to detect human faces. Identify the facial area and background.
(3) Conduct image preprocessing of the obtained facial area to facilitate eye localization and eye data extraction.
(4) Conduct an in-depth study of the various fatigue detection methods. Then apply the circle Hough transform method to detect the eye pupils, and determine fatigue.
(5) Use a dynamic facial fatigue system to assess fatigue: First, the human face is identified. Next, the eyes are located. A line chart is then composed to illustrate the changes of the pupil over time. Finally, the degree of fatigue is determined based on the proportion of the number of "fatigue frames" to the total number of frames. The face-based fatigue detection system is demonstrated in Fig. 2.

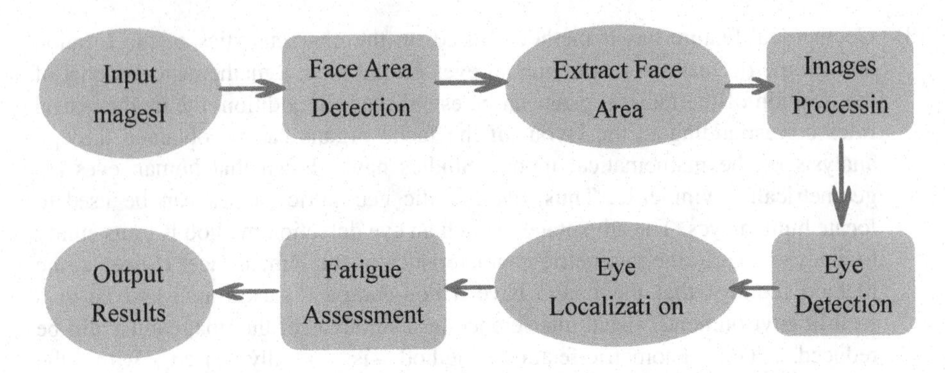

Fig. 2. Face-based fatigue detection system.

2.2 Research Data

This study constructed a dynamic-expression-recognition-based human-computer interaction system. By identifying the human face and eyes, and producing a line chart of the changes in pupil size over time, the system could determine the degree of fatigue based on PERCLOS. The facial and eye areas in the key frames of expression identification were adopted to determine the expression of fatigue. A skin-color-based technique was utilized for face detection using the HSV color model [4], which was then used to detect the skin color of any given face. The improved circle Hough transform algorithm was applied for use in eye detection. Blink rate (pupil region extraction) and PERCLOS were combined in order to detect fatigue. The binarization results of the HSV format are better than those of the Ycbcr format [5]. Image binarization is used to facilitate the extraction of information from an image. Using binary images improves the efficiency of computer identification [6]. As shown in Fig. 3, the process of converting HSV binary images into grayscale images using morphological processing is shown in Fig. 4.

Using circle Hough transformations for eye detection enables the system to detect and locate the pupils of the eye more accurately. First, a Canny operator was adopted to extract the image edge of the object to be detected [7]. The area of the pupil was next determined. Then, the edge of the pupil was defined based on an image with open eyes, and the improved circle Hough transforms were utilized to detect the quasi-circled pupil. Eye blink rate and PERCLOS were combined to detect fatigue. Specifically, when the object to be detected appeared tired, the eye blink rate was likely to increase and the degree of fatigue could be determined by comparing the proportion of "fatigue frames" to the total number of frames and the threshold value [8]. Table 1 shows that the maximum pixel value of the human eye area was 4900, and pupil region extraction (PRE) could be calculated for each frame. When *PRE* was less than 0.3, the image was defined as being in a fatigue state; otherwise, it was defined as being in a non-fatigue state.

The improved circle Hough transform algorithm was applied to eye detection. Blink rate (PRE) and PERCLOS were combined to detect fatigue. Based on image

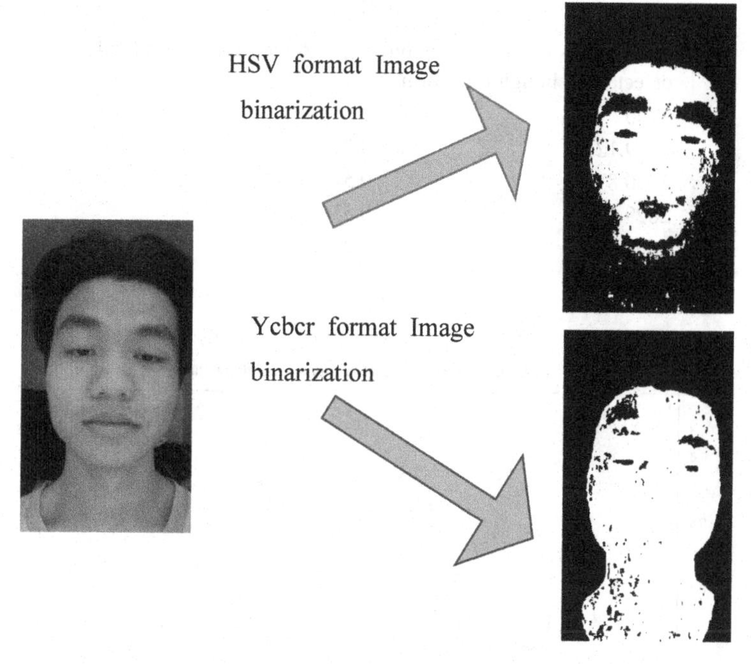

Fig. 3. Skin color detection. Comparison of the binarization results of the HSV and Ycbcr formats.

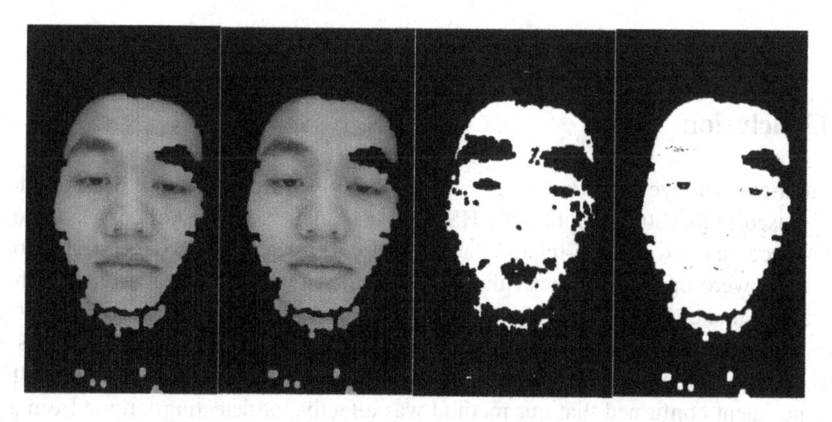

Fig. 4. Process of converting HSV binary images into grayscale images through morphological processing.

capture and data acquisition, a line chart was generated to illustrate changes in the size of drivers' eyes when in a state of fatigue. The results of the association between changes in blink rate and distribution of apparent state of fatigue are shown in Fig. 5.

Table 1.

Pixel values of the eye region detected by hough transform	Pupil Region Extraction (PRE)
4900	1
3499.12	0.7141
2520.87	0.5144
2629.87	0.5367
1264.25	0.2580

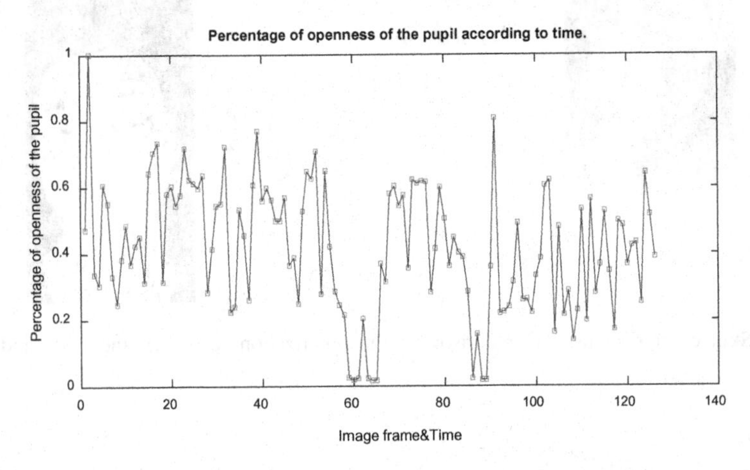

Fig. 5. Blink rate versus apparent state of fatigue.

3 Conclusion

The non-contact eye-detection-based, fatigue-early-warning, image-processing technology used in this study applied the HSV color model to detect skin color, define the human face, and extract eye-related data in order to assess fatigue. The circle Hough transforms were applied to detect the pupils in the human eye area, from which we generated a line chart of changes in the size of the pupil over time. Determining fatigue was accomplished by determining the total number of "fatigue frames." PERCLOS and blink rate were combined as a single analytical parameter of fatigue, and the results of the experiment confirmed that our method was effective at detecting fatigue from a set of dynamic facial expressions. The proposed method also succeeded in achieving a human-computer interactive early warning system that generated reminders and feedback. The system may be widely applied in the field of fatigue detection to help reduce the number of traffic accidents.

References

1. Xie, X., Sudhakar, R., Zhuang, H.: On improving eye feature extractionusing deformable templates. Pattern Recognit. **27**(6), 791–799 (1994)
2. Liu, N., Su, H., Guo, C.-H., et al.: Jiyu Hough Bianhuanyuan Jiance de Renyan Dingwei Fangfa Gaijin (Development of a Human Eye Localization Method based on Circle-Hough-Transform Detection). Comput. Eng. Des. **32**(4), 1359–1362 (2011)
3. Li, S., Zhang, L., Liao, S., et al.: A near-infrared image based face recognition system. In: Proceedings of Seventh IEEE International Conference on Automatic Face and Gesture Recognition, pp. 455–460 (2006)
4. Ganesan, P., Rajini, V.: Value based semi automatic segmentation of satellite images using HSV color space, histogram equalization and modified FCM clustering algorithm. In: 2013 International Conference on Green Computing, Communication and Conservation of Energy (ICGCE), 12–14 December 2013, pp. 77–82 (2013). doi:10.1109/ICGCE.2013.682340
5. Shaik, K.B., et al.: Comparative study of skin color detection and segmentation in HSV and YCbCr color space. Procedia Comput. Sci. **57**, 41–48 (2015)
6. Bao, H.-Q.: Jiyu Neirong de Shipin Yundong Duixiang Fenge Jishu Yanjiu (Research on Content-Based Segmentation Technology of Moving Video Objects). Shanghai University (2005)
7. Zhang, Z., Ma, S.-L., Zhang, Z.-B., et al.: Improved image edge extraction algorithm based on canny operator. J. Jinlin Univ. (Sci. Ed.) **45**(2), 224–248 (2007)
8. Song, Z.-H., Zhou, Y.-M.: PERCLOS-based recognition algorithms of motor driver fatigue. J. China Agric. Univ. **7**(2), 104–109 (2002)

Face Recognition Based on HOG and Fast PCA Algorithm

Xiang-Yu Li(✉) and Zhen-Xian Lin

School of Communication and Information Engineering,
Xi'an University of Posts and Telecommunications, Xi'an 710061, China
lxy_john_5@163.com, lzhxl26@126.com

Abstract. A new method of face recognition based on gradient direction histogram (HOG) features extraction and fast principal component analysis (PCA) algorithm is proposed to solve the problem of low accuracy of face recognition under non-restrictive conditions. In this method, the Haar feature classifier is used to extract and extract the original data, and then the HOG features are extracted from the image data and the PCA dimension reduction is processed, and the Support Vector Machines (SVM) algorithm is used to recognize the face. The experimental results of the classification recognition on the LFW face database verify the effectiveness of the method.

Keywords: Haar feature classifier · HOG · Fast PCA · SVM

1 Introduction

As one of the important research topics in the field of biometrics, face recognition has been favored by its non-contact and non-stealing characteristics [1, 2]. With the machine learning, pattern recognition, artificial intelligence and computer vision technology continue to develop, face recognition technology has made great progress. The commonly used methods of face recognition are: Support Vector Machine (SVM), Principal Component Analysis (PCA), Histogram of Oriented Gradient (HOG), and so on. Among them, the improved SVM algorithm based on Optimization of kernel function [3] and classification algorithm for face recognition [4], due to the extraction of all features of the image, the computation is large and it is difficult to achieve the rapid identification of huge amounts of data; PCA is the feature extraction of the global feature, although the dimension of the feature is reduced, its recognition accuracy is affected by the illumination [5–7]; HOG method based on feature extraction [8], although not affected by illumination and geometric deformation, the calculation is still large, difficult to achieve rapid identification [9]. At present, most of the face recognition methods are based on restrictive conditions (illumination, gesture, expression and other specific circumstances) of the data classification and identification [10], and non-restrictive conditions (illumination, gesture, expression and other real state) of the face recognition is a difficult problem [11].

In this paper, Haar feature classifier is introduced into the preprocessing process of raw data. A new method of face recognition for non-restrictive conditions is

P. Krömer et al. (eds.), *Proceedings of the Fourth Euro-China Conference on Intelligent Data Analysis and Applications*, Advances in Intelligent Systems and Computing 682, DOI 10.1007/978-3-319-68527-4_2

constructed by combining HOG feature extraction, fast PCA dimensionality reduction and SVM classification algorithm.

2 Raw Data Preprocessing

The preprocessing of raw data consists of two main parts: extraction of raw data from human faces database and detection and extraction of human faces. The flow of raw data preprocessing is shown in Fig. 1.

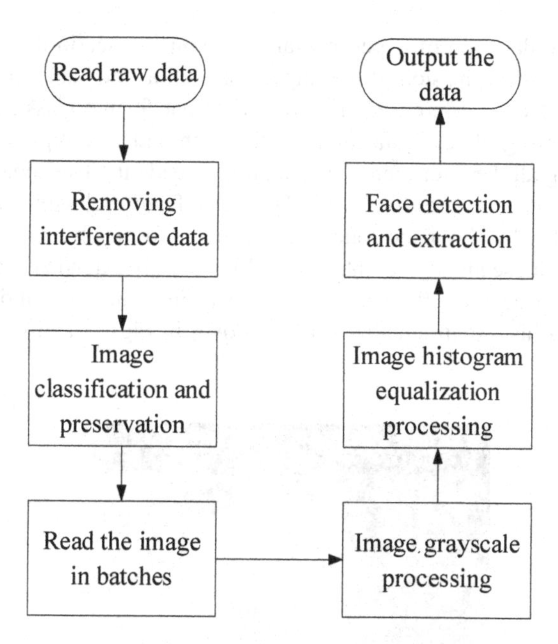

Fig. 1. Preprocessing process

2.1 Extract Raw Data

As the original data is collected under non-limiting conditions, some of the data due to its different acquisition path, resulting in a greater difference in imaging results. Such as facial mask, facial expression is too exaggerated, large deviation angle of the face, etc., as shown in Fig. 2. These data will have a greater impact on the training and final recognition of the classification model, so the interference data is first removed during the preprocessing phase. In order to facilitate the mass processing of data, the sorting of categories and names are ordered at the same time as the original data is extracted.

2.2 Face Detection and Extraction

The difficulty of face recognition in non-restrictive conditions is that there is a lot of redundant information and interference information in the original data. Figure 2 is

Fig. 2. Useless interference data

extracted from the database of experimental data, you can see in the sample not only the light, posture and expression of the impact, as well as the background of the items and other human face interference. In this paper, Haar feature classifier is introduced into the preprocessing of raw data for face detection and face region extraction. The effect of removing all the interference information outside the face area is realized, and the experimental samples are optimized. The Haar feature classifier is based on the AdaBoost algorithm by Viola and Jones [12], which is formed by cascading the trained strong classifier. Subsequently, Lienhart and Maydt [13] extended the classifier, and finally formed the Haar classifier used in this paper. First, the original data is processed by gray scale and histogram equalization, as shown in Figs. 3 and 4.

Fig. 3. Grayscale image

The effect of the two steps is to reduce the amount of data information and speed up the follow-up processing, while enhancing the contrast of the image and weakening the influence of illumination. The Haar feature classifier for face detection and face region extraction is shown in Fig. 5.

Haar feature classifier can achieve 95% detection accuracy, but its extraction of data still has a small amount of error data [14]. Delete the extracted data and save it

Fig. 4. Histogram equalization of the image

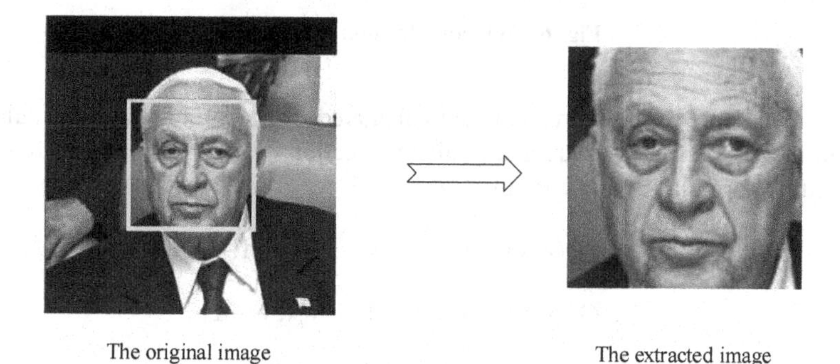

Fig. 5. Face area extraction process

again. As the HOG feature extraction has a certain requirement on the image size, the image data is normalized to a size of 64 × 128 pixels.

3 HOG Feature Extraction and PCA Dimensionality Reduction

3.1 HOG Feature Extraction

Histogram of Oriented Gradient (HOG) was proposed by Dalal et al. [15] in 2005 to detect pedestrians. The HOG feature is robust and has no sensitivity to both light and geometric changes, and the computational complexity of the HOG feature is much less than that of the original data. The main steps of HOG feature extraction are as follows:

(1) A 64 × 128 pixel window is divided by 8 × 8 pixel cell, forming 8 × 16 = 128 cells, as shown in Fig. 6. The gradient components of each pixel (x, y) in

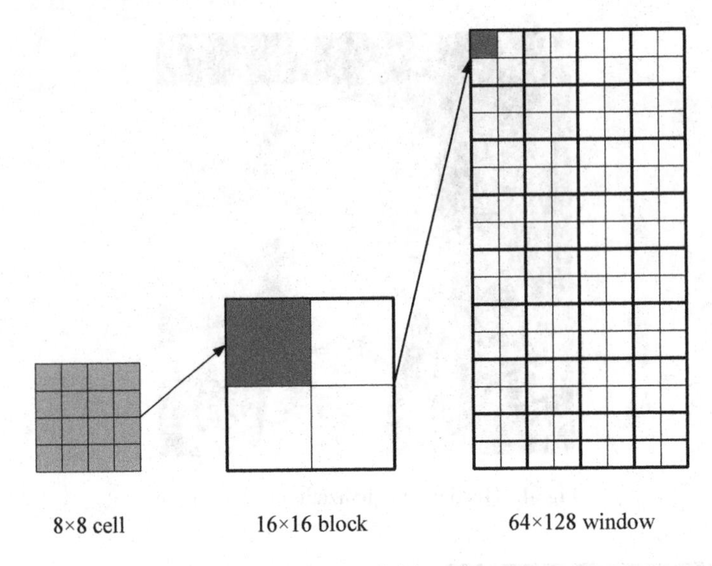

Fig. 6. Window division process

horizontal and vertical directions are calculated by formula (1) and formula (2), and the gradient magnitude and gradient direction of each pixel point are calculated by formula (3) and formula (4).

$$G_x(x, y) = I(x + 1, y) - I(x - 1, y) \tag{1}$$

$$G_y(x, y) = I(x, y + 1) - I(x, y - 1) \tag{2}$$

$$m(x, y) = \sqrt{(G_x(x, y))^2 + (G_y(x, y))^2} \tag{3}$$

$$\theta(x, y) = \arctan\frac{G_y(x, y)}{G_x(x, y)} \tag{4}$$

(2) A block of 16 × 16 pixels is composed of 2 × 2 = 4 cells, and 7 × 15 = 105 blocks are composed. The block step size of 8 pixels, the number of blocks in the horizontal direction is (64−16)/8 + 1 = 7, and the number of blocks in the vertical direction is (128−16)/8 + 1 = 15, as shown in Fig. 7.

(3) Take a histogram of 9 gradient directions for each cell, as shown in Fig. 8. Such a block has 4 × 9 = 36 feature vectors, and then 105 blocks of feature vectors are connected in series to form an image of 36 × 105 = 3780 HOG features.

3.2 PCA Dimensionality Reduction

The main principle of the PCA algorithm is to transform the original data into a set of linearly independent data by linear transformation, which can be used to extract the main feature components of the data. Therefore, it is often used for dimensionality

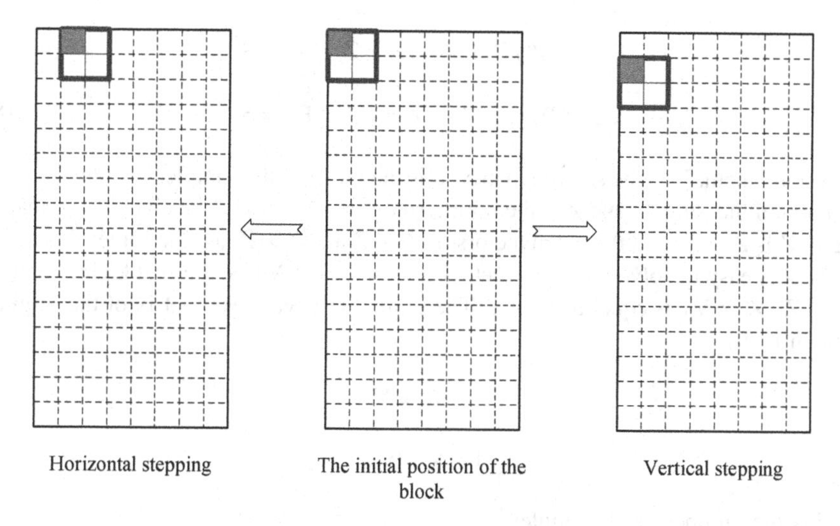

Fig. 7. Stepping process of the block

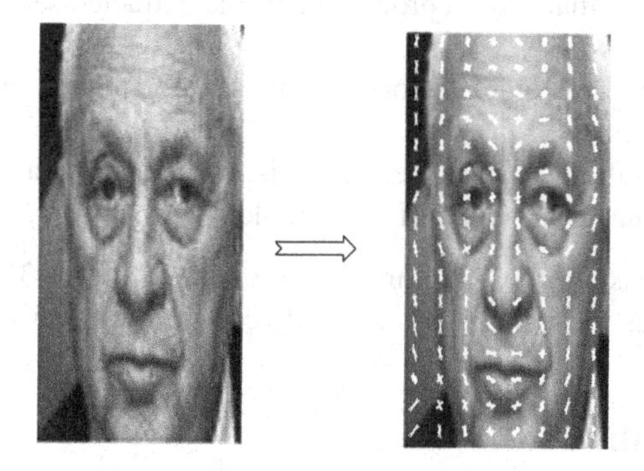

Fig. 8. HOG feature extraction display

reduction of high dimensional data. For face recognition problems, the feature dimension is often much higher than the number of samples. The original data of this paper is $64 \times 128 = 8192$, even if the extracted HOG feature is 3780 dimensions. For the solution of the covariance matrix in the PCA algorithm, the computational data is 3780×3780, which will lead to the running time, and the general computer hardware is difficult to meet the requirements. In this paper, the traditional PCA algorithm is improved by using fast-PCA method to solve the eigenvalues and eigenvectors of the sample matrix. The main principle of fast-PCA is shown in Eq. (5)–(9):

$$Z = \frac{X - \mu}{\sigma} \tag{5}$$

$$P^{-1} \times (Z \times Z') \times P = S \tag{6}$$

$$P^{-1} \times (Z')^{-1} \times Z' \times Z \times Z' \times P = S \tag{7}$$

$$(Z' \times P)^{-1} \times Z' \times Z \times (Z' \times P) = S \tag{8}$$

Among them, X is the sample matrix, μ is the mean of the sample, σ is the standard deviation of the sample, and Z is the normalized sample matrix. The eigenvector matrix of $Z' \times Z$ is $Z' \times P$, and the eigenvectors of $Z' \times Z$ and $Z \times Z'$ are the same. Z is a row matrix, it is easy to solve the eigenvectors of $Z \times Z'$. Let V1 be the first k eigenvectors of $Z \times Z'$, $Z' \times V1$ is equal to $Z' \times P$. Therefore, the covariance matrix of the original problem is (9).

$$V = Z' \times V1 \tag{9}$$

$$1 \leq k \leq N,$$

N is the number of all samples.

The main steps of PCA dimensionality reduction are as follows:

(1) The zero-mean normalized preprocessing of the extracted sample matrix is shown in Eq. (5).
(2) The eigenvalues and eigenvectors of the sample matrix are obtained by the fast-PCA method.
(3) Finally, the resulting feature vector V is processed in a modular way, thus obtaining the final PCA value of the sample.

After PCA processing, the data dimension is reduced from 3780 to k dimension, which greatly reduces the amount of data and accelerates the speed of operation.

4 Face Recognition

4.1 Data Normalization

The feature data for support vector machine classification is obtained by HOG feature extraction and PCA dimensionality reduction. However, there is a large difference in the numerical size of the different feature components in the same sample, and if there is no normalization, there will be no comparability [16]. Therefore, before the final classification and identification, it needs to data normalization. In this paper, using the median normalization method, as shown in Eq. (10), the same sample of data normalized to [−1, 1].

$$X' = 2 \times \frac{X - X_{min}}{X_{max} - X_{min}} - 1 \tag{10}$$

Where X' is the normalized data, X is the original data, X_{max} is the maximum value in the original data, and X_{min} is the minimum value in the original data.

4.2 Face Recognition Based on SVM

SVM algorithm [17, 18] has been widely used in various classification scenes because of its strong classification ability for small samples and high dimension data. Based on the statistical learning theory of structural risk minimization, the SVM algorithm separates the different categories by finding the optimal hyperplane among different samples. The optimal hyperplane can be expressed as:

$$y = \omega^T \phi(x) + b \tag{11}$$

where ω is the normal vector of hyperplane, and b is the offset vector of hyperplane.

For the problem of linear indivisibility, we need to transform the nonlinear classification problem into quadratic optimization problem. Then the Lagrangian multiplier is used to transform the classification problem into its dual problem. The final hyperplane classification function is:

$$f(x) = \mathrm{sign}\left(\sum_{i=1}^{n} \alpha_i y_i (\phi(x) \cdot \phi(x_i)) + b\right) \tag{12}$$

where sign is a sign function and α_i is a Lagrange multiplier.

The kernel function $k(x_i \cdot x)$ instead of $(\phi(x) \cdot \phi(x_i))$, on into:

$$f(x) = \mathrm{sign}\left(\sum_{i=1}^{n} \alpha_i y_i k(x_i \cdot x) + b\right) \tag{13}$$

The common kernel functions include linear kernel function, Gauss radial basis function (RBF kernel function), polynomial kernel function and Sigmoid kernel function. In this paper, the linear kernel function is used as the kernel function of SVM.

Face recognition based on SVM includes two parts: model training and sample testing. Firstly, the training samples are trained in the case of setting SVM specific parameters, and the training model is obtained. And then use the trained model to test the test samples, get the accuracy of face recognition and test time.

5 Experiment

5.1 Experimental Environment

The experimental environment is divided into two parts: hardware environment and software platform:

(1) Hardware environment: PC, Windows 7_64 bit operating system, the processor for the Intel (R) Core (TM) i5-2450 M CPU @ 2.50 GHz, memory 4.0 G
(2) Software platform: Visual Studio 2015, OpenCV 3.2.0, MATLAB R2015b

Among them, the libsvm toolkit is used in the MATLAB platform, which is a simple, easy-to-use and fast and effective SVM pattern recognition and regression software package designed by Professor Lin Chih-Jen of Taiwan University.

5.2 Experimental Database

This paper adopts the LFW face database under the condition of non restriction. The LFW face database contains 5749 people, 13233 images in total, and the image size of 250 × 250 pixels. In this experiment, 14 experimental samples were selected as the experimental samples with face images with more than 40 data volumes, as shown in Fig. 9. The model training and sample testing were carried out in the following 4 ways:

Fig. 9. Experimental sample

(1) 10 samples of each type were selected as experimental samples, of which 5 were used as training samples and 5 as test samples;
(2) 20 samples of each type were selected as experimental samples, of which 10 were used as training samples and 10 as test samples;
(3) 30 samples of each type were selected as experimental samples, of which 15 were used as training samples and 15 as test samples;
(4) 40 samples of each type were selected as experimental samples, of which 20 were used as training samples and 20 as test samples.

5.3 Comparison and Analysis of Experimental Results

In order to verify the effectiveness of the proposed method, the results of the face recognition were compared with the SVM algorithm, the PCA + SVM algorithm and the HOG + SVM algorithm under the same experimental conditions, respectively, in the above four experimental samples of different sizes, The experimental results shown in Fig. 10. At the same time, the algorithm is compared with the time of the test phase, as shown in Fig. 11.

It can be seen from Fig. 10 that the recognition accuracy of SVM algorithm is the lowest when the same number of training samples and test samples are used, and the recognition rate of PCA + SVM algorithm and HOG + SVM algorithm is improved, but their recognition rate is still low. The method proposed in this paper is superior to the other three algorithms, and the highest recognition accuracy is above 90%. At the

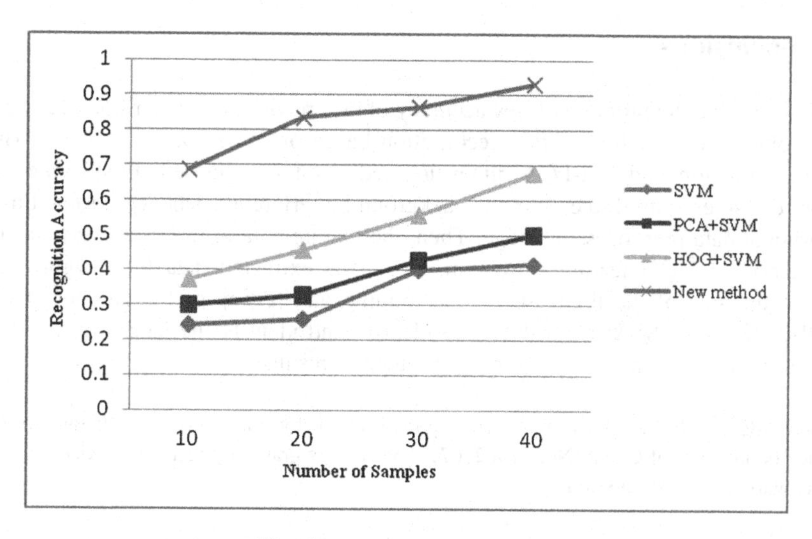

Fig. 10. Accuracy comparison

same time, it can be seen from Fig. 11 that the time of SVM algorithm increases rapidly with the increase of the number of samples at the experimental phase, although the time of HOG + SVM algorithm reduced, it also affected by experimental sample number. The method used in this paper is basically the same as the PCA + SVM algorithm, and it is not affected by the experimental sample number. The comparison of the above two experimental results fully proves that the proposed method not only uses less time, but also has higher recognition accuracy.

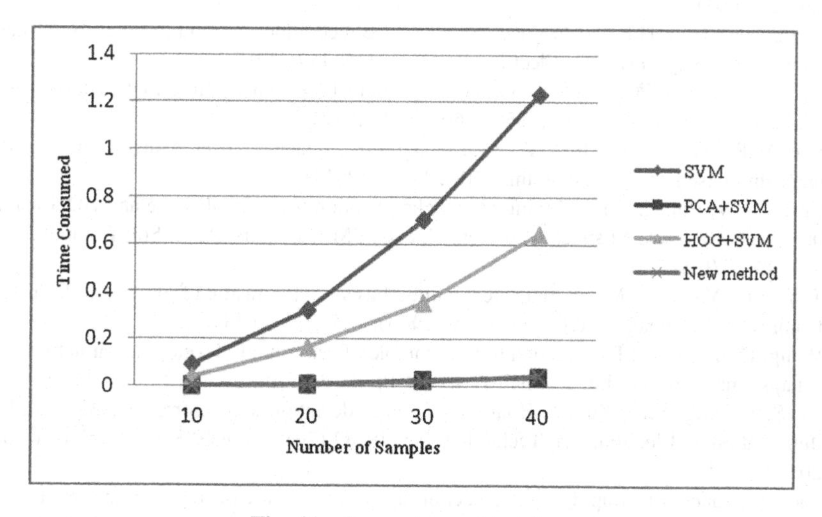

Fig. 11. Comparison of test time

6 Conclusions

In order to solve the problem of low accuracy of face recognition under non - restrictive conditions, a new method of face recognition based on Haar feature classifier, HOG feature extraction and fast-PCA dimension reduction is proposed. Firstly, the Haar feature classifier is used to extract the background interference data at the same time in the original data preprocessing stage. Then, the feature data of the face is extracted by the method of HOG feature extraction. Then, the extracted data PCA algorithm is reduced the size of the final use for the training and testing of the amount of data. Finally, the use of SVM algorithm is to identify and identify the face. It is verify the effectiveness of the method with the experimental results.

Acknowledgements. This research was supported in part by grants from the National Natural Science Foundation of China (No. 61402367). The authors gratefully thank Pro. Xiao-Qiang XI for his warmhearted discussion.

References

1. He, C.: Survey of face recognition technology. Intell. Comput. Appl. **6**(5), 112–114 (2016)
2. http://www.en.cnki.com.cn/Article_en/CJFDTotal-DLXZ201605034.htm
3. Muhammad, S., Mohsin, S., Javed, M.Y.: A survey: face recognition techniques. Res. J. Appl. Sci. Eng. Technol. **4**(23), 1–10 (2012)
4. Shu, S.B., Luo, J.R., Xu, C.D., et al.: A new method of face recognition based on support vector machine. Comput. Simul. **28**(2), 280–283 (2011)
5. Yan, Z.C., Yao, L., Han, C.: A face recognition method based on hybrid kernel function SVM. J. Sichuan Univ. Sci. Technol. (Nat. Sci. Ed.) **29**(3), 23–26 (2016). doi:10.11863/j.suse.2016.03.06
6. Zhao, L.Y., Gao, H.B., Li, Y.: Research on face recognition algorithm based on principal component analysis (PCA). Electron. World **2**, 116–117 (2017)
7. Li, G., Li, Q.: A PCA face recognition algorithm based on facial kernel features and its applications. Electron. Devices **35**(5), 607–610 (2012)
8. Sun, W.R., Zhou, X.C., Ji, Y.T.: Face recognition equalization, PCA algorithm and SVM algorithm based on the histogram. **35**(8), 11–15 (2014)
9. Liang, C.W., Juang, C.F.: Moving object classification using local shape and HOG features in wavelet-transformed space with hierarchical SVM classifiers. Appl. Soft Comput. **28**(C), 483–497 (2015)
10. Tan, H.L., Yang, B., Ma, Z.: Face recognition based on the fusion of global and local HOG features of face images. IET Comput. Vis. **8**(3), 224–234 (2014)
11. Wang, D.S., Liu, L.: Face recognition in complex background: developmental network and synapse maintenance. Int. J. Smart Home **9**(10), 47–62 (2015)
12. Lv, S.D., Song, Y.D., Xu, M., Huang, C.Y.: Face detection under complex background and illumination. J. Electron. Sci. Technol. **13**(1), 78–82 (2015). doi:10.3969/j.issn.1674-862X.2015.01.014
13. Paul, V., Jones, M.: Rapid object detection using a boosted cascade of simple features. In: Proceedings of the 2001 IEEE Computer Society Conference on Computer Vision and Pattern Recognition, CVPR 2001, p. 511. IEEE (2003). doi:10.1109/CVPR.2001.990517

14. Lienhart, R., Maydt, J.: An extended set of haar-like features for rapid object detection. In: Proceedings of International Conference on Image Processing, vol. 1, no. 1, pp. I-900–I-903. IEEE (2002). doi:10.1109/ICIP.2002.1038171
15. Zhu, X.T., Xi, W.: Design and implementation of face recognition system based on C++ and OpenCV. Autom. Instrum. **8**, 127–128 (2014)
16. Dalal, N., Triggs, B.: Histograms of oriented gradients for human detection. In: IEEE Computer Society Conference on Computer Vision and Pattern Recognition, CVPR 2005, pp. 886–893. IEEE (2005)
17. Pan, J.Q., Zhuang, Y., Fong, S.: The impact of data normalization on stock market prediction: using SVM and technical indicators. In: Soft Computing in Data Science. Springer, Singapore (2016). doi:10.1007/978-981-10-2777-2_7
18. Vapnik, V.N.: The Nature of Statistical Learning Theory, pp. 988–999. Springer, New York (1995)
19. Zheng, H.J., Zhou, X., Duyan, B.I.: Introduction statistical learning theory and support vector machines. Mod. Electron. Tech. **4**, 59–61 (2003)

Day and Night Image Stitching and Rendering

Jeng-Shyang Pan[1,2(✉)], Lei Liang[2,3], Mao-Hsiung Hung[1], and Yongjun Zhuang[3]

[1] Fujian Provincial Key Lab of Big Data Mining and Applications, Fujian University of Technology, Xueyuan Road University Town, Fuzhou, China
jspan@cc.kuas.edu.tw

[2] Department of Computer Science and Technology, Harbin Institute of Technology, HIT Campus of University Town of Shenzhen, Shenzhen, China

[3] Qihan Research, Qihan Technology Co., Ltd., Houhaibin Road, Nanshan District, Shenzhen, China

Abstract. We present a new method to day and night image stitching and rendering for exploring the space-time continuum within a two-dimensional still photograph. Our method is based on a two-scale decomposition of the input images. Homography matrix is calculated by matching feature points of detail layer pairs, which can avoid impact of illumination. Then detail layers and base layers are stitched together, respectively. We mapped the stitched base layer to a radiance map and then rendered it as time-lapse effect based on human vision system. Compared with previous method, our method is easier to implement and has a larger viewing angle.

Keywords: Local invariant feature matching · Image stitching · Rendering

1 Introduction

For many scenes, whether man-made or natural, day and night have their own beauty, which most people do not want to miss. Therefore, it has definitely a different feeling to see a picture for including both day and night scenes (Fig. 1). Photographer Stephen Wilkes crafts stunning compositions of landscapes as they transit from day to night. But this work takes a long time which depends on how long the whole process from dawn to dark and it is important that the location and angle of the shot cannot be moved in more than ten hours. That is, people cannot leave, has to wait and take hundreds of photos. In addition, after the shooting, he also spent a few months to do post-processing, and finally get the set of works [10]. In this paper, we propose a method which only needs two images which are taken at day time and at night time. The two images are stitched together to achieve the effect of time lapse, and has wide angle view. Because parallax is allowed between the two images, so there are not many

P. Krömer et al. (eds.), *Proceedings of the Fourth Euro-China Conference on Intelligent Data Analysis and Applications*, Advances in Intelligent Systems and Computing 682, DOI 10.1007/978-3-319-68527-4_3

Fig. 1. A picture of the day to night.

restrictions on shooting. After shooting the scene picture of the day, people can do other things, and then come back at night to shoot another one, so do not have to wait in place.

Our method has the advantages of wide angle view and simple implementation, but there are two problems because of the absence of the intermediate state image: First, the image stitching problem is usually modeled by a geometrical relationship between two images, known as homography, which is usually estimated by matching local invariant feature points of two images [1]. However, due to the different shooting time, we need to deal with great illumination change of the image pairs, that is, how to effectively complete the extraction and matching of the local invariant features in the case of large illumination changes. The second problem is how to render the effect of time lapse in the panoramic image in the absence of an intermediate time image.

In this paper, we proposed a method to deal with these two problems based on two-scale computation model [2]. The model decomposes the image into base layer and detail layer based on the fact that the human visual system (HVS) to local contrast is more sensitive than the global contrast. Such a model is much related to the texture illuminance decoupling technique [2], the texture and illuminance information of the image are included in the detail layer and the base layer, respectively. We perform local invariant feature extraction and matching on the detail layer to realize the perfect image stitching, which can solve the first problem. The second problem is solved by mapping the effect of the elapsed time on the base layer.

Figure 2 is the flow chart of the proposed method. Our method starts by decomposing the day and the night image into the base layer and the detail layer respectively, which are modified in different ways. In the detail layer, the feature points of the image pair are extracted and matched, so that the homography matrix is calculated. Then we can register and fuse the detail layer and the base layer. Registration only needs to transform one image and fusion using weighted average method. We render time-lapse effects on the stitching base layer by using reverse and forward tone mapping which is based on the human eye's response to realistic lighting.

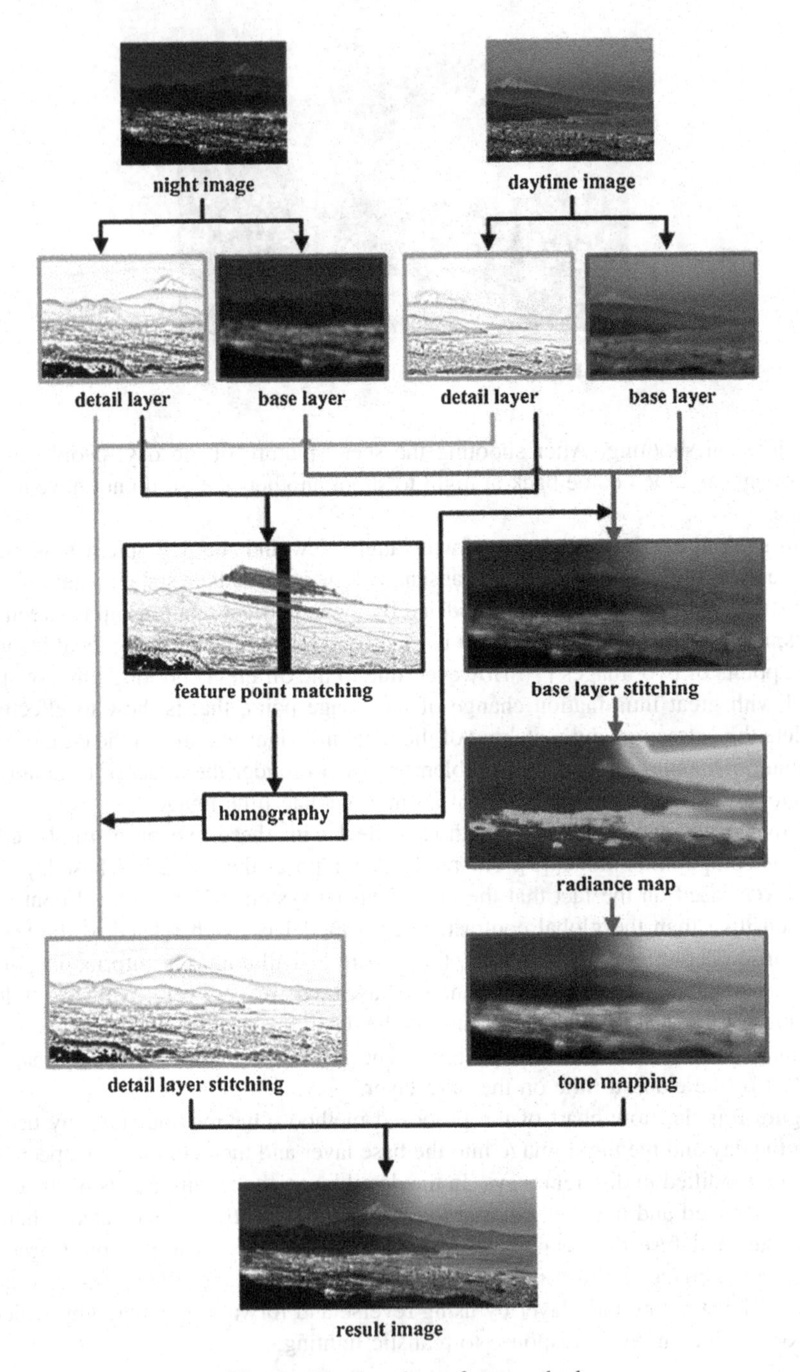

Fig. 2. The flowchart of our method.

2 Image Stitching

In order to calculate the homography matrix of image stitching, we need to match the feature points between image pairs. At present, the main method to extract the feature points of the image is Scale-invariant feature transform (SIFT) [5], which mainly addresses the problem of scale change, but it is not robust enough for large illumination changes, so it is not suitable for day and night image stitching. In the proposed algorithm, we avoid the influence of illumination by using two-scale decomposition model which is similar to Durand's [2], but the base layer is estimated by guided filtering [3] which has obvious advantages in computational efficiency and maintaining gradient near the edge compared with bilateral filtering. Directly using the SIFT algorithm to extract the feature of the detail layer will get fewer matching points, which will calculate the wrong homography matrix, resulting in dislocation stitching, as shown in Fig. 3. This is because the overlap between the image pairs is small and the SIFT detector does not extract many feature points from the detail layer image. In the proposed method, two improvements are obtained: First, in order to ensure the unity of the synthetic image style, there is no large scale change between the two images. So in order to reduce the number of error matching point, we use image pyramid instead of Difference of Gaussians (DoG) in SIFT. Second, blob detector of SIFT is replaced by corner detector [9], which can increase the number of detected feature points in detail layer.

Fig. 3. Detail layer stitching result using SIFT.

3 Time Lapse Rendering

Since the pixel value of the image I is the natural radiance value compressed by the camera's camera curve, we first restore the radiance information of the image L by the inverse camera response function (ICRF) [4].

$$I \xrightarrow{ICRF} L \tag{1}$$

The image that stores the radiance information is also called high dynamic range (HDR) image [7]. Since the HDR image cannot be displayed on the low dynamic range (LDR) medium, so we use pseudo color image representation (Fig. 4).

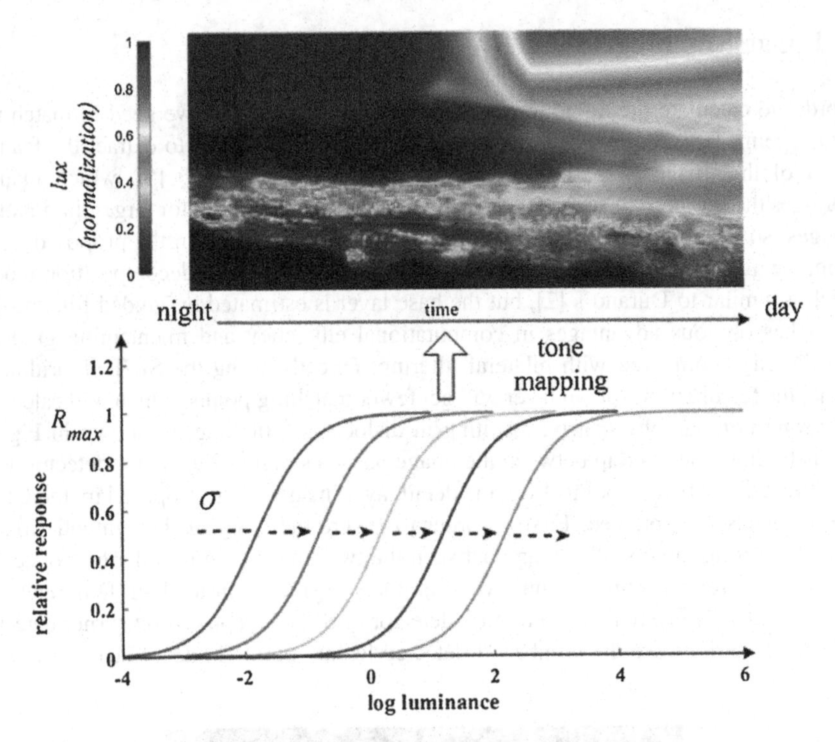

Fig. 4. Time lapse tone mapping for HDR image.

Because the current display media is LDR, we also need to compress the dynamic range of HDR image, and HVS-based mapping curve can make the compressed image more realistic. Relevant physiology study found that human eye compress dynamic range of luminance mainly depends on the photoreceptors of retina [6]. Photoreceptors' adaptive luminance curve is shown in Fig. 4. We can see that with the increase of the external luminance intensity, the response curve of photoreceptors shift right along the *X* axis. The result is that the response curve can cover different dynamic ranges continuously. The photoreceptors' response *R* as function of luminance *L* may be modelled by [6]:

$$R = \frac{L}{L + \sigma} \cdot R_{max} \tag{2}$$

$$\sigma = (k \cdot L_a)^m \tag{3}$$

where k and m are constants. Half-saturation parameter σ determines the degree of photoreceptor response curve shifting to the right along the *X* axis (Fig. 4). σ is a function of the adaptive luminance L_a. In order to render the effect of the passage of time in a day, we use the columns of the input image n to correspond to the time of day t, and let $R_{max} = 1$, we obtained:

$$R(n) = \frac{L}{L + (k \cdot L_a(n))^m} \tag{4}$$

L_a calculation can be divided into global and local algorithms. Global algorithms generally use the logarithm average of the image as the adaptive luminance value. Local algorithms determine the mapping value for a particular pixel as reference to its neighbouring pixel information. In this paper, a semi-global algorithm is proposed by using the following linear model to calculate the corresponding adaptive luminance of each column n.

$$L_a(n) = \alpha \cdot n + \beta \tag{5}$$

Like many tone mapping methods [8], we view the log-average luminance as a useful approximation to the key of the scene. This L_a is computed by:

$$L_a(day/night) = \frac{1}{N} exp(\sum_{x,y} log(\delta + L)) \tag{6}$$

where N is the total number of pixels in the image and δ is a small value to avoid the singularity that occurs if black pixels are present in the image. We would like to map this $L_a(day/night)$ to middle-column of the input day and night images. As shown in Fig. 4, the time variation of the result image is from night to day, and according to the model (5), the following equations are obtained:

$$\begin{cases} L_a(night) = \frac{W}{2} \cdot \alpha + \beta \\ L_a(day) = (2W - o - \frac{W}{2}) \cdot \alpha + \beta \end{cases} \tag{7}$$

where W is the width of each input images, o is the width of the overlapping part of the stitching. The equations are solved:

$$\begin{cases} \alpha = \frac{L_a(night) - L_a(day)}{W-o} \\ \beta = \frac{W(L_a(night) + L_a(day)) - 2oL_a(night)}{2(W-o)} \end{cases} \tag{8}$$

and L_a of each columns n in stitched image is obtained:

$$L_a(n) = \frac{L_a(night) - L_a(day)}{W - o} \cdot n + \frac{W(L_a(night) + L_a(day)) - 2oL_a(night)}{2(W - o)} \tag{9}$$

Substituting $L_a(n)$ into Eq. (4), then we can get the final effect of time-lapse rendering.

4 Experimental Results

The proposed method has two main contributions. Firstly, day and night image stitching is completed based on a two-scale decomposition framework. Image pyramid and corner detector is used for detecting feature points in detail layer, and compared with SIFT, the number of correct matching points obtained by the improved method is increased by about 10 times. Secondly, based on HVS, we propose a tone mapping method that can render the image into time-lapse effect. Figure 6 is the result of the proposed method. As shown in the figure, compared with the image without rendering (Fig. 5), the result image of the proposed method (Fig. 6) is more natural and more realistic from night to day.

Fig. 5. Image stitching without rendering.

Fig. 6. Result of the proposed method.

5 Conclusion

This paper presents a new stitching and rendering method based on two-scale decomposition, in which the stitching method can perfectly combine the day and night images. Compared with the traditional method, the proposed method is easier to implement and has a larger viewing angle. The HVS-based rendering algorithm can make the stitching image more realistic, and show the time-lapse in the final image.

References

1. Brown, M., Lowe, D.G.: Automatic panoramic image stitching using invariant features. Int. J. Comput. Vis. **74**(1), 59–73 (2007)
2. Durand, F., Dorsey, J.: Fast bilateral filtering for the display of high-dynamic-range images. ACM Trans. Graph. **21**(3), 257–266 (2002)

3. He, K., Sun, J., Tang, X.: Guided image filtering. IEEE Trans. Pattern Anal. Mach. Intell. **35** (6), 1397–1409 (2013)
4. Lin, S., Gu, J., Yamazaki, S., Shum, H.Y.: Radiometric calibration from a single image. In: Computer Vision and Pattern Recognition, CVPR, vol. 2, pp. 938–945. IEEE (2004)
5. Lowe, D.G.: Distinctive image features from scale-invariant keypoints. Int. J. Comput. Vis. **60**(2), 91–110 (2004)
6. Reinhard, E., Devlin, K.: Dynamic range reduction inspired by photoreceptor physiology. IEEE Trans. Vis Comput. Graph. **11**(1), 13–24 (2005)
7. Reinhard, E., Ward, G., Pattanaik, S., Debevec, P., Heidrich, W., Myszkowski, K.: High Dynamic Range Imaging - Acquisition, Display, and Image-Based Lighting, 2nd edn. Political Parties and the State, Princeton University Press, Princeton (2010)
8. Reinhard, E., Stark, M., Shirley, P., Ferwerda, J.: Photographic tone reproduction for digital images. ACM Trans. Graph. **21**(3), 267–276 (2002)
9. Rosten, E., Drummond, T.: Machine learning for high-speed corner detection. In: European Conference on Computer Vision, ECCV, pp. 430–443 (2006)
10. Wilkes, S.: Homepage. http://www.stephenwilkes.com/fine-art/day-to-night/. Accessed 24 June 2017

Application of Image Preprocessing in Soil Mesostructure

Suping Zheng[1] and Qingjun Fang[2,3](✉)

[1] College of Engineering, Fujian Jiangxia College, Fuzhou 350108, China
[2] School of Civil Engineering, Fujian University of Technology, Fuzhou 350118, China
89100941@qq.com
[3] Key Laboratory of Geomechanics and Embankment Engineering, Ministry of Education, Nanjing 210098, China

Abstract. Following on the development of soil shear bands, this paper studies the processing method of soil mesostructure images taken by optical microscope. The study shows that: the images fusion method based on wavelet transform can solve the problems caused by depth of field existing in the images of soil mesostructure; the image mosaic based on the module matching is a better method to solve the problems of the limited observation range for the mesostructure; the images of the soil mesostructure processed by combination of the two methods can meet the needs for the studies of the soil mesostructure parameters.

Keywords: Soil mesostructured · Image processing · Module matching · Wavelet transform · Image mosaic

1 Introduction

According to incomplete statistics, the economic loss caused by soil break amounts to 0.5 to 1 billion every year in China. Therefore, soil failure is a problem requiring further studies. A great deal of engineering practice and theoretical research showed that the failure of soil was closely related to the status and variation of its own mesostructure and to a large extent was controlled by them. In fact, various kinds of macroscopic mechanical behavior are the result of the soil mesostructure's variation. Thus, the experimental research and theoretical analysis of the development of shear bands from mesoscopic scale has great scientific and engineering significance. At present, the study of soil mesostructure remains experimental, which observes soil mesoscopic variation while loaded with the help of a magnifying device similar to a microscope. In this experimental approach, one of the most crucial steps is the processing of soil mesostructure images. The difference in observing device and research targets may result in the difference in the images taken and furthermore the difference in processing methods. This paper focuses its study on the processing methods of the soil mesostructure images taken by optical microscope.

P. Krömer et al. (eds.), *Proceedings of the Fourth Euro-China Conference on Intelligent Data Analysis and Applications*, Advances in Intelligent Systems and Computing 682, DOI 10.1007/978-3-319-68527-4_4

2 Image Processing of Soil Mesostructure

In order to understand the internal variation of the shear failure from the mesoscopic scale and the development of shear bands caused by the local deformation, there are two things need addressing regarding the image collection and processing. The soil shear failure is a process which needs a sequence of images to describe and record. Therefore the quality of each image is very important because image of good quality can minimize the interrupt of light stability, test environment and the background noise of the test equipment. Meanwhile, as the observing range is increasing gradually during the developmental process of shear bands, the image mosaic is necessary to ensure the information acquired from the images can reflect the whole process of the development of soil shear bands caused by the local deformation.

Following developmental characteristics of soil shear bands, based on the wavelet transform, this paper offers a suitable method for images fusion when collecting the mesoscopic information. Using the module matching technique, this paper offers an image mosaic method that can enlarge the observing range.

2.1 Images Fusion Based on the Wavelet Transform

When taking the mesostructure images, the depth of field is very clear because of the amplification, which makes the image partly clear and partly vague. This may cause the loss of some information in the images which can greatly affect the extraction of quantization parameters of soil mesostructure. Therefore, image processing is necessary to solve the problems caused by the depth of field.

Images fusion is effective in solving the above-mentioned problem. In this paper, we apply the multi-focus approach to process the images for clear picture in the field. The most common three ways of multi-focus images fusion are: pyramid based [1], discrete cosine transform based [2] and wavelet transform based [3]. Among these, wavelet analysis has the characteristics of good localizing quality and multi-resolution response analysis at both time domain and frequency domain, i.e. it has higher frequency resolution and lower time resolution in low frequency section. This enables the observers to focus on any small part of the analysis object, which serves precisely for the images fusion of the soil mesostructure. Thus, we use the wavelet transform based multi-focus images fusion to solve the problems caused by the depth of field. The following images processing of the soil mesostructure can prove the applicability of wavelet transform in our study.

In Fig. 1, Image 1 to Image 3 are the result of different focuses on the same field and Image 4 is the fusion of the above three images. After the fusion of Image 1 and Image 2, we fuse it with Image 3 and get Image 4. The quality of Image 4 is good and the whole field is very clear, which effectively solves the problems caused by the depth of field. The final result of Image 4 shows that the wavelet transform based multi-focus image fusion can satisfy our needs of image processing in this study. Therefore, we apply it in our processing of the images to ensure the accuracy of data analysis.

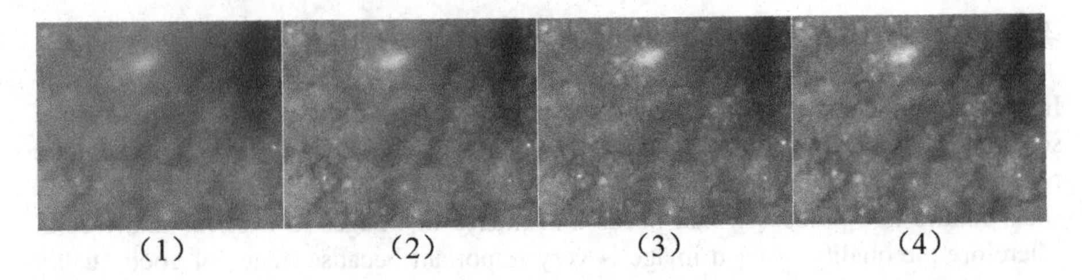

Fig. 1. Images processed based on wavelet transform

2.2 Image Mosaic Based on the Module Matching

The measurement unit of soil mesostructure image in one field is usually mm2. However, in such a small field, the fracture will quickly enlarge beyond the field. Which is harmful to the observation of the development of soil shear bands and would in turn affect the analysis of failure mechanism. An effective solution to the problem is the enlargement of the observing range.

Reducing magnification can directly enlarge the observing range, but it lead to the observation loss of some parts of the mesostructure in the original field. To solve this dilemma, we find another way - image mosaic, which is applied in this study to enlarge the observing range.

Image mosaic is mainly composed of three parts, image preprocessing, image registration and image fusion. The fusion in this part is essentially the same as the image fusion generally mentioned. The only difference lies that we generally fused the clear parts of images of the same observing field with different focus and finally get a clear and complete filed; while the fusion in this part is the fusion of the overlaps between two images of different fields.

In recent years, with the broadening of its application scope, the studies on image mosaic have increased greatly, which focus mainly on the image registration and image fusion. Image registration is a registration and superposition process which uses two or more images for the same scene obtained from different angle, different imaging modalities and different timing [4]. Due to difference in the light and field of vision, the images taken are different which would lead to the appearance of the seam after splicing. Image fusion is effective in erasing the splicing seam [5]. In the following, after the detailed study of image registration and image fusion, we put forward an image mosaic method that can be applied to the studies of soil mesostructure.

Selection of Registration Method. There are three common methods of registration: gray level information based registration, frequency domain based [6] and features based registration [7]. Generally speaking, the last two methods have more superiority over the first one and have wider application. Actually, the selection of registration method is closely related to image acquisition and testing environment. According to the characteristics of image acquisition of soil mesostructure, we find that there are only relation of pure translation existing between these images. For example, in Fig. 2, Image 2 is taken in the central position; Image 1 is taken after upward translation of the camera; Image 3 is taken after the right translation of the camera. It can be found that

Fig. 2. Distribution of mosaic images

there has no rotation or affine among these images. Besides, the quality of these images are greatly improved after wavelet transform based image fusion and the calculation speed of modern computers also increase substantially. Considering the above factors, we intent to adopt the gray level information based registration.

Selection of Matching Module. After the selection of registration method, we need to choose the matching module. The following are the most commonly used modules: module based matching, ratio based matching and grid structure based fast multiple template matching.

The module based matching algorithm is the most mature one. Its principle is relatively simple. First, we need to select a module sized $m \times n$ ('m' and 'n' are the number of pixels) from the overlap area in the first image. Second, we match this selected module with the overlap areas in other images at the same size of $m \times n$ each time. Third, we calculate the error according to certain formula. When the error is minimum, it is thought that the optimal matching region is found. This can further identify the size of the overlap area. Finally, we get the seamlessly tiled images with the help of effective method of images fusion. When the matching module is big enough, the application of this module based matching algorithm would guarantee the accuracy though the amount of calculation is relatively large [8].

Ratio based matching is the optimal algorithm in module matching and grid structure based matching can reduce the amount of computation. However, when considering the accuracy, the module based matching is more suitable for this research, though it has larger amount of computation. With the development of IT industry, the update of hardware can meet the need of computation.

Selection of Registration Formula. After the selection of the matching module, we need to choose the registration formula. The most commonly used formulas include: absolute error formula, function of minimum average absolute difference, maximum registration pixel statistics, minimum mean-squared error function, normalized cross correlation function and so on [9].

We assume that the size of selected module is $m \times n$, $A(x, y)$ is the pixel value for any point in the module and $B(x_1, y_1)$ stands for pixel value of any point in the overlap areas between images. Meanwhile, pixels in the module correspond one by one with that in the overlap areas between images. As shown in Fig. 3, Image 2 is the original image; Image 1 is the image taken after upward translation of the camera; Image 3 is taken after the left translation of the camera. When stitching with Image 1, we need to

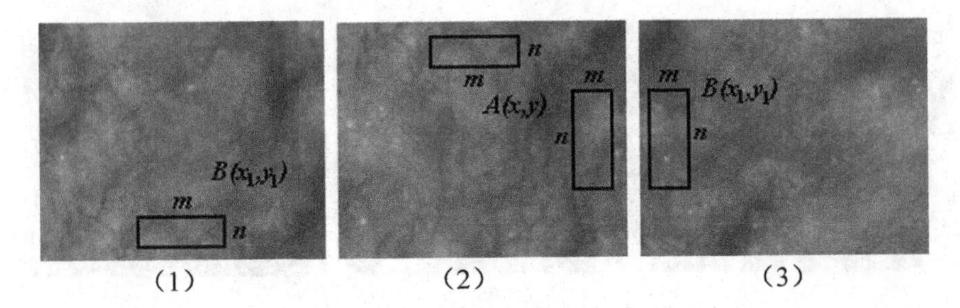

Fig. 3. Design of registration module

select a module sized $m \times n$ like the black rectangle given in the upper part of Image 2 and then search it in the overlap areas between images; finally we select the optimal matching points by a certain registration formula. The image mosaic process between Image 2 and Image 3 is almost the same with that between Image 2 and Image 1; the only difference lies in that the selected module is the black rectangle given in the right of Image 2.

The commonly used formulas are as follows:

Minimum Mean-squared Error Function

$$= \min\left(\frac{1}{m \times n}\sum_{x=1}^{m}\sum_{y=1}^{n}(A(x,y) - B(x_1,y_1))^2\right) \tag{1}$$

Minimum Average Absolute Difference Function

$$= \min\left(\frac{1}{m \times n}\sum_{x=1}^{m}\sum_{y=1}^{n}|((A(x,y) - B(x_1,y_1))|\right) \tag{2}$$

Maximum Registration Pixel Statistics

$$= \max\left(\frac{1}{m \times n}\sum_{x=1}^{m}\sum_{y=1}^{n}T\right) \text{ where } T = \begin{cases} 1 & |A(x,y) - B(x,y)| \leq t \\ 0 & others \end{cases} \tag{3}$$

Normalized Cross Correlation Function ①

$$= \frac{\sum_{x=1}^{m}\sum_{y=1}^{n}A(x,y) \times B(x,y)}{\sqrt{\sum_{m=1}^{m}\sum_{n=1}^{n}[A(x,y)]}\sqrt{\sum_{m=1}^{m}\sum_{n=1}^{n}[B(x_1,y_1)]^2}} \tag{4}$$

Normalized Cross Correlation Function ②

$$= \frac{\sum_{x=1}^{m} \sum_{y=1}^{n} A(x,y) \times B(x,y)}{\sqrt{\sum_{m=1}^{m} \sum_{n=1}^{n} [A(x,y)]^2} \sqrt{\sum_{m=1}^{m} \sum_{n=1}^{n} [B(x_1,y_1)]^2}} \tag{5}$$

From the formulas given above, we can find that the computation by minimum mean-squared error, minimum average absolute difference and maximum registration pixel statistics are relatively simple and clear. However, these three formulas are easily influenced by gray variation, i.e. when one gray value of one image linearly changes, these three formulas are completely useless. Meanwhile, the selection of threshold is also a difficult problem. Different modules, different images and different search windows would require different thresholds. Therefore, choosing these three formulas as registration formulas are not suitable.

As a registration formula, normalized cross correlation function is more accurate and suitable compared with the above three. Besides, it is not affected by the linear variation of gray value. Hereby, we choose normalized cross correlation function as registration formula in this paper. In fact, the formulas of normalized cross correlation function and number operation are identical. In reference to others' research results, we intent to use the following as the registration formula:

Normalized Cross Correlation Function ③

$$= \frac{(\sum_{x=1}^{m} \sum_{y=1}^{n} A(x,y) \times B(x,y))^2}{\sum_{m=1}^{m} \sum_{n=1}^{n} [A(x,y)]^2 \sum_{m=1}^{m} \sum_{n=1}^{n} [B(x_1,y_1)]^2} \tag{6}$$

The third normalized cross correlation function has the advantages of the no. 1 and no. 2 normalized cross correlation functions. Besides, when programming, involution operation is easier than that of development.

Searching Method. When doing the matching module search in the images, scholars have proposed many methods to reduce the registration time and improve the accuracy, such as pyramid search method, genetic algorithm search method and so on. Considering the characteristics of the images in our studies and the simple translations of up and down, left and right, we use direct search method.

If it is up and down translation, the direct search will only search the matching rectangles of same abscissa in the images after the selection of module. Besides, because the overlap area is unknown, the search would start from the bottom of the module which is shown in Fig. 4. The searching method of left and right translation are identical to that of up and down with a slight difference in direction which is shown in Fig. 5.

Images Fusion. Images fusion is one of the most important steps of image mosaic which directly affect the image mosaic. The commonly used images fusion algorithms are: gray-scale-based images, color-space conversion and transformation domain.

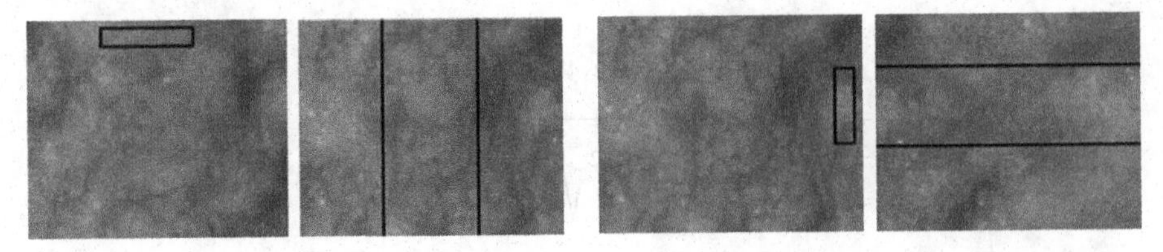

Fig. 4. Up and down search diagram **Fig. 5.** Left and right search diagram

Among these three, gray-scale-based images fusion is most widely used at present. It can be further divided into: weighted average method, fusion of region of interest, Toet method, contrast modulation fusion and so on. Because the images are taken only by the translation of camera, weighted average method has its unique advantages compared with others. Therefore, we adopt the weighted average method in our studies.

In Szeliski's weighted average, each of the corresponding pixel points between two images multiply a weight first; then the addition of these two figures would be taken as the pixel value for the fused image. Assuming $f(x, y)$ is the image after fusion, $f(x_1, y_1)$ $f(x_2, y_2)$ are images to be fused, α_1 and α_2 are weight values, then image after fusion can be represented as follow:

In which, $\alpha_1 + \alpha_2 = 1$, and $0 < \alpha_1 < 1$, $0 < \alpha_2 < 1$. To eliminate the seam after splicing, α_1 is a gradient value, i.e. based on the width of d in the overlap, the value of α_1 begins with 1 and each pixel minuses $1/d$ until equals to 0. Meanwhile, the value of α_2 starts from 0, with increase by degrees to the value of 1.

The program of image mosaic is based on the vb.net and matlab. The size of the module is measured by the coordinate values of two corners. It is not fixed which can be adjusted to the conditions. Besides, the matching module can also be a red rectangle made by the press the right mouse button. Searching range is also flexible which can be entered according to the size of the image. When the image sequences are taken at the same time, camera's up translation and down translation are equidistant. The program is designed as follow: after one image mosaic, it can calculate the number of pixels of the width of the overlap areas for next time's image mosaic. If the numbers of pixels are known, it can be directly used for image mosaic.

The following sequences of images of soil mesosturcture are to be stitched to illustrate the reliability of adopted algorithm and program. In order to analyze the mesostructure of shear band, generally nine images are needed to be stitched. As shown in Fig. 6, there are nine images of soil mesostructure which are to be stitched.

After the confirmation of registration algorithm, module form, registration formula, images fusion method and searching method, the size of the module becomes the most important factor affecting the images fusion. The background of the soil mesostructure image is usually complicate. If the size of the module is too large, the operation speed will be affected; if it is too small, it will cause wrong registration easily. Therefore, the size of the module should be reasonable. The calculation of overlap area during image stitching of image 1 and image 4 in Fig. 6 is shown in Tables 1 and 2. As shown in Tables 1 and 2, the height of module does not have much influence on the calculation result because the overlap area is small. As image mosaic is considered, the search of the whole image is vertically during the calculation. When the module is too small, a

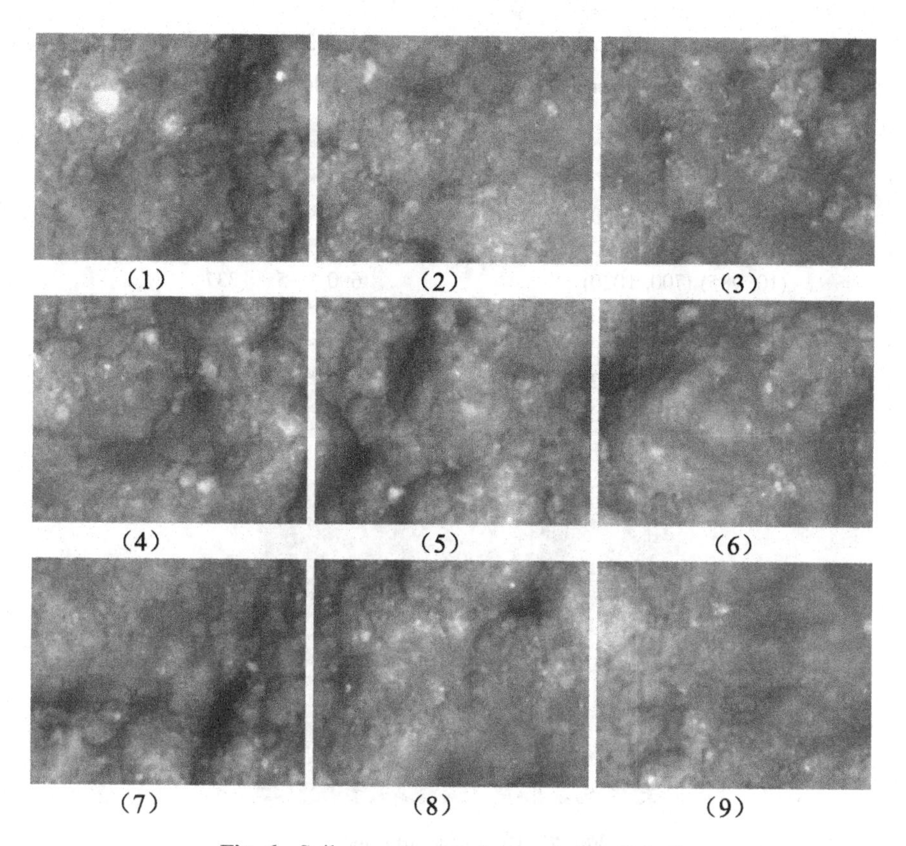

Fig. 6. Soil mesostructure images to be stitched

Table 1. Calculation of overlap area with height of 30 and different width

Number	Coordinates of upper and lower angular point of a module	Module size	Pixels of an overlap area's height
1	(10, 990) (300, 1020)	290 * 30	348
2	(10, 990) (400, 1020)	390 * 30	983
3	(10, 990) (500, 1020)	490 * 30	981
4	(10, 990) (600, 1020)	590 * 30	419
5	(10, 990) (700, 1020)	690 * 30	339
6	(10, 990) (800, 1020)	790 * 30	339
7	(10, 990) (900, 1020)	890 * 30	90
8	(10, 990) (1000, 1020)	990 * 30	90
9	(10, 990) (1100, 1020)	1090 * 30	90

mismatch will easily occur. When the width of module reaches 890 pixels, an accurate match can be acquired. Enlarging the module over 890 will not make the match more accurate. The other image mosaic calculation is basically the same as this.

Table 2. Calculation of overlap area with height of 35 and different width

Number	Coordinates of upper and lower angular point of a module	Module size	Pixels of an overlap area's height
1	(10, 985) (300, 1020)	290 * 35	350
2	(10, 985) (400, 1020)	390 * 35	985
3	(10, 985) (500, 1020)	490 * 35	981
4	(10, 985) (600, 1020)	590 * 35	423
5	(10, 985) (700, 1020)	690 * 35	337
6	(10, 985) (800, 1020)	790 * 35	337
7	(10, 985) (900, 1020)	890 * 35	90
8	(10, 985) (1000, 1020)	990 * 35	90
9	(10, 985) (1100, 1020)	1090 * 35	90

Fig. 7. The image after mosaic

After the determination of the size of module, a proper module is chosen and nine images in Fig. 6 are stitched. The result is shown in Fig. 7. As shown in Fig. 7, the result of image mosaic is ideal, there is no obvious trace of image mosaic and the transition region appears natural and smooth.

3 Conclusions

The conclusions of our studies are as follows: the images fusion based on wavelet transform can solve the problems caused by the depth of field in the soil mesostructure images; the image mosaic based on the module matching is a better method to solve the problems of the limited observation range for the mesostructure; furthermore, to select the representative parameters of soil mesostructure, we still need to address the problems of image segmentation, definition and extraction of parameters.

Acknowledgments. This work is supported by the Science and technology project in Fujian Province Education Department (No. JAT160549, No. JA15339), Development Fund of Key Laboratory of Geomechanics and Embankment Engineering, Ministry of Education (No. GH201405), and startup funds of Fujian University of Technology (No. GY-212081).

References

1. Yu, Z.M., Gao, F.: Laplacian pyramid and contrast pyramid based image fusion and their performance comparison. Appl. Res. Comput. **10**, 128–130 (2004). (in Chinese)
2. Li, G., Wang, G., Wang, R., Zhang, L.: Multi-focus image fusion based on automatic focus algorithm. Appl. Res. Comput. **3**, 166–168 (2005). (in Chinese)
3. Wang, H., Jing, Z., Li, J.: Multi-focus image fusion using image black segment. J. Shanghai Jiaotong Univ. **11**, 1743–1746 (2003). (in Chinese)
4. Yang, Z.: Research on image mosaic algorithm based on block matching and key point matching. Southwest Jiaotong University, Cheng Du (2009). (in Chinese)
5. Kuglin, C.D.: The Phase Correlation Image Alignment Method (1975)
6. Li, H., Manjunath, B.S., Mitra, S.K.: A contour-based approach to multisensor image registration. IEEE Trans. Image Process. **4**(3), 320–334 (1995)
7. Schenk, T., Jin-Cheng, L., Toth, C.: Towards an autonomous system for orienting digital stereopairs. Photogram. Eng. Remote Sens. **57**(8), 1057–1064 (1991)
8. Shao, X.: Study on key algorithm of digital image mosaic. JiLin University, Changchun (2010). (in Chinese)
9. Szeliski, R.: Video mosaics for virtual environments. IEEE Comput. Graph. Appl. **16**(2), 22–30 (1996)
10. Li, J., Chen, Z., Huang, X.: CTtriaxial test for collapsibility of undisturbed Q3 loess. Chin. J. Rock Mech. Eng. **29**(6), 1288–1296 (2010). (in Chinese)
11. Yao, Z.H., Chen, Z.H., Zhu, Y.Q., et al.: Meso-strutural change of remolded expansive soils during wetting-drying cycles and trixial soaking tests. Chin. J. Geotech. Eng. **32**(1), 68–76 (2010). (in Chinese)
12. Zhu, G.P., Chen, Z.H., Wei, C.F., et al.: Microstrcture evolution of red clay during wet-dry cycles. J. Water Resour. Archit. Eng. **14**(4), 42–49 (2016). (in Chinese)
13. Fang, X., Shen, C., Chen, Z., et al.: Triaxial wetting tests of intact Q2 loess by computed tomography. Chin. Civil Eng. J. **44**(10), 98–106 (2011). (in Chinese)

Application of Image Proccessing in Soil's Shear Zone Mesostructure

Qingjun Fang[1,2(✉)]

[1] School of Civil Engineering, Fujian University of Technology, Fuzhou 350118, China
89100941@qq.com

[2] Key Laboratory of Geomechanics and Embankment Engineering, Ministry of Education, Nanjing 210098, China

Abstract. Appropriate method of processing the mesostructure image of soil's shear zone is the basis for accurate analysis of the mesostructure parameters of shear zone. In this paper, the segmentation of shear zones, the extraction of mesostructure quantization parameters, the determination of study area of mesostructure variation and the method of tracking observation are studied. The results show that image segmentation technique based on hue separation is suitable for segmentation of soil's shear zone mesostructure image. According to fractal theory, the study area of the shear zone's mesostructure variation can be determined. The tracking points in the study area should be selected outside the shear zone and be matched by digital image correlation method.

Keywords: Image processing · Soil's shear zone · Mesostructure

1 Introduction

With the deepen research of soil mesostructure, scholars found that the processing of soil mesostructure image play an important role in soil study. Suitable image processing method can more truthfully reflect the variation of soil mesostructure while loaded, and more effectively extract mesostructure parameters which are to be analyzed, and further explain the macro performance of soil. Therefore, it is of great significance to study image processing of soil mesostructure. The research on the pretreatment of soil mesostructure images has been discussed in the other paper. In this paper, the mesostructure parameters extraction method of soil's shear zone mesostructure images which are preprocessed by image fusion and image mosaic is studied.

2 Image Segmentation Based on Hue Separation

Image segmentation is one of the most important steps in soil mesostructure parameters analysis. The segmentation results directly affect the quantization analysis results of mesostructure parameters. Therefore, image segmentation is very significant. In article [1], the method of average gray value and the method of maximum variance automatic

P. Krömer et al. (eds.), *Proceedings of the Fourth Euro-China Conference on Intelligent Data Analysis and Applications*, Advances in Intelligent Systems and Computing 682, DOI 10.1007/978-3-319-68527-4_5

threshold are adopted to segment the microstructure images of rock and soil materials, and the results are good. But the images processed in article [1] did not appear large cracks and the pores were relatively well-distributed [1]. In this paper, the variation of shear zone mesostructure is studied. The images contain larger cracks, if the methods provided in the article [1] are adopted, the results could not be guaranteed. In order to find the best method, the suitability of various methods is studied.

2.1 The Method of Average Gray Value and the Method of Maximum Variance Automatic Threshold

The principles of the method of average gray value and the method of maximum variance automatic threshold can be found in article [1]. The applicability of the two methods is discussed in this paper. As shown in Fig. 1, there are two mesostructure images of soil's shear zone. The two images are respectively processed by the method of average gray value and the method of maximum variance automatic threshold, as shown in Figs. 2 and 3. The black part of the picture represents the pores and fractures, and the white part represents the particles. The subsequent sections will share the same representation without special explanation.

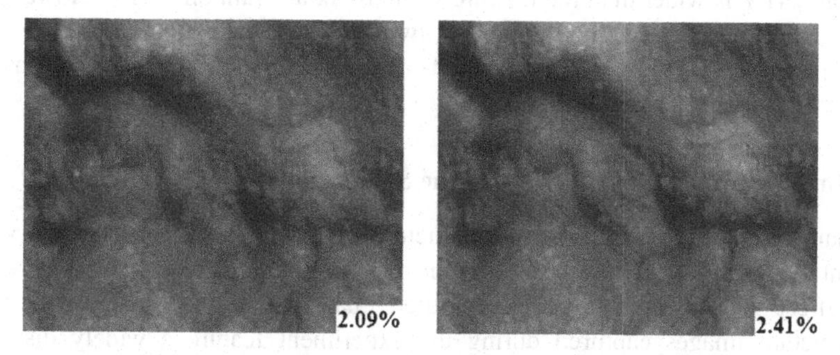

Fig. 1. The mesostructure image of soil's shear zone with axial strain of 2.09% and 2.41%

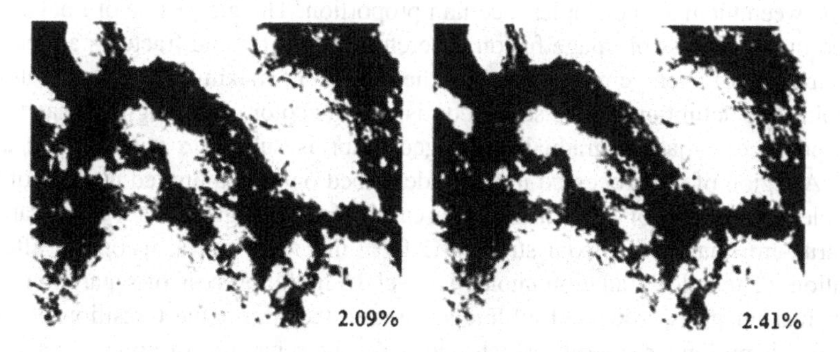

Fig. 2. Image processed by the method of average gray value

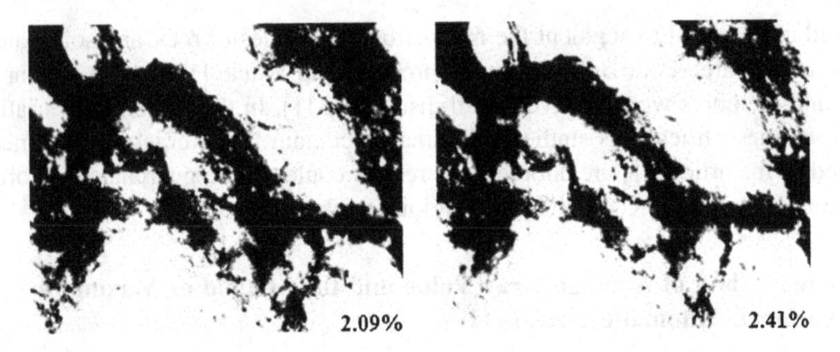

Fig. 3. Image processed by the method of maximum variance automatic thresholding

As shown in Figs. 2 and 3, it can be found that the images processed by the method of average gray value and the method of maximum variance automatic threshold are very similar. Yet compared with the original image, it can be found that the segmentation effect is not ideal. After segmentation, the area of pores and fractures are obviously enlarged, but the segmented image can not reflect the development of fracture. It is easy to observe from the original image that the fracture width of axial strain on 2.41% is wider than the fracture width of axial strain on 2.41%. However, the binary images processed by the above two methods reflect the opposite result. Therefore, these two segmentation methods can not be applied to the segmentation of soil's shear zone mesostructure image.

2.2 Image Segmentation Based on Hue Separation

The main research object is the mesostructure of soil's shear zone, the purpose of segmentation is to reflect the variation of the mesostructure fracture. In order to achieve better effect with segmentation, the captured images should be analyzed first. Soil mesostructure images captured during the experiment feature a widely distributed image gray value. Under this circumstance, there are many obvious differences in gray values between the area of pores and fractures and the area of particles. But the gray value between them also occupies a certain proportion. The gray value of image is also affected in the process of image fusion. Therefore, the pores and fractures are enlarged by the method of average gray value and the method of maximum variance automatic threshold. The definition of hue separation is that the color of an original image which is composed of adjacent gradually changed color is substitute with several abrupt colors. A region of the processed image is described only by a limited number of hues, and a clear histogram of strip distribution can be found. Figure 4 is a histogram of a mesostructure image with axial strain of 2.09% in Fig. 1 and a histogram after hue separation. (The color gradation choice is level 16 in the process of separation.) After the original image is processed by hue separation technique, the transition of middle gray value is no longer smooth, which will make the separatrix of pores, fractures and particles more obvious in soil's mesostructure images. After images are processed by hue separation, the average gray value method or the direct threshold method is used to segment the image, and better results will be achieved. Since the change of color

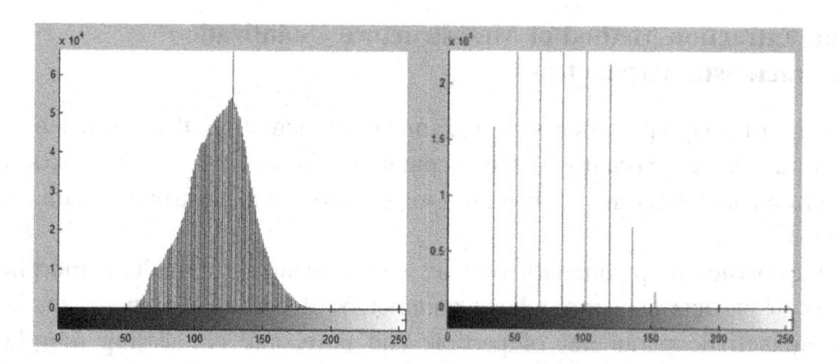

Fig. 4. The histogram of the mesostructure image and the histogram after hue separation with axial strain of 2.09%

gradation value will greatly affect the segmentation effect, it is critical to select the appropriate color gradation value. There is certain difference when the color gradation value is different. After the color gradation value reaches 16, the segmentation is almost the same as the original image processed by the method of average gray value and the method of maximum variance automatic threshold. As easily observed by comparing with the original image, when the color gradation value is 4, the final segmentation result is best and it can best reflect the characters of fracture in the original image.The following images can explain the applicability of image segmentation based on hue separation

As shown in Fig. 5 a sequence of soil mesostructure variation images following the continuous increase of axial strain of soil sample are collected. And Fig. 6 is a sequence of binary images after Fig. 5 is segmented based on hue separation. As shown in Figs. 5 and 6, we can see that the variation of the fracture in the segmented images is consistent with that of the original image, which shows that the method is applicable to the segmentation of soil's shear zone mesostructure image.

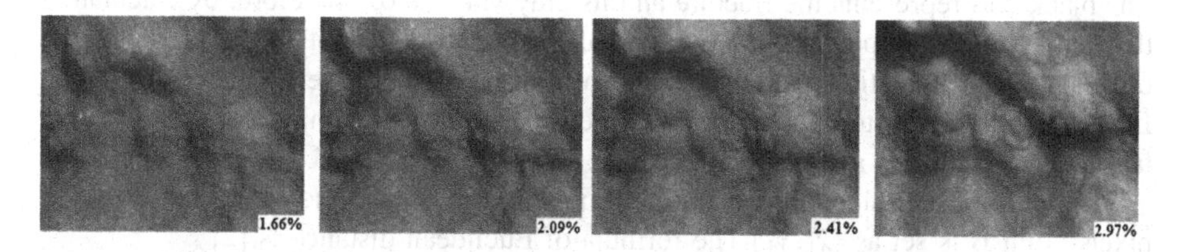

Fig. 5. A sequence of original mesostructure images of soil's shear zone

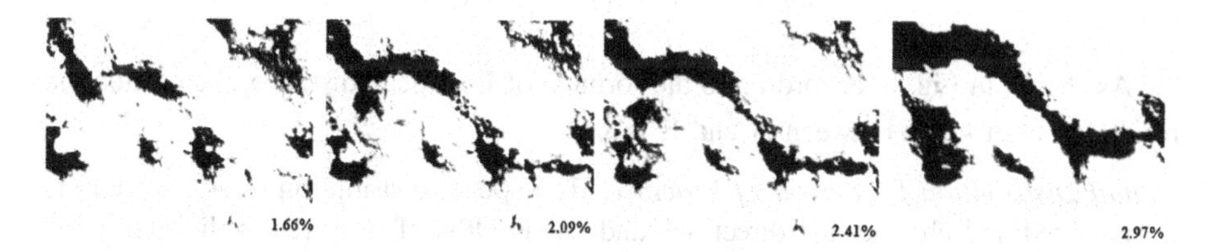

Fig. 6. A sequence of mesostructure images of soil's shear zone after segmentation

2.3 The Extraction Method of Mesostructure Quantization Characteristic Parameters

After a series of image preprocessing and image segmentation, the outlines of particles and fractures have basically been separated. According to the definition of mesostructure parameters the quantization information of mesostructure parameters can be extracted.

The mesostructure parameters outside soil's shear zone include morphological parameters of particle and pore, shape parameters of particle and pore, spatial distribution characteristic parameter of particle and pore, and connecting morphological parameters. The specific parameters of particle mesostructure include particle area, particle perimeter, particle roundness, particle orientation and particle distribution fractal dimension. The specific parameters of pore mesostructure parameters are similar to that of particles. The morphological parameters can be expressed in terms of Euler numbers. The definition and extraction methods of the structural parameters which is discussed above have been described in detail in article [3]. In this paper, the extraction of quantization information of mesostructure parameters in shear zone is studied, and the problems related to the selection of research areas are discussed.

Quantization Extraction of Characteristic Parameters of Soil's Shear Zone Mesostructure

Calculation of Fracture Area and Perimeter. Area and perimeter can be used to characterize the size of the area. The statistical method can be used to calculate the area in the image. When analyzing the images of shear zones mesostructure, the main research problem is the variation of fracture. A universal image of soil's shear zone mesostructure is segmented by image segmentation technique based on hue separation. And then finish the image binaryzation, as shown in Fig. 7.

It is relatively easy to calculate the fracture area in the shear band by applying statistical method, which is to calculate the number of pixels of the same gray value in the region. As shown in Fig. 7, there are only two kinds of gray value after processing. The black part represents the fracture and its gray value is 0. Therefore, by calculating the total number of pixels with a gray value of 0, the area of the fracture can be obtained. The calculation of the perimeter of the shear band is relatively complicated. First of all, we need to choose the measurement of distance which is Euclidean distance. The formula is as follows.

The coordinate value of pixels point A is set as (x, y) and the coordinate value of pixels point B is set as (x_1, y_1).The formula of Euclidean distance is [2]

$$D_O = \sqrt[\frac{1}{2}]{(x - x_1)^2 + (y - y_1)^2}.$$

As shown in Fig. 8, according to the formula of Euclidean distance, the calculation result of the distance between A and B is $\sqrt{2}$.

Spatial Distribution Parameter of Fracture. The spatial distribution of soil fracture is complicated and chaotic, the directions and bandwidths of fractures in different parts are different. Therefore, the classical geometry theory can not describe them. The

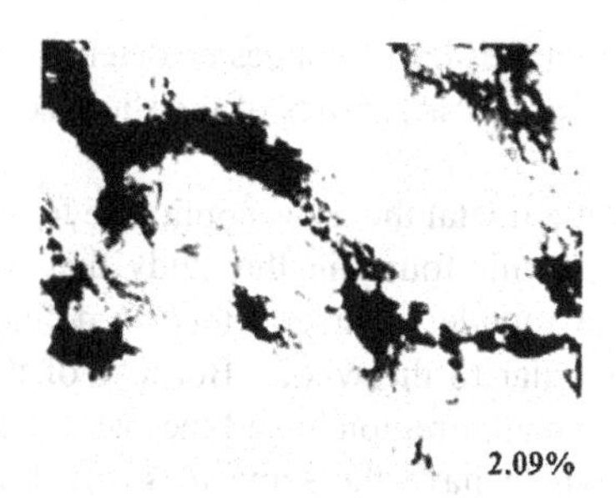

Fig. 7. Local meso-morphology

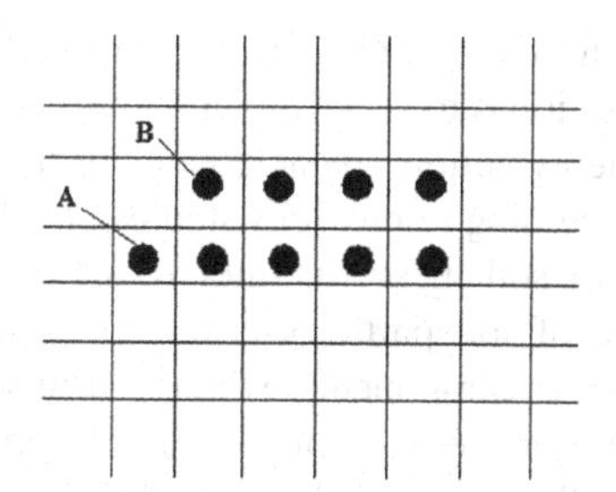

Fig. 8. Schematic diagram of fracture of shear zone perimeter

fractal theory created by Mandelbort has its unique advantages in the study of irregular, unstable and discontinuous phenomena in nature. The theory has been widely applied to many fields. Many scholars also applied the theory to the study of geotechnical engineering in order to study the fractal characteristics of fracture in geotechnical engineering. The results show that the fractal theory can better describe the complex spatial distribution characteristics of fracture.

For quantization description of things and phenomena with fractal characteristics, we must understand the concept of dimension. Generally dimensions has several common definitions, such as capacity dimension, Hansdorff dimension, spectral dimension, topological dimension, similarity dimension and box counting dimension, etc. The definition of dimension is not universal; some definitions of dimension are meaningless in certain situations. The definition of dimension should be chosen according to research objects. In this paper, the box dimension method is used to calculate the dimension of fracture. A soil's shear zone mesostructure image which contains fracture is divided into squares with a side length of r1, r2, r3…etc. The total number of squares occupied by fracture after each dividing was calculated respectively. The total number of squares is represented by N. According to previous studies, the relationship between R and N can be represented by $N - r^{-K}$. In the formula, K is the fracture dimension. According to statistical results, the relation between N and R is plotted in the double logarithmic coordinate system, and then the $\lg N(r) - \lg(r)$ curve can be obtained. The curve is linear, and its slope is K. Theoretically, more r value means smaller step size, and the result is more accurate. Because of the limitation of pixel values, we choose integer-valued r in the fractal dimension calculation of the image.

The Determination of Soil's Mesostructure Study Area. In order to make the selected characteristic parameters accurately reflect the information of the shear zone during the whole process of initiation, evolution and failure because of soil local deformation. The selection of study areas will influence the value of characteristic parameters. The purpose of this paper is to study the mesostructure of shear zone during the process of initiation and the variation of shear zone mesostructure during the process of evolution. Not all parts of image are useful after stitching. Therefore, the image needs to be clipped. It is a primary problem that the size of images be determined because it has to show the initiation region's structural characteristics of the shear zone and the evolution characteristics of the shear zone.

According to previous studies, it is known that the fractal theory is applicable to the study of soil mesostructure. As early as in 1985, Avnir found in the study that the surface of soil particles is fractal. Mandelbrot, the founder of fractal theory, defines fractals as components, which are in some way similar to the whole. Because of the fractal characteristics of soil particles, if we select a smaller region in soil mesostructure image, then expand the area slowly, two stages which have the same mesostructure parameters will appear. The selected smaller areas can be considered as the basic form that is similar to the whole. In this paper, the small area in the two stages is used as the study area for the initiation of shear zone, while the large area is used as an study area for the evolution of shear zone.

Tracking Observation Method. When the size of the study area is determined, it is necessary to consider how to track the study area. Among the existing techniques, point tracking is mostly applied [4]. If a random point in the observation region is selected to be tracked, automatic tracking can be achieved, but if the point selected is located in the shear zone, it is impossible to track the full test process. For example, If you select point A in Fig. 9, we can see from a sequence of the soil mesostructure image that point A is just on the shear zone, and when the strain of sample is 2.09%, the point no longer exists, which means the tracking cannot be achieved. Considering the above problem, in this paper, each time the image is taken, a sequence of soil mesostructure image is obtained by referring to the previous image. Then, a point outside the shear zone is selected to track according to the characteristics of the sequence images. As shown in Fig. 9, point B is chosen based on two considerations. First, the point is in the specified region and it can explain the variation of shear zone; second, the point is located outside the shear zone.

Since the sequence of image has been acquired, a smaller search region can be selected when the strain of sample is 2.09% for tracking matching, such as the rectangular region in Fig. 9. Then search matching is performed according to the digital

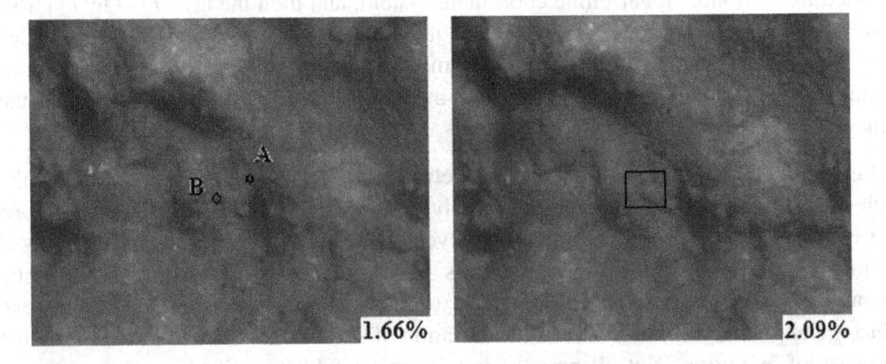

Fig. 9. Schematic diagram of tracking point

correlation method. It should be noted that the matching points should be selected from the last captured image, that is, the matching template needs to be dynamic.

3 Conclusion

The image segmentation technique based on hue separation is suitable for segmentation of soil's shear zone mesostructure image. According to the fractal theory, the study area of shear zone's mesostructure variation can be determined. The tracking points in the study area should be selected outside the shear zone and be matched by digital image correlation method.

Acknowledgments. This work is supported by the Development Fund of Key Laboratory of Geomechanics and Embankment Engineering, Ministry of Education (No. GH201405), Science and technology project in Fujian Province Education Department (No. JA15339, No. JAT 160549), and startup funds of Fujian University of Technology (No. GY-212081).

References

1. Fricke-Begemann, T., Gülker, G., Hinsch, K.D., et al.: Mural inspection by vibration measurements with TV-holography. Opt. Lasers Eng. **32**(6), 537–548 (2000)
2. Gonzalez, R.C.: Digital Image Processing, 3rd edn.
3. Hu, X.: The study of mechanism of microstructure and damage model of structural soil localization deformation. HoHai University, NanJing (2008). (in Chinese)
4. Wang, W.: Research and application of tracking test method for rock and soil microstructure deformation. Southeast University (2007). (in Chinese)
5. Xie, H., Yu, G., Yang, L., Zhang, Y.: Research on the fractal effects of crack network in overburden rock stratum. Chin. J. Rock Mech. Eng. **2**, 29–33 (1999). (in Chinese)
6. Dal Ferro, N., Charrier, P., Morari, F.: Dual-scale micro-CT assessment of soil structure in a long-term fertilization experiment. Geoderma **204–205**, 84–93 (2013)
7. Liu, Z., Liu, F., Ma, F., Wang, M., Bai, X., Zheng, Y., et al.: Collapsibility, composition, and microstructure of loess in China. **686**(October2015), pp. 673–686 (2016)
8. Deng, J., Wang, L., Zhang, Z., Bing, H.: Microstructure characteristics and forming environment of late quaternary period loess in the loess plateau of China. Environ. Earth Sci. **59**(8), 1807–1817 (2009)
9. Fang, X., Shen, C., Chen, Z., et al.: Triaxial wetting tests of intact Q2 loess by computed tomography. China Civ. Eng. J. **44**(10), 98–106 (2011). (in Chinese)
10. Li, J., Chen, Z., Huang, X.: CTtriaxial test for collapsibility of undisturbed Q3 loess. Chin. J. Rock Mech. Eng. **29**(6), 1288–1296 (2010). (in Chinese)
11. Yao, Z.H., Chen, Z.H., Zhu, Y.Q., et al.: Meso-strutural change of remolded expansive soils during wetting-drying cycles and trixial soaking tests. Chin. J. Geotech. Eng. **32**(1), 68–76 (2010). (in Chinese)
12. Zhu, G.P., Chen, Z.H., Wei, C.F., et al.: Microstrcture evolution of red clay during wet-dry cycles. J. Water Resour. Archit. Eng. **14**(4), 42–49 (2016). (in Chinese)
13. Lloret, A., Villar, M.V., Sanchez, M., et al.: Mechanical behaviour of heavily compacted bentonite under high suction changes. Geotechnique **53**(1), 27–40 (2003)

14. Penumadu, D., Hazen, I.: Geo material characterization using digital micrograph analysis. In: Proceedings of 2nd International Conference on Imaging Technologies: Techniques and Applications in Civil Engineering (1997)
15. Luo, D., Leng, X., Xue, X.: Recent studies in geotechnical image analysis. In: Proceedings 2nd International Conference on Imaging Technologies: Techniques and Applications in Civil Engineering, pp. 56–65 (1997)
16. Stock, S.R., Naik, N.K., Wilkinson, A.P., et al.: X-ray microtomography (microCT) of the progression of sulfate attack of cement paste. Cem. Concr. Res. **32**(10), 1673–1675 (2002)

Protocols for Data Security and Processing

An Improved Scheme of Secure Access and Detection of Cloud Front-End Device

Xiao-Bao Yang[1(✉)], Yan-Ping Chen[2], and Yue-Lei Xiao[1]

[1] Institute of IOT and IT-Based Industrialization,
Xi'an University of Posts and Telecommunications, Xi'an 710061, China
y78h11b09@xupt.edu.cn, xiao_yuelei@163.com
[2] School of Computer, Xi'an University of Posts and Telecommunications,
Xi'an 710121, China
chenyp@xupt.edu.cn

Abstract. Security of accessing cloud services is very crucial problem for front-end devices in network. In the research literature, the typical methods aim for certificates and key mutual authentication of devices. However, in this paper, we propose a new efficient design scheme, the key idea of the scheme is to adopt the elliptic curve cryptography (ECC) algorithm for authentication, combined with attributes information of front-end device using smart card, and use the high security Advanced Encryption Standard (AES) algorithm to encrypt data instead of the conventional DES and 3DES algorithms. Especially, in the process of data transmission, the authentication server regularly detects the legitimacy identifier of access devices and synchronously update the share key of session to resist the key hijacking crack. Thus, the front-end device with the secure modular of smart card not only becomes trusted, but also the device's information and data are well protected in the accessing cloud network.

Keywords: Cloud service · Smart card · Authentication · Device attribute · ECC · AES

1 Introduction

The development of microelectronics' technology and network technology has caused turmoil in the study of the Internet of Things (IOT) [1], the sensing layer [1], and cloud computing [2]. This development changes the interactive mode among governments, enterprises, and individuals. However, due to the wide variety of front-end devices, the lack of effective authenticating methods, appropriate supervision mechanisms, Viruses, Trojans, and malicious intrusion attacks spread within the network. This serious situation creates a negative impact on the legality of users and their terminal operations; thus, the popularization of cloud computing [3] is hindered. To improve the safety access and the supervision mechanism of the networks' front-end devices is now a topic of importance and high priority within the cloud service network.

In reference to the safety problem and accessing efficiency, many enterprises, organizations, and researchers have put forward many various solutions. For example, the Cisco System proposed a technology named Network Admission Control

P. Krömer et al. (eds.), *Proceedings of the Fourth Euro-China Conference on Intelligent Data Analysis and Applications*, Advances in Intelligent Systems and Computing 682, DOI 10.1007/978-3-319-68527-4_6

(NAC) [4] for the users. Microsoft developed a Network Access Protection (NAP) [5] technology. The Trusted Computing Group proposed a Trusted Network Connection (TNC) [6] technology. In these technologies, the authentication method of IEEE 802.1X is used by the users and the network of cloud computing, but is not included with device access. Moreover, Ref. [7] puts forward an access method based on the password and challenge/response for mobile cloud devices: The authentication server issues a challenge and waits for the access device to communicate with other devices in the system; thus, the device accepts if it passes the authentication of the website or application. This soft protection has low authentication efficiency. Reference [8] proposes a simple authentication framework that works for a specific class device and authorization mechanism. The device refers to various kinds of information equipment (not computer) that are kept in a ready state in the network that causes many limitations. For example, the authentication of digital broadcast television to its terminal is realized by using Wired Common Access Card (CAC) [9], and it only guarantees the credibility of the initial access terminal but does not guarantee the durable consistent of the device. Reference [10] develops a Trusted Network Equipment Access Authentication Protocol (TNEAAP), which uses the BAN logic system to prove that TNEAAP is secure and credible. However, the schemes or methods mentioned above are either difficult to realize by theoretical research, or are unable to adapt to the diversity of cloud front end device access.

This paper proposes a scheme that has higher security and less limitation to ensure the front-end device's initial access and the durable credibility, which has four mainly contributions: (1) The key idea of our scheme is to embed the smart card in the front-end device as a security-modular, which supplies with an intelligent hard encryption function using ECC algorithm and AES algorithm. (2) The attributes information of the device and the keys created for authentication and session are saved security into the smart card, the device is guaranteed by ECC algorithm while raw data is encrypted by AES algorithm. (3) The authentication server periodically detects the valid identity of front-end device and updated the share key of the session in the process of data transmission. (4) The transmission security of cipher-text data using AES algorithm is higher than that of using DES and 3DES algorithms.

The construction of this paper is organized as follows: the concept and framework of cloud service, the front-end device accession to cloud service, and the existing security problems are explained and analyzed in Sect. 2. Section 3 describes the improved scheme using smart card in detail, including of identifier generation, the process of accessing session, and the authentication using ECC algorithm and regular detection. Section 4 mainly describes the cipher-text session with AES algorithm. Section 5 is analyzed the security of improved scheme. The paper is concluded in Sect. 6.

2 Access of Device to Cloud Service

In this section, the basic concept and framework of cloud service, the front-end device accession to cloud service, and the existing security problems are explained and analyzed.

2.1 Concept and Framework of Cloud Service

Cloud computing is the further development and application of distributed processing, parallel processing, grid computing, and virtual network storage. The data collected through peripheral devices is distributed in a large scale cluster after completing the network calculation. In cloud service, the enterprise or the individual can access the data, realize the application, and form its unitary construction and management to the information service. The tenant and its customer do not need to construct and maintain the hardware environment; they maintain the traditional mode of service after paying the inexpensive fee that is proportional to the data's quality. The function of cloud service is rich in content and user friendly. Cloud service [3, 11] is divided into three groups: (1) Infrastructure as a Service (IaaS); (2) Platform as a Service (PaaS); (3) and Software as a Service (SaaS). The cloud service system includes front-end devices, sensor and internet networks, virtual servers, resource pools, and applications and services. The typical framework of a cloud service system is shown in Fig. 1.

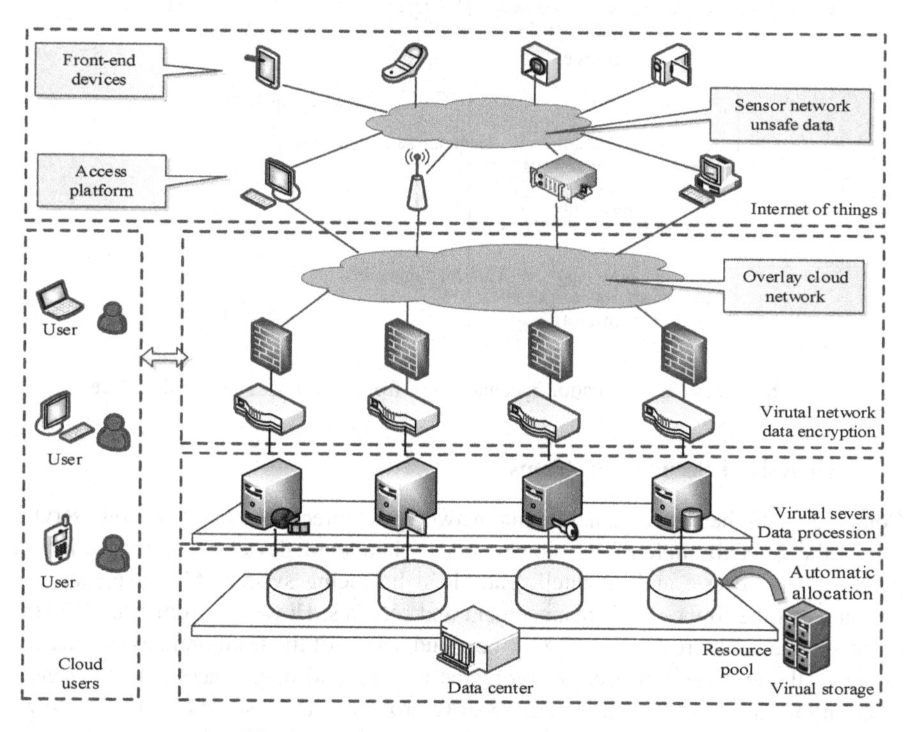

Fig. 1. Typical framework of cloud service

2.2 Front-End Device Access Cloud Service

The network of cloud service includes three layers: (1) The sensor layer; (2) the cloud computing layer; and (3) the application layer. Communication, which is based on the sensing layer between the front-end and the access platform of cloud service, includes device registration, access authentication, and data communication. The access authentication equipment includes the front-end devices, the accessing control devices, and the certificating authenticated servers. The specific process is shown in Fig. 2: The front-end device sends its identifier or self-contained information certificate to the access device of the cloud service. The access device will then return a query result or forward the digital certificate to the authentication server to verify whether the query result is true or false. If the result is true, then access is allowed; if false, then access is denied [12].

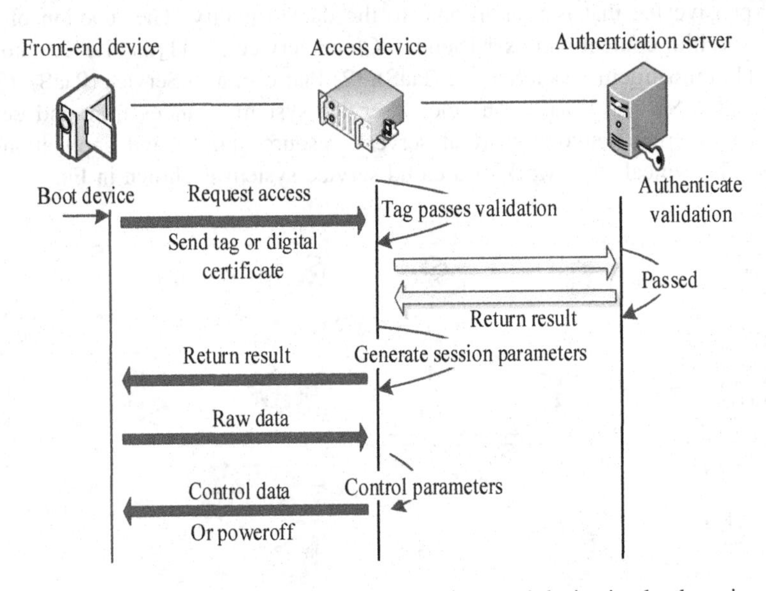

Fig. 2. Access authentication mechanism of front-end device in cloud-service

2.3 Analysis of Security Problems

Compared with the conventional private network, the three advantages of cloud service are in the breakthrough of the constraints of hardware devices; the realization of portable data access; and an intelligence load balancing system. Moreover, another advantage is the low costs of management and services. However, from Sect. 2.1 (the cloud service structure) and Sect. 2.2 (the cloud access of the front-end device), we are aware of the existence of security problems in front-end device access to the sensor layer and to raw data collection. The security problem can be summarized as follows:

(1) The diversity of the front-end devices and the complications of the network environment: The device for the cloud access is no longer just the traditional desktop computer, i.e., the mobile phone, notebook, IPAD, and other mobile devices are used for cloud access. Therefore, the complexity of the system will increase. The host network is no longer the pure internet; it has evolved into the superposition of a mobile network and a sensor network as well as the traditional

internet. These makes the connection environment complicated and negatively affects cloud system to be safety and security.

(2) The device authentication is too simple: The device in the front-end cloud does not have the security module or have a lower safety measure, e.g., certificate processing is only the query access in the gateway or the access equipment. The device information is in an open state, e.g., the camera of the tenant, the temperature controller, and the mobile device's identification, model, and power consumption.

(3) The raw data is transmitted as plaintext: The default transmission mode in most cloud front-end devices is using plaintext to transmit raw data and the state of plaintext data is synchronized with the cloud platform, thus, cannot be effectively protected. Also the attacker can capture the transmission of the device's plaintext information as well as hijack the raw data stream, thus, obtaining the enterprise tenant's private location information which further threatens key information.

(4) The security problems from the characteristics of interconnection and resource limitations of cloud front-end devices: Because of the superposition of device communication and the traditional internet, access possibilities are diverse and the link and bandwidth are not stable. Device constraints of computation power, storage power, and battery power cause failure in traditional encryption methods and access control measures for front-end devices in the cloud environment.

3 Improved the Scheme of Accessing Using the Smart Card

This section describes the improved scheme using smart card in detail, including of the some notations and their meaning using in this paper, the identifier generation, the process of accessing session, and the authentication using ECC algorithm and regular detection.

3.1 Preliminaries

In order to solve these security problems mentioned in Sect. 1.3, we propose a scheme of embedding a smart card as the security modular in the front-end device. Thus, the identifier information, the authentication key, and the connection information of the device are stored in the smart card. When the device accesses the cloud service, the smart card provides the device's identification attributes and is combined with the hash abstract calculation and the ECC algorithm. Thus, the access and routine test verification are accomplished with the cloud tenant authentication server. The raw data is translated into cipher-text to transmit and the key of encryption is regularly updated. Table 1 refers to the notations that are used in the following content.

The smart card in our scheme contains electrically erasable programmable read-only memory (EEPROM), read-only memory (ROM), and random access memory (RAM). The operating system of the smart card is written into ROM with mask. The EEPROM is used to store the identifiers of the front-end devices, the public and private keys for encryption and decryption, and the digital signature keys. The smart

Table 1. The notations and their meaning

Notations	Description
FD_ID	Front-end Device Identifier
SC_N	Smart Card Serial Number
FD_A_ID	Front-end Device Access Identifier
H(x)	The one-way hash function
a‖b	a and b splicing operation
Key	The Session Shared key
EK(M, Key)	The AES Encryption Function
DK(C, Key)	The AES Decryption Function
*	Multiply (times) Operation
[,]	Represents the range
d	The private key
Q	The public key

card has the key algorithm co-processor, which can support many key algorithms, such as AES algorithm, data encryption standard (DES), and 3DES, these algorithms are symmetric. Rivest Shamir Adlemen (RSA) [13] and ECC are asymmetric; and the Hash algorithm [14] does not belong to either of these two groups. In this scheme, the ECC algorithm is used to authenticate the device access and to generate the shared key for encryption and decryption. The AES algorithm is used to encrypt the raw data. The Hash algorithm is used to compute the information abstract.

3.2 Identifier Generation

Although the front-end device of the cloud service varies, they can always be described by their property [15]. Suppose a front-end device has the attributes: A_1, A_2 ... A_n, then the unique identifier of this device can be expressed as:

$$FD\_ID = H(A_1 \| A_2 \| \ldots \| A_n) \tag{1}$$

where n is an integer that is greater than 2. Equation (1) can give a unique identifier to different types of devices in a framework. This unique identifier is stored in the smart card, the access device, and the authentication server. When accessing the cloud service network, a front-end device needs to register in the authentication server of the cloud tenants and is bounded with the embedded smart card in the front-end deice. The identifier of the smart card is its chip serial number SC_N and it connects with the FD_ID to generate the unique identifier:

$$FD\_A\_ID = H(FD\_ID \| SC\_N) \tag{2}$$

And it is stored in the authentication server for future verification.

3.3 Improved Process of Access Session

The processing of the access authentication is to build a safe session of transmitting messages within the smart card of the front-end device, the access device, and the authentication server of cloud tenants. The detail of the access processing is shown in Fig. 3.

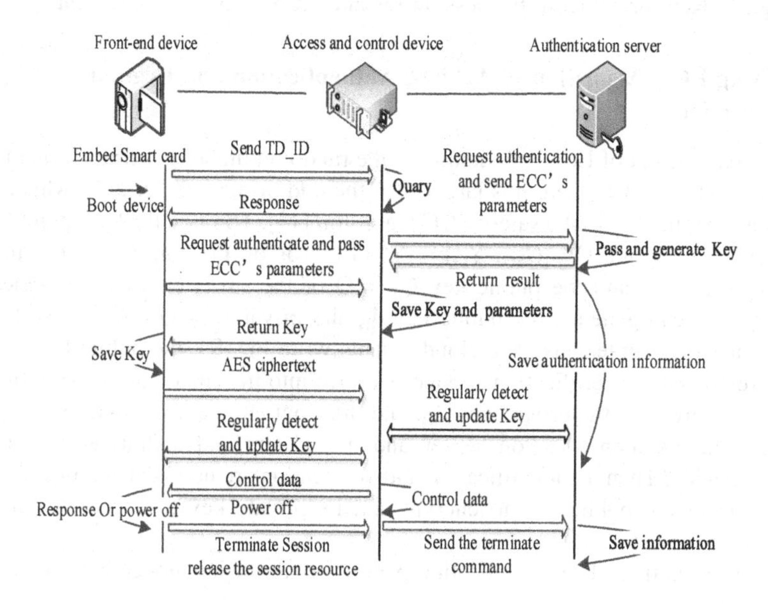

Fig. 3. Process of the access session

(1) Embed the smart card into the front-end device; boot it to send FD_ID to the access and control device of the cloud service.
(2) The access device queries FD_ID. If the result is ok, then respond to the request, otherwise deny.
(3) The device manages the returned result and calls for the stored SC_N. The ECC algorithm is activated as the computing connector FD_A_ID and the generated random number k. They are sent to the access device in cipher-text and forwarded to the authentication server of the cloud tenants to request ECC for verification.
(4) After passing the ECC certification, the authentication server of the cloud tenants calculates the temporary session shared key, Key = H (k), returns to the access device with the verification results, and then records the authentication information and the random number k in cipher-text.
(5) The cloud access device processes and verifies the returned results and saves the temporary key Key = H (k), and the access results will return to the front-end device.
(6) The smart card processes the verification results and saves the temporary session key Key = H (k).

(7) The Key = H (k) is used to perform the encryption and decryption sessions of the raw data and the control data.
(8) The front-end device and the platform device of the cloud access responds to the service termination or to shutdown operation, then notifies the authentication server of the cloud tenants to retreat from the network. The front-end device, the platform device, and the authentication server remove the recording information, e.g., k, Key, and release the session resource to interrupt communication.

3.4 Using ECC Algorithm to Achieve Authentication and Regular Detection

After the parameters of ECC are identified, the random number generator generates an integer $d \in [1, n - 1]$, where n is larger than the odd prime number 2^{191}, which is the certification requirement of a safety ECC algorithm [13]. Through the base point G, one can calculate the point $Q = (X_G, Y_G) = d * G$ that is on the elliptic curve. Then the private key $d_A = d$ and the public key $Q_A = Q$ of the smart card can be generated. Similarly, one can generate the public key d_B and private key Q_B of the certification services of the cloud tenants. The cloud tenants write FD_ID, the public key Q_B, and private key d_A of the certification service platform into the smart card. The public key Q_A of the smart card, the private key d_B, and the front-end device identifier FD_A_ID are stored in the authentication server and it registers and authorizes to bind the front-end device. Then the identification and verification are now realized in cipher-text form by using the public key to encrypt and its private key to decrypt. A detailed description follows:

The smart card sends the verification parameters to the authentication server of the cloud tenants.

(1) Read SC_N and calculate: FD_A_ID = H (TD||SC_N);
(2) Use the random number generator to generate a random number $k \in [1, n - 1]$ and keep k.
(3) Computing: $C_1 = k * G$;
(4) Computing: $C_2 = FD\_A\_ID + k * Q_B$;
5) Send cipher-text point pair C: $\{C_1, C_2\}$;

The authentication server of the cloud service tenants uses its private key d_B to decrypt the authentication parameters:

(1) Computing: $C' = C_2 - d_B * C_1 = FD\_A\_ID + k * Q_B - d_B * (k * G)$;
(2) Take $Q_B = d_B * G$ into C';
(3) The verified parameters:

$$\begin{aligned} C' &= FD\_A\_ID + k * Q_B - d_B * (k * G) \\ &= FD\_A\_ID + k * (d_B * G) - d_B * (k * G) = FD\_A\_ID; \end{aligned}$$

(4) Compare C' with the access identifier of the registration record in the authentication server. If they are the same, the device with the smart card is legal. The

verification information is recorded and the random number k is saved as cipher-text. Otherwise access is denied.

Regular checking means to periodically verify the device's legitimate access. Its certification processing is the same as the raw access certification. However, the random number k', which does not need to regenerate, is obtained as $k' = k + 1$, where k is the first random number. This regular checking mechanism ensures the certainty and continuity of the verification and prevents the attacker from using pseudo devices to do replaying attack [16, 17].

4 Cipher-Text Transmission of AES

Section 4 mainly consists of the basics of AES cryptography, the generation of the shared key in session, the encryption and decryption of data, and the key updated in detail.

4.1 AES Cryptography and the Session Shared Key

After the front-end device passes the authenticated access and connects to the cloud service network, the raw data collected and the control data are transferred as cipher-text. To improve communication efficiency, one can choose AES algorithm [18, 19, 21] that is supported by the smart card. The AES is symmetric cryptography and is also referred to as single key cryptography. In this encryption process the receiver and the sender must agree upon a single secret key, such as the session shared key.

When authentication is valid, the authentication server in cloud service uses a random number k to produce the session shared key Key = H (k) by the Hash algorithm, and a 128 bit length is required. The Key is returned and saved in the smart card and access device.

4.2 Encryption and Decryption

The raw data M as the length of Key is encrypted to produce cipher-text data C = EK (M, Key) and the encryption process needs many rounds. Each round, except for the last, consists of four transformations called ByteSub, ShiftRow, MixColumn, and AddRoundKey [19, 21]. In the last round the transformation MixColumn is omitted. The cipher-text C has also the same length of M, which transfers to the access device of the cloud service. The receiver of the access device performs M = DK(C, Key), and the receiver performs the compression processing in the next step. Decryption, which is the reverse of encryption, uses the same Key as encryption. Figure 4 shows the procession of AES encryption and decryption.

4.3 Updating the Session Shared Key

The session shared key is updated regularly with each authentication accessing of the front-end device. When $k' = k + 1$ changes, the key also changes by Key = H (k'); then the key is protected synchronously by the smart card and the access device of the

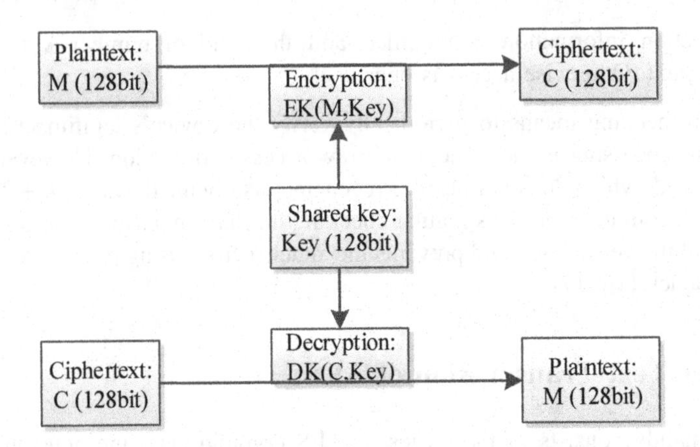

Fig. 4. AES encryption and decryption process

cloud service. It is almost impossible for the attacker to access the information of this session shared key, thus, the data stream transmission is guaranteed to be effective and safe.

5 Analysis of Security and Efficiency

In this section, Security refers to the protection of the front-end device's information, the access device and servers for the cloud service tenants, and the transmitted data among them. Security includes the authenticity, privacy, and integrity of data. Efficiency means that the raw services remain unchanged while security increases. The following is the analysis of the security and efficiency of the scheme.

5.1 Analysis of Security

Compared with the raw scheme mentioned by the Sect. 2, the proposed scheme have the following securities for the front-end devices to access the cloud service.

(1) Smart card as the secure modular is authenticated based on the device's attributes.

The unified identifiers of the device's attributes stored in the help to solve the previous problem of needing to verify multiple access ways, thus, the access environment is simplified. The front-end device and the smart card are bound together for verification in the form of "one-machine; one-card" to ensure that the access device is the authorized device of cloud service tenants.

(2) Prevent privacy leakage and malicious attacks.

In the scheme, the FD_ID is word strings calculated by the Hash algorithm. The attacker must decode it before obtaining the relevant access information of the front-end device. However, it is almost impossible for the attacker to do so because the Hash abstract calculation is irreversible. Also, the raw data cannot be analyzed and cracked by a middleman attacker because the data is encrypted by the regularly updating the session shared key. Furthermore, the front-end device's information, including of FD_ID, FD_A_ID, and public and private keys, are stored in the smart card. The front-end device's access verification and periodic detection mechanism effectively prevent the intervention of a pseudo-device.

(3) Strengthen authentication and cryptographic algorithm.

Accord to [19], compared ECC with RSA under the same security condition, the Table 2 shows the same security level in the different key size of RSA and ECC. In another words, in the network environment of the front-end devices accessing the cloud service, with increasing the key length, ECC security is much higher than RSA in the smart card, e.g., a 240 bit key length of ECC is safer than a 2048 bit RSA.

Table 2. Different key sizes of RSA and ECC

RSA Key Size (in bits)	ECC key size (in bits)
1024	160
2048	224
3072	256
7680	384
15360	512

At the same time, the majority smart card uses the DES or 3DES algorithms to encrypt data. DES [20] only uses a 56 bit key. One bit in each of the 8 octets is used for odd parity on each octet. This is a weakness and allows attacks and other known methods to exploit the DES weaknesses, which makes the DES an insecure block cipher. 3DES [21] is called Triple Data Encryption Algorithm which is a block cipher, and which applies DES cipher three times to each block of data, encryption – decryption – encryption using DES. Yet, AES, known as the Rijndael (pronounced as Rain Doll) algorithm encrypts data blocks of 128 bits using symmetric keys 128, 192, or 256 bits [20, 21], Each round of AES uses permutation and substitution network, and which is suitable for both hardware and software implementation [22]. Thus, AES is introduced to replace DES and 3DES to improve security of information and raw data in our scheme.

5.2 Analysis of Efficiency

In our scheme, the smart card embedding method to ensure the security of the front-end device leads to more verification and response session. This scheme's absolute efficiency is lower than that of the no protection scheme. However, when the smart card scheme is compared to the Digital Certificate and other safely schemes, its security service occupies very little computation and storage resources: This increases the efficiency and security of access certification. Moreover, as a hardware protection, the smart card method is more flexible. Compared with RSA, the smart card's ECC algorithm needs less space for storing the key and has a higher efficiency of encryption and decryption [23]. The AES algorithm requires less processing time for encryption and decryption and performs better than DES, and 3DES on the smart card.

6 Conclusions

The safety access of the front-end device is fundamental to ensure cloud service security. In our scheme, the front-end device with embedded smart card is reliable and safe. It provides first security protection for cloud service by (1) efficiently and securely completing the access authentication; (2) the cipher-text session; and (3) the periodical checking mechanism; thus, protecting the data from its collecting sources. Compared with currently existing schemes, our scheme is safer and more efficient than the current schemes.

In future research, we will design the corresponding access standard and interface as well as to prepare for the realization of the controllability and security of the cloud service front-end device in the Wisdom City. Our scheme needs the smart card as described in this paper and adjustments to the front-end devices, which will cause a small increase in the production cost of the system.

Acknowledgements. This research was supported in part by grants from the National Natural Science Foundation of China (No. 61402367), the Science and Technology Project in Shaanxi Province of China (No. 2016GY-092), and the Project of Education Department of Shaanxi Province (No. 16JK1701). The authors gratefully thank Pro. Xiao-Qiang XI for his warmhearted discussion.

References

1. Zanella, A., Bui, N., Castellani, A., Vangelista, L.: Internet of things for smart cities. IEEE Internet Things J. **1**(1), 22–32 (2014). doi:10.1109/JIOT.2014.2306328
2. Botta, A., Donato, W.D., Persico, V., Pescapé, A.: Integration of cloud computing and internet of things: a survey. Future Gener. Comput. Syst. **56**(C), 684–700 (2016). doi:10.1016/j.future.2015.09.021
3. Kashif, M., Palaniappan, S.: Framework for secure cloud computing. Adv. Int. J. Cloud Comput.: Serv. Archit. (IJCCSA) **3**(2), 21–35 (2013). http://airccse.org/journal/ijccsa/papers/3213ijccsa02.pdf
4. Fong, P.W.L.: Relationship-based access control: protection model and policy language. In: Proceedings of the First ACM Conference on Data and Application Security and Privacy, 191–201. ACM (2011). doi:10.1145/1943513.1943539
5. Hurst, R.M., Manaktala, E.H., Mayfield, P.G., et al.: Network access protection: U.S. Patent 7, 793, 096[P], 2010-9-7
6. Zhang, H., Chen, L., Zhang, L.: Research on trusted network connection. J. Chin. J. Comput. **33**(4), 706–717 (2010). (in Chinese)
7. Alizadeh, M., Abolfazli, S., Zamani, M., et al.: Authentication in mobile cloud computing: a survey. J. Netw. Comput. Appl. **61**(3), 59–80 (2016). doi:10.1016/j.jnca.2015.10.005
8. Hirano, M., Okuda, T., Yamaguchi, S.: Application for a simple device authentication framework: device authentication middleware using novel smart card software. In:

International Symposium on Applications and the Internet Workshops, SAINT Workshops 2007, p. 31. IEEE, Hiroshima (2007). doi:10.1109/SAINT-W.2007.26
9. Chen, X.H.: The study of multi-business authentication access is based on the technology of DHCPv6. J. Cable Technol. **6**, 28–30 (2014)
10. Lai, Y.X., Chen, Y.N., Zou, Q.C., et al.: Design and analysis on trusted network equipment access authentication protocol. Simul. Model. Pract. Theory **51**, 157–169 (2015). doi:10.1016/j.simpat.2014.10.011
11. Balkhi, M.: A view of cloud computing. Int. J. Innov. Sci. Res. **4**(1), 54–60 (2014)
12. Mohamed, A., Grundy, J., Müller I.: An analysis of the cloud computing security problem. arXiv preprint arXiv:1609.01107 (2016)
13. Rivest, R.L., Shamir, A., Adleman, L.: A method for obtaining digital signatures and public-key cryptosystems. Commun. ACM **21**(2), 120–126 (1978). doi:10.1145/359340.359342
14. Liu, Y., Xiong, R., Chu, J.: Quick attribute reduction algorithm with hash. Chin. J. Comput. **32**(8), 1493–1499 (2009). doi:10.3724/SP.J.1016.2009.01493
15. Patel, V., Patel, R.: Improving the security of SSO in distributed computer network using digital certificate and one time password (OTP). Int. J. Comput. Appl. **89**(4), 10–14 (2014). doi:10.5120/15489-4227
16. Liu, C.-W., Tsai, C.-Y., Hwang, M.-S.: Cryptanalysis of an efficient and secure smart card based password authentication scheme. In: Recent Developments in Intelligent Systems and Interactive Applications, pp. 188–193. Springer, Cham (2017). doi:10.1007/978-3-319-49568-2_26
17. Dua, G., Gautam, N., Sharma, D., et al.: Replay attack prevention in kerberos authentication protocol using triple password. **5**(2), 449–457 (2013). doi:10.5121/ijcnc.2013.5205
18. Kandala, S., Sandhu, R., Bhamidipati, V.: An attribute based framework for risk-adaptive access control models. In: 2011 Sixth International Conference on Availability, Reliability and Security (ARES), pp. 236–241. IEEE (2011). doi:10.1109/ARES.2011.41
19. Olufunso, D.A., Kayode, A.B., Adebayo, A.O.: Secured cloud application platform using elliptic curve cryptography. In: Proceedings of the World Congress on Engineering and Computer Science, vol. 1 (2016)
20. Zhao, K.X., Cui, J., Xie, Z.Q.: Algebraic cryptanalysis scheme of AES-256 using Gröbner basis. J. Electr. Comput. Eng. **2017** (2017). doi:10.1155/2017/9828967
21. Patil, P., Narayankar, P., Narayan, D.G., et al.: A comprehensive evaluation of cryptographic algorithms: DES, 3DES, AES, RSA and Blowfish. Proc. Comput. Sci. **78**, 617–624 (2016). doi:10.1016/j.procs.2016.02.108
22. Li, M.G.: Study on public key infrastructure in support of public key cryptographic algorithm SM2 based on elliptic curves. Inf. Secur. Commun. Priv. **9**, 78–80 (2011)
23. Jeeva, A.L., Palanisamy, D.V., Kanagaram, K.: Comparative analysis of performance efficiency and security measures of some encryption algorithms. Int. J. Eng. Res. Appl. (IJERA) **2**(3), 3033–3037 (2012)

The Order and Failure Estimation of Redundancy System Based on Cobweb Model

Chao-Fan Xie[1], Lin Xu[2](✉), Lu-Xiong Xu[3], and Fuquan Zhang[4,5]

[1] Network Center, Fuqing Branch of Fujian Normal University, Fuqing, Fuzhou, Fujian, China
119396356@qq.com

[2] Economic Institute, Fujian Normal University, Fuqing, Fuzhou, Fujian, China
xulin@fjnu.edu.com

[3] Electronic Information and Engineering Institute, Fuqing Branch of Fujian Normal University, Fuqing, Fuzhou, Fujian, China
xlx123456@139.com

[4] Fujian Provincial Key Laboratory of Information Processing and Intelligent Control, Minjiang University, Fuzhou 350121, China
8528750@qq.com

[5] School of Software, Beijing Institute of Technology, Beijing 100081, China

Abstract. In order to improve the reliability and life of components, product providers often take the form of redundant backup. The lower the failure rate of the product itself and the more the number of redundant backups, the more reliable and stable system is, but the number of redundant backups will greatly increase the cost of the product. Moreover, the product can not be detected through dismantling products to achieve, This provides the provider with the possibility of speculation, so the provider will provide a lower degree of redundancy of products or secondary products with high-level redundancy is a huge problem faced by buyers. The study will use the method of truncation estimation to make a asymptotic estimation, provide statistic reference formula and feasible calculation method for the purchaser, which can greatly reduce the cost of the demand and the reliability of the system operation.

Keywords: Reliability · Redundant backup · Truncation estimation

1 Introduction

Reliability theory is a discipline that can analyze characterize the products specified functions probability of occurrence of random events, which Is the sixties of last century developed new interdisciplinary subject. Thus, the reliability theory is based on probability theory, the first based on field research is machine maintenance problem [1]. At present, the main study of the reliability is the system reliability indices, as well as the reliability of the index on the basis of the optimal detection time to avoid faults, reduce the losses caused by the fault, such as the literature [2–4]. Research which has done basically is qualitative analysis or uses numerical analysis to find the

P. Krömer et al. (eds.), *Proceedings of the Fourth Euro-China Conference on Intelligent Data Analysis and Applications*, Advances in Intelligent Systems and Computing 682, DOI 10.1007/978-3-319-68527-4_7

approximation value of solution in actual system [5–7]. Fan et al. proposed numerical simulation algorithm for solving reliability problems which belongs to intelligent algorithm [7–10]. So all their researches were focusing on engineering method.

In N element exponentially distributed parallel system, Xie Chaofan and Xu Luxiong has been analyzed reliability and extreme value, obtained some relevant theoretical guidance. Without considering the economic constraints, the failure rate is equal to all of the components, the system reliability reaches a minimum, in considering the economic constraints, and when the failure rate is unit elastic, the conclusion is the same. And if you choose a good product in parallel with a poor product, the product reliability are more higher than parallel on two moderate of the failure rate. During operation, according to the actual situation, if the failure rate in the envelope line, then the whole system components should be replaced altogether, but if the failure rate is far away from the envelope, then for economic performance considerations, just to update the highest failure rate of several originals out [11]. But in this paper, it only consider parallel connection mode, it doesn't consider redundant backup connection mode. In research of redundancy's system, the reliability of the distribution of redundancy system is always greater than the parallel systems'. And the average lifetime difference of the two systems increases gradually with the increase of the number of parallel systems. The average lifetime difference of the two systems Δs_n, s order is $O(n)$. When you make economic analysis of two systems, In cases that without consideration of economic costs, redundancy system's economic benefits is definitely better than the parallel system. If funds are limited, And without considering the economic impact of the failure rate selection on system cost or the cost of the detection switch is too large. Then we should choose the optimal scheme for parallel system. In consideration of the effect of failure rate selection on the economic cost of the system, it gives a specific calculation inequality [12]. When the optimal value point of the redundancy distribution in parallel system and it is at the right side of the intersection point. Then we should select the redundancy distribution system. When the optimal value point is at the left side of the intersection point, Then we should compare the optimal values of the two systems and last choose the best one.

In the actual system, components are often redundant backup, but the number of redundant backups will greatly increase the cost of the product, the product can not be detected through the dismantling of products to achieve, which provides product providers with the possibility of speculation, the study will use the method of truncation estimation to make a asymptotic estimation, provide statistic reference formula and feasible calculation method for product consumers, so as to avoid the influence of provider's speculation to the consumer's system.

2 The Definition of the Main Indicators of Reliability

(1) Reliability

The definition of reliability *R(t)* [13]: it is the probability that product completes the required function under the specified conditions and within the prescribed time.

If the life distribution of product is $F(t)$, $t > 0$, the reliability $R(t) = P(T \geq t) = 1 - F(t)$. This is a function of time(t), so it can be called as reliability function. To the components obeying exponential distribution λ, its reliability is $e^{-\lambda t}, t \geq 0$.

(2) Failure rate

Failure rate $\lambda(t)$: It is the probability of occurring failure in the unit of time after product has worked a period of time(t). According to reliability theory, $\lambda(t) = \frac{f(t)}{1-F(t)}$, when $t > 0$, the failure rate of exponential distribution is constant λ.

(3) System parameter specification

A: represents normal working events of system.
A_i: represents normal working events of the element i.
λ_i: represents failure rate of the element i.
$R_s(t)$: represents system reliability, that is,$P(A) = R_s$.
$R_i(t)$: represents reliability of the element i, that is, $P(A_i) = R_i$.
m_s: represents the average lifetime of the system, $m_s = \int_0^{+\infty} R_s(t)dt$

Redundant backup systems, some of the elements work, the other unit does not work, in a waiting or a standby state, When the unit the failure rate of the secondary unit in a standby period is zero, in other words, is a hundred percent reliability during standby. One work, n − 1 standby redundant system framework is shown below Fig. 1, where, K is detected and switches.

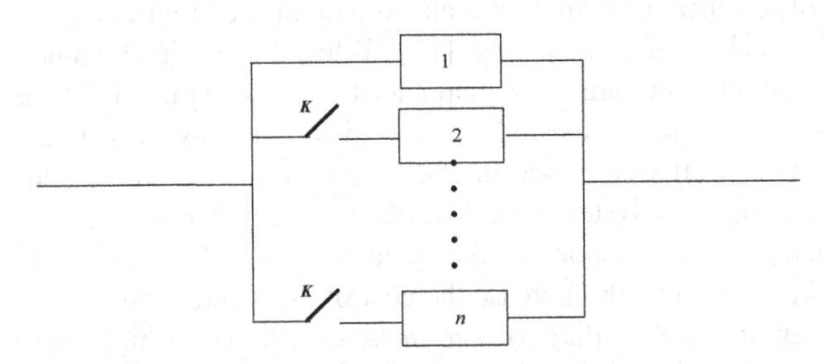

Fig. 1. The logical block diagram of redundancy distribution system.

The following lemma is assumed that when the switch is absolutely reliable:

Lemma 1.1: $X_1, X_2, \cdots X_n$ is mutually independent random variables, subject to the same parameter λ exponential distribution, then $X = X_1 + X_2 + \cdots + X_n$ obey redundancy distribution of order n, the probability density function is:

$$b_n(u) = \begin{cases} \lambda e^{-\lambda u} \frac{(\lambda u)^{n-1}}{(n-1)!}, & u \geq 0 \\ 0, & u < 0 \end{cases} \tag{1}$$

Proof: Let $p(u)$ is a probability density function of random variables, Since X_i are independent and identically distributed random variables exponentially distributed, then $p(u)$ is exponential distribution probability density function, to make $\varphi(t) = \int_{-\infty}^{+\infty} e^{itu}p(u)du$ is a characteristic function of random variable X_i then $\varphi(t) = (1 - \frac{it}{\lambda})^{-1}$. Because of $X = X_1 + X_2 + \cdots + X_n$, then X shows the probability density function of X_i do the n-fold convolution, Since the Fourier transform can become convolution to multiplication. Therefore, the characteristic function is:

$$\varphi_n(t) = \varphi(t)^n = (1 - \frac{it}{\lambda})^{-n}$$

The probability density function of X is, when $u \geq 0$, $b_n(u) = \frac{1}{2\pi} \int_{-\infty}^{+\infty} e^{-itu}\varphi_n(t)dt = \frac{1}{2\pi}\int_{-\infty}^{+\infty} e^{-itu}(1 - \frac{it}{\lambda})^{-n}dt$, points can be obtained from the Division:

$$\frac{1}{2\pi}\int_{-\infty}^{+\infty} e^{-itu}(1-\frac{it}{\lambda})^{-n}dt = \frac{1}{2\pi}\left[\frac{\lambda}{(n-1)i}e^{itu}(1-\frac{it}{\lambda})^{-n+1}\Big|_{-\infty}^{+\infty} + \frac{\lambda u}{n-1}\int_{-\infty}^{+\infty} e^{-itu}(1-\frac{it}{\lambda})^{-n+1}dt\right]$$

$$= \frac{1}{2\pi}\frac{\lambda u}{n-1}\int_{-\infty}^{+\infty} e^{-itu}(1-\frac{it}{\lambda})^{-n+1}dt = \cdots = \frac{1}{2\pi}\frac{(\lambda u)^{n-1}}{(n-1)!}\int_{-\infty}^{+\infty} e^{-itu}(1-\frac{it}{\lambda})^{-1}dt = \lambda e^{-\lambda u}\frac{(\lambda u)^{n-1}}{(n-1)!}$$

So for the $n - 1$ redundancy elements and system of component life exponentially distributed, its distribution follows n-order redundancy distribution, proof.

As the product life is generally longer, so the life test to do all the failure to test samples will take a long time. Especially for long-life products will take longer, it is all difficult to achieve that. Therefore, the use of censored life test method has become the best choice. Here we use the non-replacement censored life test, assuming that there are N samples to participate in life test, the test to the prior provisions of r failure to stop the test, and the failure time of failure samples are:

$$t_1 \leq t_2 \leq \cdots \leq t_r$$

Lemma 1.2: In the case of no replacement fixed number of truncated samples, and the life of the product obeys the nth-order Irish distribution, the failure time of r failure products are r order statistics: $t_1 \leq t_2 \leq \cdots \leq t_r$, their joint distribution density are:

$$f(t_1, t_2, \cdots, t_r) = \begin{cases} \frac{N!}{(N-r)!}\lambda^r e^{-\lambda\sum_{i=1}^{r} t_i} \bullet \prod_{i=1}^{r}\frac{(\lambda t_i)^{n-1}}{(n-1)!} \bullet \{\sum_{k=0}^{n-1}\frac{(\lambda t_r)^{k-1}}{(k-1)!}e^{-\lambda t_r}\}^{N-r}, & 0 < t_1 \leq t_2 \leq \cdots \leq t_r \\ 0, & \text{other} \end{cases} \quad (2)$$

It is a likelihood function of truncated r sample failure of N samples.

Proof: From Lemma 1.1 we can see that the probability density function of system life $X_i, (i = 1, 2, \cdots, N)$ is:

$$b_n(t) = \begin{cases} \lambda e^{-\lambda t} \frac{(\lambda t)^{n-1}}{(n-1)!}, & t \geq 0 \\ 0, & t < 0 \end{cases}$$

The i-th order statistics $X_{(i)}$ fall into the infinitesimal interval statistics $(t_i, t_i + \Delta t_i]$, $i = 1, 2, \cdots, r$, This event is equivalent to "the capacity of the sub-sample $X_1, X_2, \cdots, X_N$ of N each have a component fall into the interval $(t_i, t_i + \Delta t_i]$, while the remaining *N-r* components fall into the interval $(t_r + \Delta t_r, +\infty]$", According to the form of its composition, as shown in Fig. 2, set the probability of this event is equal to $f(t_1, t_2, \cdots, t_r)\Delta t_1 \Delta t_2 \cdots \Delta t_r$, The probability of each sub-component falling into the interval $(t_i, t_i + \Delta t_i]$ is $b_n(t_i)\Delta t_i + o(\Delta t_i)$.

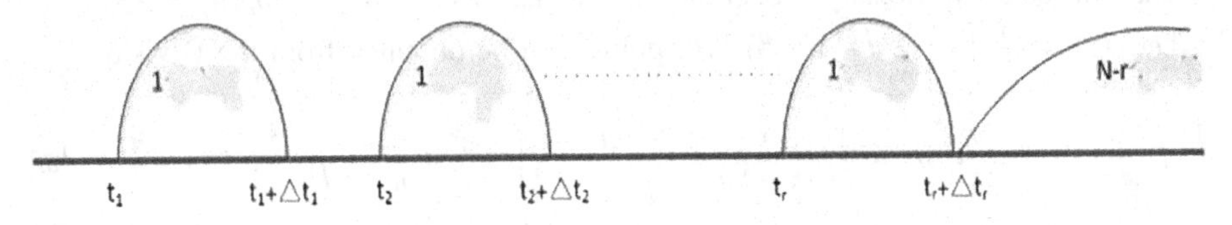

Fig. 2. The sample component distribution in the interval

The probability of falling $(t_r + \Delta t_r, +\infty]$ is $1 - \int_{t_r}^{+\infty} \lambda e^{-\lambda t} \frac{(\lambda t)^{n-1}}{(n-1)!} dt$. And the N components are divided into r + 1 groups, From the first group to the rth group each have a component, and the last group has N-r components. There are $\frac{N!}{1!1!\cdots 1!(N-r)!}$ possibilities for this grouping. So when $0 < t_1 \leq t_2 \leq \cdots \leq t_r$,

$$f(t_1, t_2, \cdots, t_r)\Delta t_1 \Delta t_2 \cdots \Delta t_r = \frac{N!}{1!1!\cdots 1!(N-r)!} \lambda^r e^{-\lambda \sum_{i=1}^{r} t_i} \bullet \prod_{i=1}^{r} \frac{(\lambda t_i)^{n-1}}{(n-1)!}$$

$$\bullet \left(1 - \int_{t_r}^{+\infty} \lambda e^{-\lambda t} \frac{(\lambda t)^{n-1}}{(n-1)!} dt\right)^{N-r} \Delta t_1 \Delta t_2 \cdots \Delta t_r + o(\Delta t_1 \Delta t_2 \cdots \Delta t_r)$$

$$= \frac{N!}{(N-r)!} \lambda^r e^{-\lambda \sum_{i=1}^{r} t_i} \bullet \prod_{i=1}^{r} \frac{(\lambda t_i)^{n-1}}{(n-1)!} \bullet \left\{\left[1 + \lambda t_r + \frac{(\lambda t_r)^2}{2!} + \cdots + \frac{(\lambda t_r)^{n-1}}{(n-1)!}\right] e^{-\lambda t_r}\right\}^{N-r}$$
$$\Delta t_1 \Delta t_2 \cdots \Delta t_r + o(\Delta t_1 \Delta t_2 \cdots \Delta t_r)$$

Order $\Delta t_i \to 0$ can get that:

$$f(t_1, t_2, \cdots, t_r) = \begin{cases} \frac{N!}{(N-r)!} \lambda^r e^{-\lambda \sum_{i=1}^{r} t_i} \bullet \prod_{i=1}^{r} \frac{(\lambda t_i)^{n-1}}{(n-1)!} \bullet \left\{\sum_{k=0}^{n-1} \frac{(\lambda t_r)^{k-1}}{(k-1)!} e^{-\lambda t_r}\right\}^{N-r}, & 0 < t_1 \leq t_2 \leq \cdots \leq t_r \\ 0, & \text{other} \end{cases}$$

we can get the conclusion, proof.

3 Approximate Estimation of Order and Failure Rate of Redundancy System

Lemma 1.3: The asymptotic estimation of the order of the redundancy system is $\hat{x}_{k+1} = \frac{(q^2)\hat{x}_k}{c^2}$, the failure rate is estimated as $\hat{\lambda} = \frac{\hat{x}_k}{T_1}$

Proof: From Lemma 1.2, we can know that the likelihood function of the r truncated samples in the N samples is:

$$L(t_1, t_2, \cdots, t_r; \lambda, n) = \frac{N!}{(N-r)!}\lambda^r e^{-\lambda \sum_{i=1}^{r} t_i} \bullet \prod_{i=1}^{r} \frac{(\lambda t_i)^{n-1}}{(n-1)!} \bullet \{\sum_{k=0}^{n-1} \frac{(\lambda t_r)^{k-1}}{(k-1)!} e^{-\lambda t_r}\}^{N-r}$$

According to the necessary conditions for the extremum, the partial derivative of its logarithmic function λ can be obtained:

$$\frac{\partial InL}{\partial \lambda} = \frac{r}{\lambda} - T_1 + \frac{(n-1)r}{\lambda} - (N-r)t_r \frac{\frac{(\lambda t_r)^{n-1}}{(n-1)!}}{\sum_{k=0}^{n-1} \frac{(\lambda t_r)^{k-1}}{(k-1)!}} = 0 \tag{3}$$

Among $T_1 = \sum_{i=1}^{r} t_i$, the failure rate of the general products are quite small, and λt_r is relatively small, then the Eq. (3) approximates:

$$\frac{\partial InL}{\partial \lambda} = \frac{r}{\lambda} - T_1 + \frac{(n-1)r}{\lambda} = 0 \tag{4}$$

Get that:

$$\lambda = \frac{n}{T_1} \tag{5}$$

For the order n,

$$\begin{aligned} \frac{L(t_1, t_2, \cdots, t_r; \lambda, n+1)}{L(t_1, t_2, \cdots, t_r; \lambda, n)} &= \frac{\lambda^r \prod_{i=1}^{r} t_i}{n^r} \bullet \left(\frac{\sum_{k=0}^{n+1} \frac{(\lambda t_r)^{k-1}}{(k-1)!}}{\sum_{k=0}^{n} \frac{(\lambda t_r)^{k-1}}{(k-1)!}} \right)^{N-r} \\ &= \frac{\lambda^r \prod_{i=1}^{r} t_i}{n^r} \bullet \left(1 + \frac{\frac{(\lambda t_r)^n}{n!}}{\sum_{k=0}^{n} \frac{(\lambda t_r)^{k-1}}{(k-1)!}} \right)^{N-r} \approx \frac{\lambda^r \prod_{i=1}^{r} t_i}{n^r} \bullet \left(1 + \frac{(\lambda t_r)^n}{n!}\right)^{N-r} \end{aligned} \tag{6}$$

According to the independent variables for discrete cases to obtain the necessary conditions for the extreme:

$$\begin{cases} \dfrac{L(t_1, t_2, \cdots, t_r; \lambda, n+1)}{L(t_1, t_2, \cdots, t_r; \lambda, n)} < 1 \\ \dfrac{L(t_1, t_2, \cdots, t_r; \lambda, n)}{L(t_1, t_2, \cdots, t_r; \lambda, n-1)} \geq 1 \end{cases} \tag{7}$$

Order $T_2 = \prod_{i=1}^{r} t_i$ Substituting Eq. (5) into Eq. (6), Then (7) is formulated as:

$$\begin{cases} \dfrac{(\frac{t_r}{T_1}n)^n}{n!} < \sqrt[N-r]{\frac{T_1^r}{T_2}} - 1 \\ \dfrac{[\frac{t_r}{T_1}(n-1)]^{n-1}}{(n-1)!} \geq \sqrt[N-r]{\frac{T_1^r}{T_2}} - 1 \end{cases} \tag{8}$$

By Stirling's approximation, when the n is large, n factorial calculation is very large, so the Stirling's formula is very easy to use, and, even when n is small, the value of calculation has been very accurate formula too. Substituting $n! \approx \sqrt{2\pi n}(\frac{n}{e})^n$ into formula (8) available, the same as $(n-1)! \approx \sqrt{2\pi n}(\frac{n-1}{e})^{n-1}$ can get that:

$$\begin{cases} \dfrac{(\frac{t_r}{T_1}e)^n}{\sqrt{2\pi n}} < \sqrt[N-r]{\frac{T_1^r}{T_2}} - 1 \\ \dfrac{[\frac{t_r}{T_1}e]^{n-1}}{\sqrt{2\pi(n-1)}} \geq \sqrt[N-r]{\frac{T_1^r}{T_2}} - 1 \end{cases} \tag{9}$$

From the form of the inequality set (9), it can be seen that solving this inequality group corresponds to the intersection of the exponential function $q^n (q \ll 1)$ and the power function $cn^{\frac{1}{2}}$. The shape of the two curves is shown below (Fig. 3):

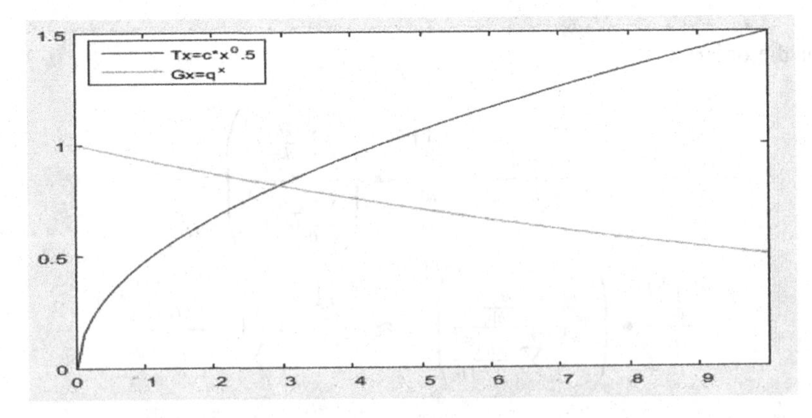

Fig. 3. The cross curve of exponential and power function

It can be seen from the figure that there is only one intersection x^*, but because it is transcendental equation, using the spider-web model iterative solution, order $q = \frac{t_r}{T_1} e\,(q \ll 1), c = \sqrt[N-r]{\frac{T_1^r}{T_2} - 1}\sqrt{2\pi}$. Structure mapping is as follows:

$$\begin{cases} Tx = cx^{\frac{1}{2}} \\ Gx = q^x \end{cases}$$

Iterative Algorithms: $x_{k+1} = T^{-1}(Gx_k) = \frac{(q^2)^{x_k}}{c^2}$, now prove $x_k \to x^*$:

When $x_{k-1} < x^* < x_k$, $q^{x_k} = q^{\frac{(q^2)^{x_{k-1}}}{c^2}} > q^{\frac{(cx_{k-1}^{\frac{1}{2}})^2}{c^2}} = q^{x_{k-1}} > cx_{k-1}^{\frac{1}{2}}$

$$|Tx_k - Gx_k| = Tx_k - Gx_k = q^{x_{k-1}} - q^{x_k} < q^{x_{k-1}} - cx_{k-1}^{\frac{1}{2}} = |Tx_{k-1} - Gx_{k-1}|$$

The same as $x_k < x^* < x_{k-1}$,

$$\lim_{k\to\infty} x_{2k+1} = x', \lim_{k\to\infty} x_{2k} = x'', x' = \lim_{k\to\infty} x_{2k+1} = \lim_{k\to\infty} \frac{(q^2)^{x_{2k}}}{c^2} = \frac{(q^2)^{x''}}{c^2},$$
$$q^{x^*} = cx^{*\frac{1}{2}}, x' = x^* = x''.$$

4 Conclusion

The components of many systems are often redundant backup, but the number of redundant backups will greatly increase the cost of the product, the product can not be detected through the dismantling of products to achieve, This paper evaluates the order and failure rate of the redundancy distributed redundant backup system, thus providing consumers with a computable and measurable product reliability calculation method. By using the irreplaceable censored life test, Providing a statistical significance of the estimation formula. We first derive the likelihood estimation function of the redundant system, use the Stirling formula to smooth the factorization, and finally obtain the asymptotic estimation formula through the spider-web model: The asymptotic estimation of the redundancy order is $\hat{x}_{k+1} = \frac{(q^2)\hat{x}_k}{c^2}$, and the failure rate estimate is: $\lambda = \frac{\hat{x}_k}{T_1}$. To provide consumers with a scientific computing method, not only can greatly reduce costs, but also can improve the reliability of the system.

Acknowledgments. This research was partially supported by the School of Mathematics and Computer Science of Fujian Normal University, by the Institute of Innovative Information Industry of Fujian Normal University, by the School of Economic of Fujian Normal University.

References

1. Luss, H.: An inspection policy model for production facilities. Manag. Sci. **29**, 101–109 (1983)
2. Leung, K.N.F., Lai, K.K.: A preventive maintenance and replacement policy of a series system with failure interaction. Optimization **61**(2), 223–237 (2012)
3. Sung, C.K., Sheu, S.H., Hsu, T.S., Chen, Y.C.: Extended optimal replacement policy for a two-unit system with failure rate interaction and external shocks. Int. J. Syst. Sci. **44**(5), 877–888 (2013)
4. Lai, M.T., Yan, H.: Optimal number of minimal repairs with cumulative repair cost limit for a two-unit system with failure rate interactions. Int. J. Syst. Sci. **47**(2), 466–473 (2014)
5. Merle, G.: Algebraic Modeling of Dynamic Fault Trees, Contribution to Qualitative and Quantitative Analysis. ENS de Caehan, France (2010)
6. Liang, X., Hong, Y., Zhang, Y., et al.: Numerical simulation to reliability analysis of fault-tolerant repairable system. J. Shanghai Jiaotong Univ. **15**(5), 526–534 (2010)
7. Ye, Z.S., Wang, Y., Tsui, K.L., et al.: Degradation data analysis using wiener processes with measurement errors. IEEE Trans. Reliab. **62**(4), 772–780 (2013)
8. Fan, J.J., Yung, K.C., Michael, P.: Lifetime estimation of high-power white LED using degradation data driven method. IEEE Trans. Device Mater. Reliab. **12**(2), 470–477 (2012)
9. Laurenciu, N.C., Cotofana, S.D.A.: A nonlinear degradation path dependent end-of-life estimation framework from noisy observations. Microelectron. Reliab. **53**, 1213–1217 (2013)
10. Zhai, G., Zhou, Y., Ye, X., et al.: A method of multi-objective reliability tolerance design for electronic circuits. Chin. J. Aeronaut. **26**(1), 161–170 (2013)
11. Xu, L., Xie, C.F., Xu, L.X.: Reliability envelope analysis. J. Comput. **25**(4), 26–34 (2014)
12. Xu, L., Xie, C.F., Xu, L.X.: Reliability envelope analysis. Adv. Intell. Syst. Comput. (535), 35–42 (2017)
13. Zong-shu, W.: Probability and Mathematical Statistics. China Higher Education Press, Beijing (2008)

Comments on Yu et al's Shared Data Integrity Verification Protocol

Tsu-Yang Wu[1,2](✉), Yueshan Lin[3], King-Hang Wang[4], Chien-Ming Chen[3], and Jeng-Shyang Pan[1,2]

[1] Fujian Provincial Key Lab of Big Data Mining and Applications, Fujian University of Technology, Fuzhou 350118, China
wutsuyang@gmail.com, jengshyangpan@fjut.edu.cn

[2] National Demonstration Center for Experimental Electronic Information and Electrical Technology Education, Fujian University of Technology, Fuzhou 350118, China

[3] School of Computer Science and Technology, Harbin Institute of Technology - Shenzhen, Shenzhen 518055, China
243994545@qq.com, chienming.taiwan@gmail.com

[4] Department of Computer Science and Engineering, Hong Kong University of Science and Technology, Clear Water Bay, Kowloon, Hong Kong
kevinw@cse.ust.hk

Abstract. Recently, Yu et al. proposed a secure shared data integrity verification protocol called SDVIP^2 to ensure the integrity of outsourced file in the cloud. Unfortunately, we exploit the vulnerability of their protocol in this paper. We demonstrate an active adversary can modify the outsourced file such that the auditor is unable to detect in the auditing process. At the end of this paper, we also provide two suggestions to fix the proposed attack.

Keywords: Cloud storage security · Data integrity verification · Bilinear pairings · Cryptanalysis

1 Introduction

With the rapid growth of cloud computing, Storage-as-a-Service brings a convenience way for users. Though they can access their data over the Internet at any time and any place, the new security and privacy issues appears where users handover the physical control to data to the cloud. The data integrity verification problem appears when a user wish to assert the whole set of data is kept on cloud without incurring too much local storage and computation effort. To address this problem, auditing protocols based on the framework that involving a trusted third party (TPA) audit protocol were proposed [2,5,8,10,13]. They focus on checking the integrity of outsourced data in cloud computing. Some other auditing protocols [6,7] achieve further to preserve the privacy of a group of users at the same time assert the integrity of the cloud storage.

P. Krömer et al. (eds.), *Proceedings of the Fourth Euro-China Conference on Intelligent Data Analysis and Applications*, Advances in Intelligent Systems and Computing 682, DOI 10.1007/978-3-319-68527-4_8

In 2014 Ni et al. presented an attack [4] on an auditing protocol [10]. We refer this attack as a *data-corruption attack* where an adversary corrupts the storage at a cloud server and creates a fake-proof in an auditing protocol to convince the TPA that the file is intact. Later the same group of authors [12] proposed a data integrity verification protocol called SDIVIP^2 that allows a secure data sharing amount a group of users over a cloud. In this paper, however, we found that SDIVIP^2 was in fact vulnerable against the data-corruption attack. As a result the integrity of the file storage is compromised while the TPA cannot perform its auditor role properly. We provide two suggestions to prevent this kind attack.

2 Review the SDIVIP^2 Protocol

2.1 Settings

We assume there is a group of mobile users in the system wish to share a file over a cloud server. These users agree on a group key and one of these users commits a file together with a set of tags associated with this file to the cloud server. Each tag is generated based on the group key and the file. A TPA, holding the public key of the group, assists the mobile users to audit the file integrity on the cloud server. It randomly generates a set of challenge and send it to the cloud server. The cloud server shall computes a proof responding the challenge based on the file content and the tags. The TPA will be able to verify if the file content are securely stored at cloud without accessing the file content. The flowchart of SDIVIP^2 protocol is depicted in Fig. 1.

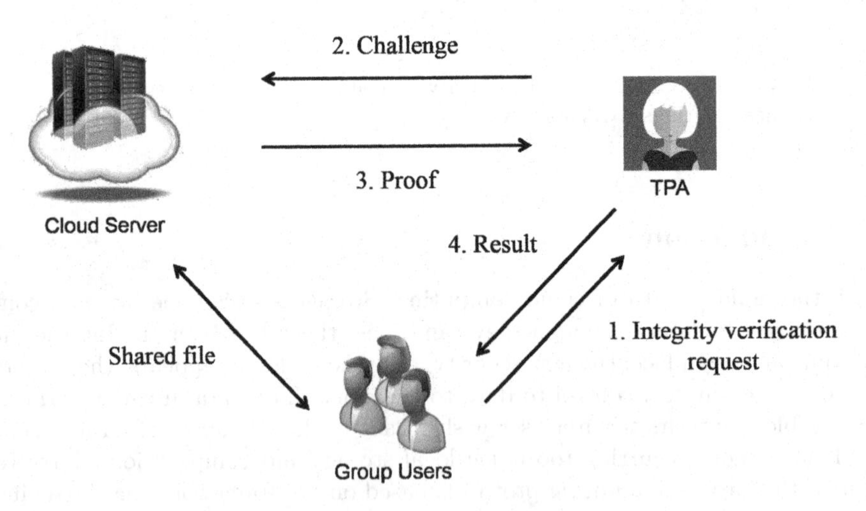

Fig. 1. The framework of SDIVIP^2 protocol

2.2 Preliminaries and Notation

Let e be a bilinear map $e : G \times G \rightarrow G_T$, where G and G_T are two multiplicative cyclic groups with a large prime order p so that e satisfying the properties.

1. *Bilinear.* For $g_1, g_2 \in G$ and $a, b \in \mathbb{Z}_p^*$, $e(g_1^a, g_2^b) = e(g_1, g_2)^{ab}$.
2. *Non-degenerate.* There exists an identity $1_G \in G$ such that $e(1_G, 1_G) = 1_{G_T}$, an identity of G_T.
3. *Computational.* There exist several efficient algorithms to compute bilinear map e.

For the details about bilinear pairings, readers can refer to [1,3,9,11] for a full descriptions.

Some other notations are used in the protocol and they are summarized in Table 1.

Table 1. Notation

Notations	Meanings
H	A cryptographic hash function. $H : \{0,1\}^* \rightarrow \mathbb{Z}_p^*$
H_1	A cryptographic hash function. $H_1 : \{0,1\}^* \rightarrow G$
g	Generator of G
U_i	A mobile user
$\mathbf{S}$	A group of mobile users who wish to share a file
n	Number of mobile users in the group $\mathbf{S}$, thus $\mathbf{S} = \{U_1, U_2, \ldots, U_n\}$
F	The file being shared and divided into $t \times s$ blocks $F = \{m_1, \ldots, m_t\}$ and $m_j = \{m_{j1}, \ldots, m_{js}\}$
pk_i	Public key of user U_i
$Chal$	A challenge sent from the TPA to the cloud server
P	A proof responded by the cloud server to the auditor TPA

2.3 Description of the Protocol

SDIVIP^2 can be described with the following six phases:

1. **Key generation.** User U_i selects its secret key value $x_i \in_R \mathbb{Z}_p^*$ and computes his/her public key $pk_i = g^{x_i}$.
2. **Group key generation.** Mobile users in $\mathbf{S} = \{U_1, U_2, \ldots, U_n\}$ formed a cycle, i.e., $U_n = U_0$, $U_{n+1} = U_1$ and share a group secret/public key pair (Gsk, Gpk) where $Gpk = g^{Gsk}$. The secret key is assumed only known within the group.

3. **Tag generation.** Given the shared file F which is divided into $t \times s$ blocks We select $b_k \in_R \mathbb{Z}_p^*$ and computing $u_k = g^{b_k}$ for $k = 1, 2, \ldots, s$. For each block m_j, U_i generates its tag $T_j = (H_1(F_{id}||j) \cdot \prod_{k=1}^{s} u_k^{m_{jk}})^{Gsk}$, where F_{id} denotes the unique identifier of F. Finally, U_i outputs $\mathbf{T} = \{T_1, T_2, \ldots, T_t\}$ and outsources $\{F, \mathbf{T}\}$ to the cloud server.
4. **Challenge.** The TPA randomly generates a challenge set $\mathbf{Q} \subset \{1, 2, \ldots, t\}$ and selects $v_j \in \mathbb{Z}_p^*$ for each m_j. Then, it sends $Chal = \{(j, v_j)|j \in \mathbf{Q}\}$ to the cloud server.
5. **Proof generation.** Upon receiving $Chal$ from the TPA, the cloud server computes $\mu_k = \sum_{j \in \mathbf{Q}} v_j \cdot m_{jk} \in \mathbb{Z}_p$ for $k = 1, 2, \ldots, s$ and $\sigma = \prod_{j \in \mathbf{Q}} T_j^{v_j}$. Finally, the cloud server sends the proof $\mathbf{P} = \{\mu_1, \mu_2, \cdots, \mu_s, \sigma\}$ to the TPA.
6. **Proof checking.** Upon receiving $\mathbf{P}$, the TPA verifies $e(\sigma, g) \stackrel{?}{=} e(\prod_{j \in \mathbf{Q}} H(F_{id}||j)^{v_j} \cdot \prod_{k=1}^{s} u_k^{\mu_k}, Gpk)$. If the verification holds, the TPA returns "true". Otherwise, it returns false.

3 Cryptanalysis on SDIVIP²

Here, we find some active adversary $\mathcal{A}$ may exist in their protocol such that it can modify the outsourced file sent by the mobile user and then generates a fake proof to pass the proof verification. In other words, the adversary can cheat the TPA and the file owner who believes the file are well maintained in the cloud. We assume that a user U_i wants to outsource $\{F, \mathbf{T}\}$ to a cloud server, where $\{F, \mathbf{T}\}$ is mentioned in the previous section. The details descriptions are described as follows.

1. An adversary $\mathcal{A}$ corrupts the file F on a cloud server by replacing each block m_{jk} as m''_{jk}, where $m''_{jk} = m_{jk} + n_{jk}$.
2. Upon receiving a challenge $Chal = \{(j, v_j)|j \in \mathbf{Q}\}$ from the TPA, the cloud server computes $\mu''_k = \sum_{j \in \mathbf{Q}} v_j \cdot (m''_{jk})$ for $k = 1, 2, \ldots, s$ and $\sigma = \prod_{j \in \mathbf{Q}} T_j^{v_j}$. Then, the server sends a proof $\mathbf{P}'' = \{\mu''_1, \mu''_2, \ldots, \mu''_s, \sigma\}$ to the TPA.
3. $\mathcal{A}$ intercepts $\mathbf{P}''$ and then modifies μ''_k to $\mu_k = \mu''_k - \sum_{j \in \mathbf{Q}} v_j \cdot n_{jk}$. Finally, $\mathcal{A}$ sends the modified proof $\mathbf{P} = \{\mu_1, \mu_2, \ldots, \mu_s, \sigma\}$ to the TPA.
4. Upon receiving $\mathbf{P}$, the TPA verifies $e(\sigma, g) \stackrel{?}{=} e(\prod_{j \in \mathbf{Q}} H(F_{id}||j)^{v_j} \cdot \prod_{k=1}^{s} u_k^{\mu_k}, Gpk)$. It is easy to see that the verification holds and the TPA returns "true".

After executing the above four steps, the cloud server believes U_i outsourcing $\{F'', \mathbf{T}\}$ and the TPA believes the proof $\mathbf{P}$ generated by the cloud server. It demonstrates that Yu et al.'s SDIVIP² protocol has a security flaw.

4 Possible Quick Fix

Ni et al. [4] presented their solution to the data corruption attack by introducing a digital signature on the proof $\mathbf{P}$. The cloud server digitally signed on the proof and so no adversary described above may amend the challenge protocol. This solution rests on the assumption that a trustworthy public key of a cloud server is available. Yet, if cloud data can be corrupted by an adversary, it is also reasonable to assume the corresponding private key can be exploited by the adversary too.

The other way round is to encrypt the proof using the TPA's public key. Since the proof **P** is encrypted using the TPA's public key and the adversary has no idea which TPA the users will choose to run the protocol, it is impossible for the adversary to corrupt a TPA's public key.

5 Conclusion

In this paper, we have revisited SDIVIP^2 protocol and pointed out that ironically this protocol is suffering the data-corruption attack proposed by the same group of authors. In the future, we are exploring the possibility to further optimize the protocol so that a better security can be achieve without involving TPA or the cloud server's credentials.

Acknowledgments. The authors would thank anonymous referees for a valuable comments and suggestions. The work of Chien-Ming Chen was supported in part by the Project NSFC (National Natural Science Foundation of China) under Grant number 61402135 and in part by Shenzhen Technical Project under Grant number JCYJ20170307151750788.

References

1. Boneh, D., Franklin, M.: Identity-based encryption from the Weil pairing. In: Proceedings of the 21st Annual International Cryptology Conference on Advances in Cryptology, pp. 213–229. Springer, Heidelberg (2001)
2. Kim, D., Kwon, H., Hahn, C., Hur, J.: Privacy-preserving public auditing for educational multimedia data in cloud computing. Multimedia Tools Appl. **75**(21), 13077–13091 (2016)
3. Li, C.T., Wu, T.Y., Chen, C.L., Lee, C.C., Chen, C.M.: An efficient user authentication and user anonymity scheme with provably security for IoT-based medical care system. Sensors **17**(7), 1482 (2017)
4. Ni, J., Yu, Y., Mu, Y., Xia, Q.: On the security of an efficient dynamic auditing protocol in cloud storage. IEEE Trans. Parallel Distrib. Syst. **25**(10), 2760–2761 (2014)
5. Wan, C., Zhang, J., Pei, B., Chen, C.: Efficient privacy-preserving third-party auditing for ambient intelligence systems. J. Ambient Intell. Humaniz. Comput. **7**(1), 21–27 (2016)
6. Wang, B., Li, B., Li, H.: Oruta: privacy-preserving public auditing for shared data in the cloud. IEEE Trans. Cloud Comput. **2**(1), 43–56 (2014)
7. Wang, B., Li, H., Li, M.: Privacy-preserving public auditing for shared cloud data supporting group dynamics. In: 2013 IEEE International Conference on Communications (ICC), pp. 1946–1950. IEEE (2013)
8. Wang, C., Chow, S.S., Wang, Q., Ren, K., Lou, W.: Privacy-preserving public auditing for secure cloud storage. IEEE Trans. Comput. **62**(2), 362–375 (2013)

9. Xu, Y., Zhong, H., Cui, J.: An improved identity-based multi-proxy multi-signature scheme. J. Inf. Hiding Multimedia Sig. Process. **7**(2), 343–351 (2016)
10. Yang, K., Jia, X.: An efficient and secure dynamic auditing protocol for data storage in cloud computing. IEEE Trans. Parallel Distrib. Syst. **24**(9), 1717–1726 (2013)
11. Yin, S.L., Li, H., Liu, J.: A new provable secure certificateless aggregate signcryption scheme. J. Inf. Hiding Multimedia Sig. Process. **7**(6), 1274–1281 (2016)
12. Yu, Y., Ni, J., Xia, Q., Wang, X., Yang, H., Zhang, X.: SDIVIP^2: shared data integrity verification with identity privacy preserving in mobile clouds. Pract. Exp. Concurr. Comput. **28**(10), 2877–2888 (2016)
13. Zhang, J., Zhao, X.: Efficient chameleon hashing-based privacy-preserving auditing in cloud storage. Cluster Comput. **19**(1), 47–56 (2016)

Efficient Anonymous Password-Authenticated Key Exchange Scheme Using Smart Cards

Tsu-Yang Wu[1,2](✉), Weicheng Fang[3], Chien-Ming Chen[3], and Eric Ke Wang[3]

[1] Fujian Provincial Key Laboratory of Big Data Mining and Applications, Fujian University of Technology, Fuzhou 350118, China
wutsuyang@gmail.com

[2] National Demonstration Center for Experimental Electronic Information and Electrical Technology Education, Fujian University of Technology, Fuzhou 350118, China

[3] Harbin Institute of Technology Shenzhen Graduate School, Shenzhen, China
626558837@qq.com, chienming.taiwan@gmail.com, wk_hit@hit.edu.cn

Abstract. Anonymous authentication is designed to hide the user's identity from any verifiers during an authentication session. Since passwords prevail in many authentication systems, anonymous password-authenticated key exchange (APAKE) has become a candidate technique for privacy-enhancing applications. Recently, Shin and Kobara proposed an improved APAKE protocol using general devices such as public directories. However, we find that their scheme is vulnerable to a credential forgery attack. Then, we propose an efficient protocol using tamper resistant smart cards. The security and efficiency analysis shows that our protocol obtains high security and efficiency.

Keywords: Password authentication · Anonymity · Security · Smart card

1 Introduction

Anonymous authentication allows registered users to authenticate themselves without revealing their identities. It is desirable when users concern about their privacy. For example, users may be reluctant to vote or comment if the privacy is compromised. In fact, as the growth and development of the Internet becomes faster, many privacy-enhancing technologies have been invented to protect user's anonymity.

Recently, many researchers focus on the design of anonymous password-authenticated key exchange (APAKE) protocol. It not only achieves authentication and key exchange based on a low-entropy password, but also preserves the client's privacy. In 2006, Chai et al. [1] found the previous APAKE protocol

P. Krömer et al. (eds.), *Proceedings of the Fourth Euro-China Conference on Intelligent Data Analysis and Applications*, Advances in Intelligent Systems and Computing 682, DOI 10.1007/978-3-319-68527-4_9

proposed by Viet et al. [2] not efficient enough, and thus proposed a new protocol using smart cards. Since then, related schemes [3–5] have been proposed to additionally support traceable client anonymity, but they all failed to resist known attacks. In 2009, a protocol using general devices such as public directories was first proposed by Yang et al. [6] and later improved in [7]. In 2012, to obtain better performance, Qian et al. [8] also utilized general devices in the design of a new APAKE protocol. However, in 2016, Yang et al. [9] pointed out that their scheme suffers from the credential sharing problem. Also, Shin and Kobara [10] observed that their protocol is vulnerable to an active attack, and proposed an improved protocol (S2APA). Despite their improvement, we find that the S2APA protocol cannot resist a credential forgery attack.

Our contribution is twofold. Firstly, we find that S2APA is vulnerable to a credential forgery attack due to the use of general devices. Secondly, to overcome such attacks, we propose an efficient APAKE protocol using tamper-resistant smart cards. Although our scheme resorts to the dedicated devices such as smart cards, according to the analysis, it achieves higher efficiency and resists various known attacks. The rest of the paper is organized as follows: Sect. 2 shows the weakness of Shin and Kobara's protocols. Our proposed APAKE scheme using smart cards is presented in Sect. 3. In Sect. 4, we perform the security and efficiency analysis on the proposed protocol. Finally, we conclude the paper in Sect. 5.

2 Security Weakness in the S2APA Protocols

In this section, we briefly review Shin and Kobara's S2APA protocol, and show that their protocol is vulnerable to a credential forgery attack. S2APA protocol aims to provide unconditional anonymity for users during the authentication, where the server does not care the users' real identities as long as they have registered and provided correct passwords. It utilizes general devices that only guarantee integrity protection, such as external hard drives, software smart cards, and public directories. Also, the protocol employs a homomorphic public key encryption scheme (Gen, E, D) instantiated by ElGamal encryption system:

- $Gen(1^k)$ generates the public key and private key for the scheme.
- $E_{pk}(\cdot)$ returns the ciphertext according to the message m and a random coin r. Also, for messages m_1 and m_2, the homomorphic property states that

$$E_{pk}(m_1 \cdot m_2; r') = E_{pk}(m_1; r_1) \odot E_{pk}(m_2; r_2), \quad (1)$$

  where $\cdot$ and $\odot$ represents the group operations in plaintext and ciphertext, and r', r_1, r_2 are some random coins.
- $D_{sk}(\cdot)$ returns the plaintext according to the input ciphertext.

2.1 Review of the S2APA Protocol

The S2APA protocol consists of three phases: the setup phase, the registration phase, and the authentication phase. In the setup phase, the server S initializes the system with some public parameters $\{G, p, g, h, H_1, H_2, H_3, Gen, E, D\}$, where G is a cyclic group generated by g of prime order p, h is another generator of G, H_1, H_2 and H_3 are one-way hash functions, and (Gen, E, D) is a public key encryption scheme described as above. S also invokes Gen to generate keys (pk, sk), and randomly selects a master secret element $N \in G$. Figure 1 shows the registration phase and the authentication phase in details.

C_i		S
Registration		
Select pw_i		
	$\xrightarrow{pw_i}$	
		$T_i = E_{pk}(N \cdot h^{-pw_i})$
	$\xleftarrow{T_i}$	
Store T_i in general device		
Authentication		
$x \leftarrow_R \mathbb{Z}_p^*, X = g^x$		
$\hat{X} = E_{pk}(X \cdot h^{pw_i}; s) \odot T_i$		
	$\xrightarrow{C, \hat{X}}$	
		$X' = D_{sk}(\hat{X})/N$
		$y \leftarrow_R \mathbb{Z}_p^*, Y = g^y$
		$K' = (X')^y$
		$trans' = C\|S\|\{T_i\}\|pk\|\hat{X}\|Y\|X'\|K'$
		$V_S = H_1(trans')$
	$\xleftarrow{S, Y, V_S}$	
$K = Y^x$		
$trans = C\|S\|\{T_i\}\|pk\|\hat{X}\|Y\|X\|K$		
Check V_S		
$V_C = H_2(trans)$		
	$\xrightarrow{V_C}$	
		Check V_C
$SK = H_3(trans)$		$SK = H_3(trans')$

Fig. 1. The registration phase and the authentication phase of the S2APA protocol

2.2 A Credential Forgery Attack

Although the S2APA protocol was proved to be secure against various known attacks, we find it vulnerable to a credential forgery attack. Note that a client C_i's credential T_i is issued by the Server S, and then stored in a general device. In our attack, we assume that an adversary $\mathcal{A}$ has registered with S, and thus obtained a credential $T_{\mathcal{A}} = E_{pk}(N \cdot h^{-pw_{\mathcal{A}}})$ after the registration phase, where $pw_{\mathcal{A}}$ is the password chosen by the adversary. Since the credential is stored in a general device, $\mathcal{A}$ can extract T_i easily. Then, to forge a credential, $\mathcal{A}$ computes

$$T_{\mathcal{A}}^* = E_{pk}(h^{-pw_{\mathcal{A}}}) \odot T_{\mathcal{A}} \equiv E_{pk}(N). \tag{2}$$

In fact, according to the homomorphic property shown in Eq. (1), the newly forged credential becomes S's master secret N encrypted under the public key pk. To use it in the authentication phase, $\mathcal{A}$ follows the steps as shown in Fig. 1 except that the value $\hat{X}$ is computed as follows:

$$\hat{X} = E_{pk}(X; s) \odot T^*_{\mathcal{A}}. \tag{3}$$

The modification on $\hat{X}$ with the forged credential will have no influence on S's computation of X'. Therefore, S will accept the authentication request. Note that the authentication phase no longer involves the input of passwords when using the forged credential. So, $\mathcal{A}$ can secretly shares or publishes $T^*_{\mathcal{A}}$ to those who have not registered without leaking the adversary's own password $pw_{\mathcal{A}}$. In a long run, such a credential forgery attack launched by many other registered adversaries will undermine the system's registration.

The credential forgery attack against the S2APA protocol can be applied to similar protocols that merely depend on a general device to store its secrets. The SAPAKE protocol proposed by Qian et al. [8], the SAP protocol proposed by Shin and Kobara [11], and the protocol proposed by Son et al. [12] also suffer from such an attack if they insist on using general devices. As a simple but effective countermeasure, one replaces the general devices by tamper resistant dedicated devices such as smart cards, from which the securely stored secrets cannot be extracted by an adversary any more. The replacement trades usability for security. However, we observe that once using tamper proof smart cards, their schemes can be further improved in terms of efficiency.

3 The Proposed APAKE Protocol

In this section, we propose a new APAKE using smart cards. We assume our scheme uses such a tamper resistant smart card that no information can be extracted by an adversary. The scheme includes three phases as well. In the setup phase, the Server S chooses a random secret n and a secure one-way hash function h, computes $N = h(n)$ sets $N, G, g, p, h, H_1, H_2, H_3, H_4\}$ as public parameters, G is a cyclic group generated by g of order p, and H_1, H_2, H_3 and H_4 are secure one-way hash functions.

Figure 2 shows the rest of the scheme. We describe the registration phase and the authentication phase in details.

In the registration phase, a client C_i select a password pw_i and a random nonce r, computes hashed password hpw_i and sends it to the server S. The server computes the client's credential T_i and issues a smart card. After that, the smart card computes an off-line password verifier R, and appends r and R to itself. Then, C_i hold a tamper resistant smart card $\{T_i, r, R\}$.

In the authentication phase, C_i inputs the password to complete an off-line verification. If it is not verified, the smart card rejects the login request immediately. Otherwise, it sends an authenticator A and a group element X to the server S. S checks the validity of A, selects a group element Y, forms the Diffie-Hellman key K, and computes another authenticator V_S. Then, it sends

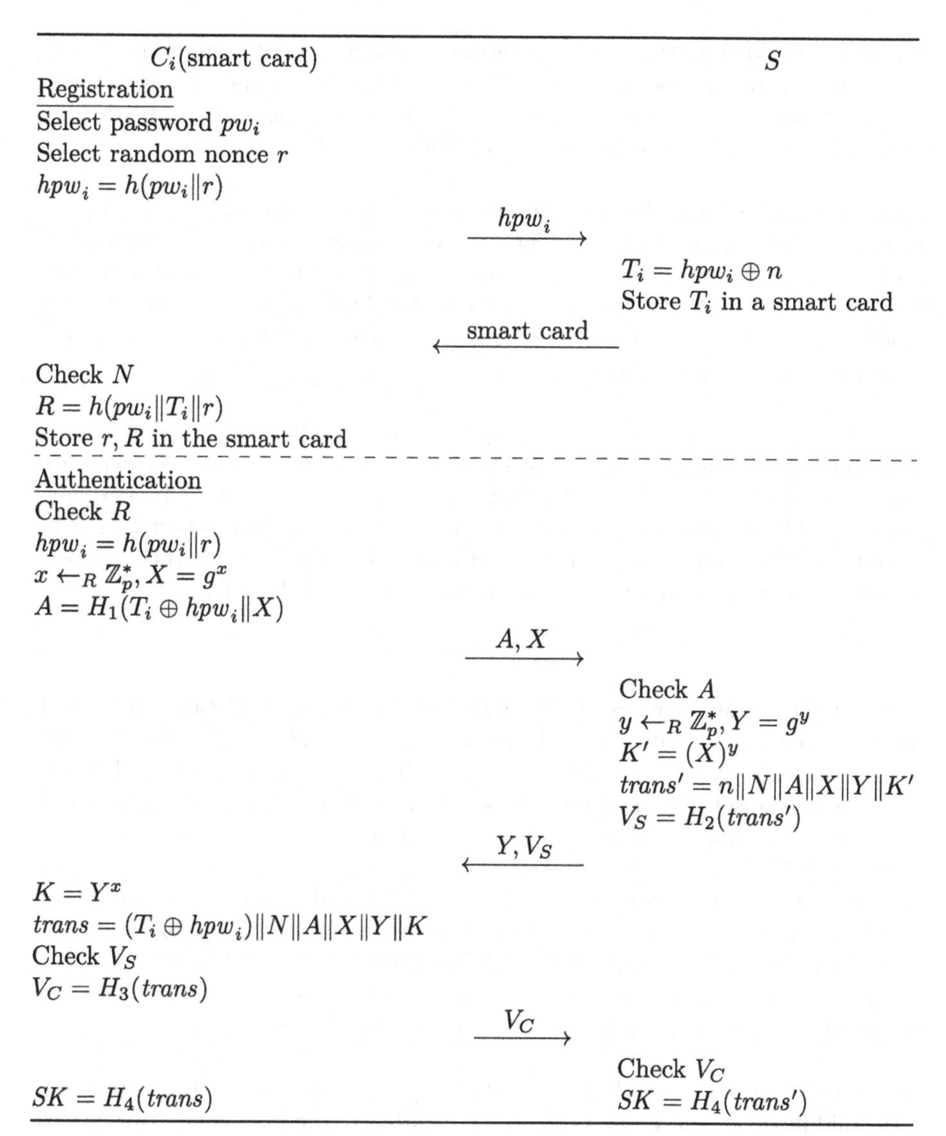

Fig. 2. The proposed APAKE protocol

$\{Y, V_S\}$ back to C_i. After that, C_i checks V_S and sends a respond V_C to S to achieve mutual authentication. Finally, S checks V_C and accepts the session. Both will compute the correct session key SK given successful verifications.

4 Security and Efficiency Analysis

For the past decade, various authenticated key exchange protocols have been proposed buy many of them have been proven insecure [13–17]. Therefore, in this section, we first demonstrate that our protocol satisfies security requirements and withstands known attack. Then, we evaluate computation and communication cost of our scheme. Trough analysis and comparison, though less user-friendly, our scheme proves to be more secure and efficient.

Mutual Authentication. In the authentication phase, both client C_i and the server S verifies each other to complete mutual authentication. If an adversary $\mathcal{A}$ attempts to impersonate either side, $\mathcal{A}$ has to compute the correct authenticators that consists of the master secret x and a Diffie-Hellman key K. However, it is impossible for $\mathcal{A}$ to obtain x. Thus, the adversary cannot generate a correct authenticator to achieve impersonation.

Unconditional Anonymity. The proposed scheme provides unconditional anonymity for the clients. During each protocol run, a client C_i sends a randomized authenticator A to the server S. So, S cannot track the user by checking A since it differs among protocol runs. Note that a malicious server trying to break the user's anonymity can modifies the credential in the registration phase. Specifically, it picks a unique random nonce n_i for each client C_i:

$$T_i' = hpw_i \oplus n_i \tag{4}$$

Then, in the authentication phase, S may check the authenticator A against different master secrets until it finds a match, i.e., $A =?H_1(n_i \| X)$. However, the proposed scheme can detect malicious servers. This is because S have publicized $N = h(n)$ in the setup phase, and the smart card will check the master secret in the registration phase. Therefore, our scheme guarantees user anonymity.

Forward Secrecy. The security requirement of forward secrecy guarantees that even if the secrets are compromised, previous session keys will not break. In our scheme, the session key consists of a Diffie-Hellman key $K = X^y = Y^x$. Owing to the hardness of computational Diffie-Hellman problem, as long as the ephemeral exponents are not compromised, the session keys will stay secure.

Resistance to Man-in-the-Middle Attack. Suppose an adversary $\mathcal{A}$ tries to stand in the middle of a client C_i and the server S, $\mathcal{A}$ has to modifies some transmitted messages. If the value X in A, X sent by C_i is modified, $\mathcal{A}$ must compute a correct authenticator A. However, this is impossible since the adversary cannot compute the master secret. Therefore, our scheme can resist man-in-the-middle attacks.

Resistance to Replay Attack. In this attack, an adversary $\mathcal{A}$ tries to replay some transmitted messages in different protocol runs to achieve malicious ends. Since all messages does not contain an identity, $\mathcal{A}$ can replay a message from other sessions. However, our scheme employs a challenge-response design pattern so that responses not corresponding to the challenge will be rejected. Therefore, our scheme can resist replay attacks.

Efficiency. Our protocol achieves high efficiency in terms of both computation and communication cost. The computation overhead in the authentication phase includes a single exclusive-or (XOR), several hash and two exponentiation operations. Compared with lightweight operations such as XOR and hash, although exponentiations are expensive, in order to achieve forward secrecy, at least two exponentiations have to be performed. Regarding to communication overhead, the authentication phase is a three-round protocol. The largest size of the messages transmitted in a single round consists of only a group element and a hash digest.

We compare our scheme with other APAKE protocols in Table 1. The computation cost is evaluated by the operations performed by a clients or the server, including paring (P), exponentiation (E), and hash (H) operations. In our analysis, the symmetric encryption/decryption is viewed as a hash operation for simplicity. The communication cost comes from the transmitted messages, including the size of any group elements ($|G|$), nonces, and authenticators. We treat all those other than the group elements as hash digests ($|H|$). In some schemes, the costs involves the number of registered clients, which is denoted by n.

Table 1. Efficiency comparison of APAKE protocols

Protocols	Computation overhead		Comm.	Usability	Revoc.
	Client	Server			
NAPAKE [3]	$4E + 2H$	$(n+3)E + 2H$	$(n+3)\|G\| + \|H\|$	(3)	✓
VEAP [4]	$2E + 5H$	$(n+2)E + nH$	$3\|G\| + (n+2)\|H\|$	(3)	✓
Yang et al. [7]	$2P + 16E + 2H$	$4P + 13E + 3H$	$8\|G\| + 6\|H\|$	(2)	✓
Yang et al. [6]	$7E + 3H$	$5E + 3H$	$3\|G\| + 5\|H\|$	(2)	✓
S2APA [2]	$5E + 3H$	$3E + 3H$	$3\|G\| + 2\|H\|$	(2)*	×
SAPAKE [5]	$6E + 3H$	$3E + 3H$	$3\|G\| + 2\|H\|$	(2)*	×
Chai et al. [1]	$2E + 5H$	$2E + (2n+2)H$	$(n+1)\|G\| + 3\|H\|$	(1)	✓
Ours	$2E + 3H$	$2E + 3H$	$2\|G\| + 3\|H\|$	(1)	×

These protocols varies in computation and communication overhead. On the one hand, they apply different level of usability. As shown in the table, there are mainly three levels related to device usage in APAKE protocols: (1) dedicated device-assisted, e.g. tamper resistant smart cards, (2) general device-assisted, e.g., public directories, and (3) password-only. The higher the level is, the more user-friendly the protocol will be in practice. Note that both the S2APA [10] protocol and the SAPAKE [8] protocol suffer from the credential forgery attack due to the use of general devices. One possible solution is to downgrade its usability to 1. Thus, among those schemes of least usability, our proposed scheme is the most efficient. On the other hand, some schemes emphasize unconditional anonymity, and thus do not take into account the client revocation, while others can be extended to support revocation efficiently. Although our scheme cannot be easily adapted to revoke user's smart card, it achieves high efficiency.

5 Conclusions

In this paper, we show that Shin and Kobara's S2APA protocol is insecure against a credential forgery attack. The attack lies in the use of general devices. A simple but effective solution is to replace with dedicated devices such as tamper proof smart cards. To improve its efficiency, we propose another APAKE protocol using smart cards. Through analysis, we show that although our protocol requires a dedicated device, it is secure against various attacks and efficient at protecting user's anonymity.

Acknowledgement. The work of Chien-Ming Chen was supported in part by the Project NSFC (National Natural Science Foundation of China) under Grant number 61402135 and in part by Shenzhen Technical Project under Grant number JCYJ20170307151750788. The work of Eric Ke Wang was supported in part by National Natural Science Foundation of China (No. 61572157), grant No. 2016A030313660 from Guangdong Province Natural Science Foundation, JCYJ20160608161351559 from Shenzhen Municipal Science and Technology Innovation Project.

References

1. Chai, Z., Cao, Z., Lu, R.: Efficient password-based authentication and key exchange scheme preserving user privacy. In: International Conference on Wireless Algorithms, Systems, and Applications, pp. 467–477. Springer (2006)
2. Viet, D.Q., Yamamura, A., Tanaka, H.: Anonymous password-based authenticated key exchange. In: International Conference on Cryptology in India, pp. 244–257. Springer (2005)
3. Kim, S., Rhee, H.S., Chun, J.Y., Lee, D.H.: Anonymous and traceable authentication scheme using smart cards. In: International Conference on Information Security and Assurance, ISA 2008, pp. 162–165. IEEE (2008)
4. Liu, Y., Zhao, Z., Li, H., Luo, Q., Yang, Y.: An efficient remote user authentication scheme with strong anonymity. In: 2008 International Conference on Cyberworlds, pp. 180–185. IEEE (2008)
5. Shao, S., Li, H., Niu, X., Yang, Y.: A remote user authentication scheme preserving user anonymity and traceability. In: 5th International Conference on Wireless Communications, Networking and Mobile Computing, WiCom 2009, pp. 1–4. IEEE (2009)
6. Yang, Y., Zhou, J., Weng, J., Bao, F.: A new approach for anonymous password authentication. In: Annual Computer Security Applications Conference, ACSAC 2009, pp. 199–208. IEEE (2009)
7. Yang, Y., Zhou, J., Wong, J.W., Bao, F.: Towards practical anonymous password authentication. In: Proceedings of the 26th Annual Computer Security Applications Conference, pp. 59–68. ACM (2010)
8. Qian, H., Gong, J., Zhou, Y.: Anonymous password-based key exchange with low resources consumption and better user-friendliness. Secur. Commun. Netw. **5**(12), 1379–1393 (2012)

9. Yang, Y., Lu, H., Liu, J.K., Weng, J., Zhang, Y., Zhou, J.: Credential wrapping: from anonymous password authentication to anonymous biometric authentication. In: Proceedings of the 11th ACM on Asia Conference on Computer and Communications Security, pp. 141–151. ACM (2016)
10. Shin, S., Kobara, K.: Simple anonymous password-based authenticated key exchange (SAPAKE), reconsidered. IEICE Trans. Fundam. Electron. Commun. Compu. Sci. **100**, 639–652 (2017)
11. Shin, S., Kobara, K.: A secure anonymous password-based authentication protocol with control of authentication numbers. In: 2016 International Symposium on Information Theory and its Applications (ISITA), pp. 325–329. IEEE (2016)
12. Son, K., Han, D.G., Won, D.: Simple and provably secure anonymous authenticated key exchange with a binding property. IEICE Trans. Commun. **98**(1), 160–170 (2015)
13. Chen, C.M., Li, C.T., Liu, S., Wu, T.Y., Pan, J.S.: A provable secure private data delegation scheme for mountaineering events in emergency system. IEEE Access **5**, 3410–3422 (2017)
14. Chen, C.M., Fang, W., Wang, K.H., Wu, T.Y.: Comments on an improved secure and efficient password and chaos-based two-party key agreement protocol. Nonlinear Dyn. **87**(3), 2073–2075 (2017)
15. Chen, C.M., Xu, L., Wu, T.Y., Li, C.R.: On the security of a chaotic maps-based three-party authenticated key agreement protocol. J. Netw. Intell. **2**, 61–65 (2016)
16. Sun, H.M., He, B.Z., Chen, C.M., Wu, T.Y., Lin, C.H., Wang, H.: A provable authenticated group key agreement protocol for mobile environment. Inf. Sci. **321**, 224–237 (2015)
17. Chen, C.M., Wang, K.H., Wu, T.Y., Pan, J.S., Sun, H.M.: A scalable transitive human-verifiable authentication protocol for mobile devices. IEEE Trans. Inf. Forensics Secur. **8**(8), 1318–1330 (2013)
18. Yang, J., Zhang, Z.: A new anonymous password-based authenticated key exchange protocol. In: International Conference on Cryptology in India, pp. 200–212. Springer (2008)
19. Shin, S.H., Kobara, K., Imai, H.: Very-efficient anonymous password-authenticated key exchange and its extensions. In: International Symposium on Applied Algebra, Algebraic Algorithms, and Error-Correcting Codes, pp. 149–158. Springer (2009)
20. Zhang, Z., Yang, K., Hu, X., Wang, Y.: Practical anonymous password authentication and TLS with anonymous client authentication. In: Proceedings of the 2016 ACM SIGSAC Conference on Computer and Communications Security, pp. 1179–1191. ACM (2016)

Digital Certificate Based Security Payment for QR Code Applications

Jing Zhang(✉), Shi-Jian Liu, Jeng-Shyang Pan, and Xiao-Rong Ji

Fujian Provincial Key Laboratory of Big Data Mining and Applications,
College of Information Science and Engineering,
Fujian University of Technology, Fuzhou, China
jing165455@126.com

Abstract. A rapidly growing social interaction technology is the use of Quick Response (QR) codes as physical entrances (i.e. the encoded URL address) to Internet resources. However, there exist various kind of security problems, for example, the URL may point to some phishing web sites. A digital certificate based signature and verification QR code is proposed in this paper. The public and private keys are generated by the authentication mechanism, which using the Public Key Infrastructure (PKI) technology. The public key can be published with the designed scanning application (App). An authenticated QR code can be created based on the private key and the merchant information. Clients can authenticate the codes by the public key and scanning App. Test results prove that the security QR code designed by this system, which binds the digital certificate, can be verified by the decoding scanner.

Keywords: QR code · Security · Certificate · Signature

1 Introduction

In this increasingly interconnected world, it is the second nature for people to communicate, socialize and share information through a huge number of media platforms, often simultaneously. Along with the emergence of new mechanisms, technologies to support this is constantly progressing. A rapidly growing social interaction technology is the use of Quick Response (QR) codes as physical shortcuts to Internet resources. Individual can use their mobile phones to capture

J. Zhang—The authors wish to thank National Natural Science Foundation of China (Grant No: 61072080, 61572010), Natural Science Foundation of Fujian Province of China (2017J05098), The Education Department of Fujian Province science and technology project (JAT160328, JZ160461), and the science research project in Fujian University of Technology (GY-Z160066, GY-Z160130, GY-Z160138).

P. Krömer et al. (eds.), *Proceedings of the Fourth Euro-China Conference on Intelligent Data Analysis and Applications*, Advances in Intelligent Systems and Computing 682, DOI 10.1007/978-3-319-68527-4_10

the QR code quickly and visit the web site through the decoded URL. Meanwhile, attacks also gain in conveniences from the popularity of QR codes, and the public should concern about the security of the widespread used QR codes.

An ordinary usage of QR code is shown in Fig. 1 without the procedures marks with rectangles. In this typical scenario, a merchant can create a QR code which is corresponding to the payment of some goods for selling. It can be done by QR code generators as simple as the online tools. Then, the QR code will be distributed by media such as Internet, waiting for being scanned by clients. Today, markets in China are filled with such codes for payment, after scanning the QR code, user will be redirected to an intermediate payment agent or a company's web page to buy the goods for convenience, which is often referred to as "one-click" payment [1]. But along with the convenience, there are also many risks.

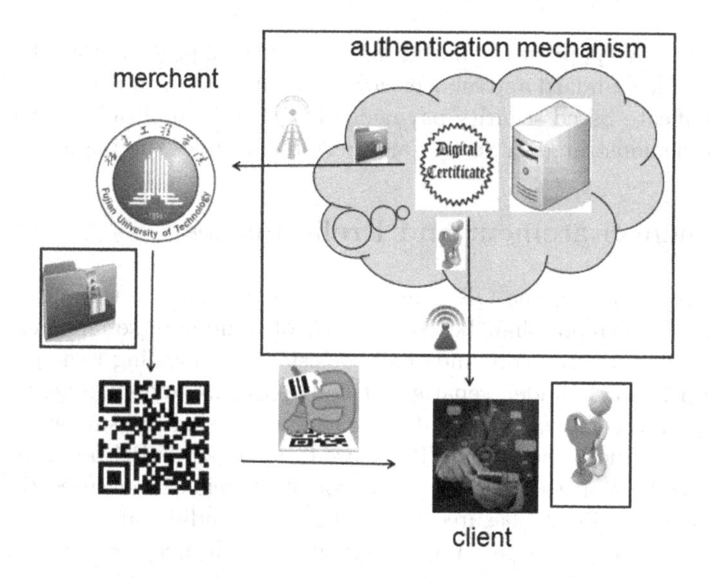

Fig. 1. Demonstration of the usage of QR code.

One of the most dangerous issue is the risk of phishing. In this case, the attacker misleads the user to a malicious site and then "phishing" for users' information such as the bank/credit card details, personally identifying information like their mothers maiden name and postal address. It is difficult to identify the authenticity of QR codes by ordinary users, because firstly the codes are not human-readable but a ultimate form of URL obscuring service. Secondly, for the unique way in which users access QR codes, there may be a illusion for people that the QR code is safer than a web link. Other risks could be the implanting of viruses such as Trojan. Since there are many extra storage spaces in QR code, attacks can inject viruses into the code. When users scanned such code, the viruses will be downloaded and installed into their mobile terminals and make some damages.

In order to deal with the problems, security techniques can offer some help. They have been widely used in applications including encoding, encryption, secure identification, verification and watermarking. Verification techniques based on digital certifications have gained more and more attention. In recent years, experts and scholars have put forward a variety of optical encryption techniques [1–3,5].

This paper aim to solve the problem for QR code based security payment. Procedures marked with red rectangles in Fig. 1 highlight differences which distinguish ours from traditional methods. The main contributions of this paper is: An authentication scheme is introduced, including the Certificate Authority based authentication mechanism. Digital signatures techniques are involved to ensure the authentication and the way of certification based on the proposed QR code.

The rest of this paper is organized as follows: Sect. 2 presents the introduction to the QR code standard and an overview of the use cases. In Sect. 3, we present digital certificate based security payment for QR code applications. Simulation results are demonstrated in Sect. 4, and Sect. 5 concludes the paper.

2 Problem Statement and Preliminaries

A brief introduction to the QR standard [10] is provided in this section. Each QR Code 2005 symbol shall be constructed of nominally square modules set out in a regular square array and shall consist of an encoding region and function patterns, namely finder, separator, timing patterns, and alignment patterns. Function patterns do not encode data. The symbol shall be surrounded on all four sides by a quiet zone border. There are forty sizes of QR Code 2005 symbol referred to as Version 1, Version 2 ... Version 40. Version 1 measures 21 modules × 21 modules, Version 2 measures 25 modules × 25 modules and so on increasing in steps of 4 modules per side upto Version 40 which measures 177 modules × 177 modules. Figure 2 illustrates the structure of a Version 7 symbol.

There are various kind of attacks based on QR code. For example, attackers use malicious QR codes to direct users to fraudulent web sites, which masquerade as legitimate web sites aiming to steal sensitive personal information such as usernames, passwords or credit card information. As illustrated in Fig. 3, differences of the two QR codes are hard to be observed especially when they are not compared side by side. In other words, attacker replaces the QR code representing a URL equals to "http://www.fjut.edu.cn" by creating a new one encoded with a malicious link "http://www.fjat.edu.cn" and pasting it over the original one. It can be seen that this attack is simple yet effective.

There are some articles designed for avoiding the attacks [4,6–9,13–16]. Among them, digital signatures have proved to be an effective way to improve security [11]. Based on the digital signatures, the QR code can be checked

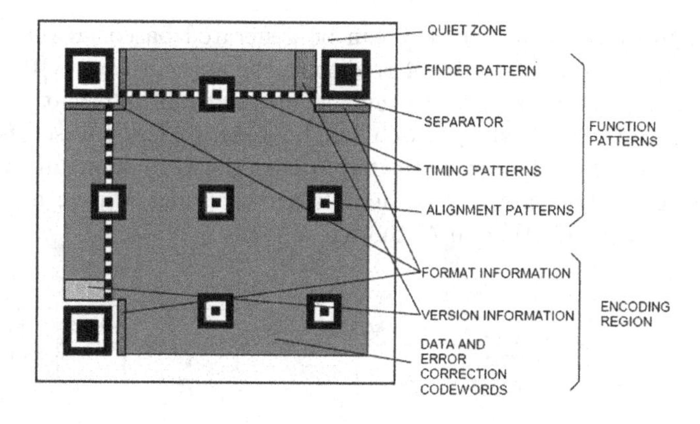

Fig. 2. QR code standard [10].

whether it has been modified or not. Shamir [12] introduced identity-based cryptography in 1984 in order to simplify the key-management procedure of traditional, certificate-based, public-key infrastructures. Shamir's approach allowed an entity's public key to be derived directly from her or his identity, such as an email address, and the entity's private key can be generated by a trusted third party. A digital certificate based security payment for QR code applications will be designed in this paper.

Fig. 3. One example of QR code trap

3 Authenticated QR Code Based on Digital Signatures

In this section, a work for QR code authentication is proposed based on techniques such as digital signature. The whole frame is shown in Fig. 4. The frame can be divided into three phases. The first phase is named the digital certificate generation (Sect. 3.1), in this phase public and private keys are generated according to a authentication mechanism, which using the Public Key Infrastructure (PKI) technology. The public key can be published with the designed scanning application (App), while private key can be generated based on the merchant information. The second phase is referred to as the authenticated QR code generation (Sect. 3.2). Using the private key and the merchant information (i.e., the signature), an authenticated QR code can be generated in the second phase. The third one is called the authenticated QR code certification. In this phase, client can authenticate the code by the designed scanning App with public key downloaded from the Certificate Authority.

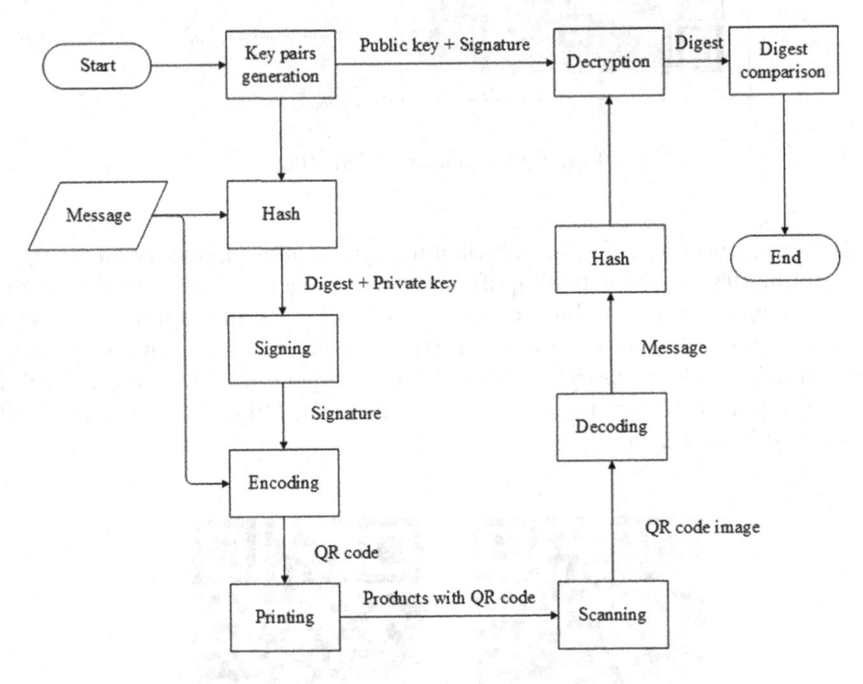

Fig. 4. The work-flow of the proposed method.

3.1 Digital Certificate Generation Phase

Our proposed scheme is closely related to the ID-based cryptography. In an ID-based cryptographic algorithms, each merchant needs to register at a CA and identify himself before joining the network. Once a merchant is accepted, the CA will generate a private key for him. The merchant's identity (e.g. merchandise's name) becomes the keyword to find the corresponding public key from CA. In order to verify a digital signature of a QR code, after the merchants associate the authenticated QR code with the merchandise, the client only needs to know the authenticated QR code of his trade partner and the public key from the CA in order to carry out the authentication process.

3.2 Authenticated QR Code Generation Phase

Before generate the authenticated QR code, basic information should be given like the ones required for ordinary QR code generation, which include the QR code version V and error correction level L for initialization. Then information from merchandise and its digital signature will be combined and treated as message for QR code encoding. If the capacity of selected version is not enough for encode the entire message, it will be promoted automatically. The algorithm for authenticated QR code generation is shown in Algorithm 1.

Algorithm 1. Authenticated QR code generation

Input: Information of merchandise M, private key K_{pri}, version V, error correction level L, authentication mode O.

Output: Authenticated QR code QR_S with authenticated information including the digital signature.

Step 1: Adopt a hash operation for M which will result in a digest $D = Hash(M)$;

Step 2: Use the K_{pri} to sign the digest D and produce the signature S;

Step 3: Let message M' be the combination of M and S with specified separators;

Step 4: Encode M' to form a QR code as the routine of standard QR code encoding procedure with V and L;

As an example shown in Fig. 5. Given the L, L as the parameters and a URL as the encoded message, then a normal QR code can be generated as shown in Fig. 5(a). While, the result of our method is the one shown in Fig. 5(b), where both the URL and generated signature are encoded into the QR code.

Fig. 5. One Authenticated QR code example

3.3 Authenticated QR Code Verification Phase

In the authentication phase, the verification scheme for the proposed QR code will be given. The basic idea for the verification is that, according to the principle of digital signature techniques, if we use the hash function, which is the same one as used for QR code generation, to produce a digest from the original information of a merchandise, then it will perfectly match the one decrypted from signature using the rightful public key from CA. Otherwise, it can be confirmed that the QR code was modified. The scheme can be described by the following Algorithm 2.

Algorithm 2. Authenticated QR code verification

Input: Authenticated QR code QR_S

Output: The certification result R of QR_S

Step 1: Decode the QR_S as the routine of standard QR code decoding procedure, which will result in a message string M';

Step 2: If M' is subject to a specified form, extract the original information of merchandise M, signature S and digital certification for public key key_{ID} from M';

Step 3: Use the hash function to generate a digest D_1 from M;

Step 4: Use the K_{id} to get the rightful public key K_{pub} from CA;

Step 5: Use K_{pub} to decrypt S which will get another digest D_2;

Step 6: Compare the two digests, and return the $R = True$ if $D_1 equal D_2$, else $R = False$;

4 Experimental Results

This section describes some experiments. Multiple system environments are used for testing as shown in Table 1.

Table 1. System testing environment

Device	Type	Param
PC	ASUS X450V	Windows 10(64); 8G (RAM); Inter Core i5-3230M; 500G
PC	Lenovo thinkpad E540	Windows 10(64); 4G (RAM); Inter Core i5-4200; 1T
Phone	HUAWEI honor 8	Android 7.0; 4G (RAM); Hisilicon Kirin 950; 32G
Phone	SAMSUNG GTI9260	Android 4.1.1; 1G (RAM); Imagination PowerVR SGX544; 4G

To validate the authenticated QR code generation, URL addresses and their digital signatures are encoded to form different authenticated QR codes. While, two of them are demonstrated in Fig. 6. Our proposed authentication scheme can identify whether the URL has been modified, as shown in Fig. 7. In other words, the truth or false of the authenticated QR code can be distinguished. If the authentication fails, entrance of the web site through the URL will be forbidden, which can make sure the security of the QR code.

Fig. 6. Test results of the proposed authenticated QR code generation.

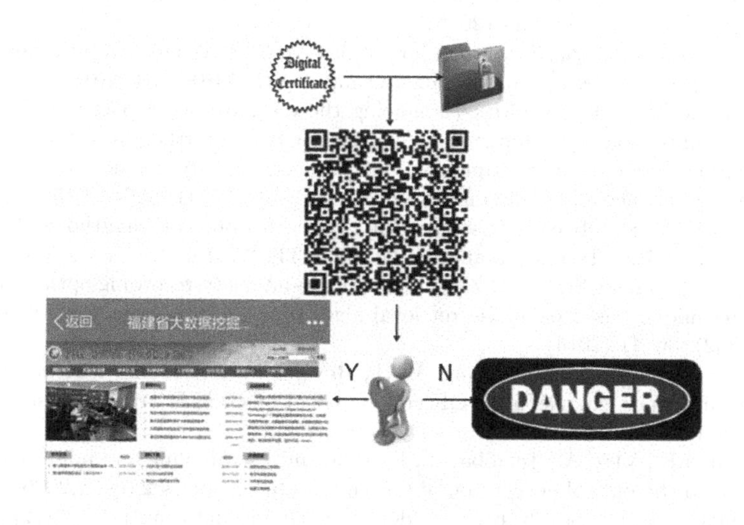

Fig. 7. Test result of the authenticate process.

5 Conclusions

Aiming for guarantee the QR Code based payment security, authenticated QR codes along with both the generating and resolving methods are proposed in this paper, digital signatures techniques are involved to ensure the certificate information of merchandise delivered by QR code. Besides, an authentication scheme is introduced, including the certificate authority based authentication mechanism and the way of certification based on the proposed QR code. If authentication fails, activities such as entrance of the given URL will be forbidden, which makes the proposed QR code secure for usage. One limitation of the proposed QR codes is that, since the signature is encoded along with the original message, a much larger capacity for QR code is required comparing with the traditional one. The problem will be solved in our future works.

References

1. Krombholz, K., Frhwirt, P., Kieseberg, P., Kapsalis, I., Huber, M., Weippl, E.: QR code security: a survey of attacks and challenges for usable security. In: Human Aspects of Information Security, Privacy, and Trust, HAS 2014. LNCS, vol. 8533, pp. 79–90 (2014)
2. Krombholz, K., Frühwirt, P., Rieder, T., Kapsalis, I., Ullrich, J.: QR code security–how secure and usable apps can protect users against malicious QR codes. In: International Conference on Availability, Reliability and Security, pp. 230–237. IEEE (2015)
3. Mary, G., Rani, M.S.: Security augmentation of visual cryptography using watermarking embedded QR code. In: International Conference on Signal Processing, Control and Data Analytics (2016)
4. Wane, A.R., Jamankar, S.P., Chandure, O.V.: An effective mechanism for ensuring security of QR code. Int. J. Adv. Res. Comput. Sci. **04**(06), 34 (2013)
5. Umaria, M.M., Jethava, G.B.: Enhancing the data storage capacity in QR code using compression algorithm and achieving security and further data storage capacity improvement using multiplexing. In: International Conference on Computational Intelligence and Communication Networks, pp. 1094–1096 (2015)
6. Kartheshwar, S.: QR code based security system to prevent unauthorized access. Int. J. Stud. Res. Technol. Manag. **3**(04), 336–339 (2015)
7. Wang, Z.P., Zhang, S., Liu, H.Z., Qin, Y.: Single-intensity-recording optical encryption technique based on phase retrieval algorithm and QR code. Opt. Commun. **332**(332), 36–41 (2014)
8. Huang, H.C., Chang, F.C., Fang, W.C.: Reversible data hiding with histogram-based difference expansion for QR code applications. IEEE Trans. Consum. Electron. **57**(2), 779–787 (2011)
9. Barrera, J.F., Vlez, A., Torroba, R.: Experimental scrambling and noise reduction applied to the optical encryption of QR codes. Opt. Express **22**(17), 20268 (2014)
10. Information technology-Automatic identification and data capture techniques-QR Code 2005 bar code symbology specification. ISO/IEC 18004 (2006)
11. Sahu, R.A., Padhye, S.: Provable secure identity-based multi-proxy signature scheme. Int. J. Commun. Sys. **28**(3), 497–512 (2015)

12. Shamir, A.: Identity-based cryptosystems and signature schemes. In: Proceedings of Advances in Cryptolog, CRYPTO 1984. LNCS, vol. 196, pp. 47-53. Springer, Heidelberg (1984)
13. Mittra, P., Rakesh, N.: A desktop application of QR code for data security and authentication. In: International Conference on Inventive Computation Technologies. IEEE (2017)
14. Purnomo, A.T., Gondokaryono, Y.S., Kim, C.S.: Mutual authentication in securing mobile payment system using encrypted QR code based on public key infrastructure. In: International Conference on System Engineering and Technology, pp. 194–198. IEEE (2017)
15. Aulya, R., Hindersah, H., Prihatmanto, A.S., Rhee, K.H.: An authenticated passengers based on dynamic QR code for Bandung smart transportation systems. In: Engineering Seminar. IEEE (2017)
16. Lu, J., Yang, Z., Li, L., Yuan, W., Li, L., Chang, C.C.: Multiple schemes for mobile payment authentication using QR code and visual cryptography. Mob. Inf. Syst. **2017**, 1–12 (2017)

New Methods for Intelligent Data Analysis

A Novel Region Selection Algorithm for Auto-focusing Method Based on Depth from Focus

J. Chen[1,2](✉), D.Q. Chen[1], and S.H. Meng[1](✉)

[1] School of Information Science and Engineering, Fujian University of Technology, Fuzhou 350118, China
jchen321@126.com, menghui@fjut.edu.cn

[2] Fujian Provincial Key Laboratory of Digital Equipment, Fujian University of Technology, Fuzhou 350118, China

Abstract. Region selection algorithm for auto-focusing method based on depth from focus plays an important role in obtaining a high-quality image. A novel algorithm is proposed after analyzing different region selection algorithms. Firstly, the gradient image is obtained through Scharr operator with improved templates. Then the region with maximum accumulation value is selected as the final auto-focusing region based on global searching with fixed step. The experiments show that the proposed algorithm is more accurate compared with other algorithms.

Keywords: Scharr operator · Region selection · Depth from focus · Auto-focusing

1 Introduction

The focusing technology plays an important part in many imaging devices. However, manual operated focusing has the disadvantages including low degree of automation, slow process of focusing and low precision affected by human factor [1]. The auto-focusing methods could solve the above problems [2]. The auto-focusing methods can be divided into the initiative controlling methods and image processing [1]. The image processing based auto-focusing can be classified as depth from defocus and depth from focus [3].

Depth from focus finds the focusing position through searching process. The method includes three parts: region selection algorithm, image definition evaluation function and searching algorithm. Region selection algorithm could improve the real-time performance of the auto-focusing method through focusing on the portion of the image instead of the whole image. Traditional fixed region selection algorithms include center region algorithm, multi-region algorithm, golden section algorithm [4–6]. With the development of image processing and performance of microcontroller unit, some automatic region selection algorithms were proposed. The first order region selection algorithm is proposed in contrast with the disadvantages of some traditional region selection algorithms [7]. An algorithm based on particle swarm optimization

P. Krömer et al. (eds.), *Proceedings of the Fourth Euro-China Conference on Intelligent Data Analysis and Applications*, Advances in Intelligent Systems and Computing 682, DOI 10.1007/978-3-319-68527-4_11

(PSO) is proposed in [8]. Entropy-based focusing region selection algorithm estimates the subject region regardless of the subject's position and the existence of high frequency component in the background [9]. Liang [10] and Lee [11] also proposed different algorithms for region selection in auto-focusing. In this study, the local gradient accumulation value based region selection algorithm is proposed based on the four directional Scharr operator with improved templates and global searching with fixed step.

This paper is organized as follows. Section 2 describes the detail of the algorithm with two stages, including obtaining gradient image and selecting focusing region. Experimental results are shown in Sect. 3. Finally, Sect. 4 concludes the paper.

2 Proposed Region Selection Algorithm

The proposed region selection algorithm can be divided into two stages. Firstly, four directional Scharr operator with improved templates is used to obtain gradient image. Then, after fixing the region size, focusing region is selected through global searching the maximum gradient accumulation value of different region with fixed step.

2.1 Gradient Image Through Scharr Operator with Improved Templates

The intense area corresponds to the detailed region in gradient image, which can be used as the focusing region in most cases. As Scharr operator owns certain advantages compared with other operators [12], the Scharr operator with improved template is used in the proposed algorithm. The proposed algorithm could down-sampled the image before using Scharr operator if the size of image is big.

Most objects in the images own multi-directional feature. Based on the 5×5 external templates with two directions in [12], the proposed algorithm introduces the 5×5 external templates with four directions (see Fig. 1).

$$\begin{bmatrix} 2 & 3 & 0 & -3 & -2 \\ 3 & 4 & 0 & -4 & -3 \\ 6 & 6 & 0 & -6 & -6 \\ 3 & 4 & 0 & -4 & -3 \\ 2 & 3 & 0 & -3 & -2 \end{bmatrix} \quad \begin{bmatrix} 6 & 3 & 2 & 3 & 0 \\ 3 & 6 & 4 & 0 & -3 \\ 2 & 4 & 0 & -4 & -2 \\ 3 & 0 & -4 & -6 & -3 \\ 0 & -3 & -2 & -3 & -6 \end{bmatrix} \quad \begin{bmatrix} 2 & 3 & 6 & 3 & 2 \\ 3 & 4 & 6 & 4 & 3 \\ 0 & 0 & 0 & 0 & 0 \\ -3 & -4 & -6 & -4 & -3 \\ -2 & -3 & -6 & -3 & -2 \end{bmatrix} \quad \begin{bmatrix} 0 & 3 & 2 & 3 & 6 \\ -3 & 0 & 4 & 6 & 3 \\ -2 & -4 & 0 & 4 & 2 \\ -3 & -6 & -4 & 0 & 3 \\ -6 & -3 & -2 & -3 & 0 \end{bmatrix}$$

(a) (b) (c) (d)

Fig. 1. 5×5 external templates with four directions. (a) The 0° directional template, (b) the 45° directional template, (a) the 90° directional template, (b) the 135° directional template.

To improve the real-time performance, the proposed algorithm calculates four directional gradient absolute values, then accumulate them, as follows:

$$G_I(i,j) = \left|\frac{\partial I(i,j)}{\partial_{0^\circ}}\right| + \left|\frac{\partial I(i,j)}{\partial_{45^\circ}}\right| + \left|\frac{\partial I(i,j)}{\partial_{90^\circ}}\right| + \left|\frac{\partial I(i,j)}{\partial_{135^\circ}}\right| \tag{1}$$

where, *I(i,j)* means the image after preprocessing, $G_I(i,j)$ means the gradient image, $\partial I(i,j)/\partial_{0^\circ}$, $\partial I(i,j)/\partial_{45^\circ}$, $\partial I(i,j)/\partial_{90^\circ}$, $\partial I(i,j)/\partial_{135^\circ}$ means four directional gradient value respectively.

Figure 2 shows the gradient image with the original two directional templates and four directional templates in the proposed algorithm. The detail information in gradient image with four directional templates is more obvious.

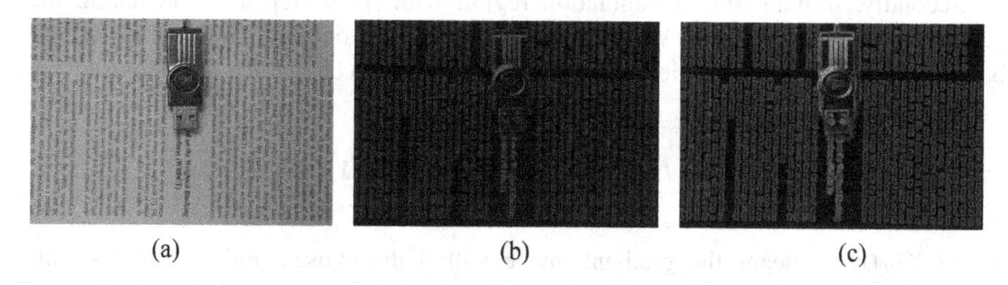

(a) (b) (c)

Fig. 2. Gradient images with different templates. (a) Original image, (b) Gradient image with two directional templates, (c) Gradient image with four directional template.

Table 1 shows the real-time performance comparison with two directional Scharr operator and four directional Scharr operator under the Matlab R2015a platform. The size of test images is 400 × 600, with 10 different images under the hardware configuration shown in Sect. 3. Two algorithms nearly own the same consuming time. Although the proposed algorithm calculates gradient value with four directional templates, the algorithm summates four directional gradient absolute values instead of square operation in two directions and one root operation in original algorithm [12].

Table 1. Real-time performance of Scharr operator with different templates

Different templates	Two directional Scharr operator	Four directional Scharr operator
Time (s)	0.0594	0.0689

2.2 Focusing Region Selection

Inspired by the concept of region division in [9], the proposed algorithm divides the gradient image into M × N regions as the first step. Then gradient value in each region is accumulated. Figure 3 shows the gradient distribution with three different block sizes (or region sizes).

In the original algorithm, if the region size is too big, more non-detailed area is introduced, while if the region size is too small, the value of image definition evaluation function could be easily affected by noise. At the same time, the subjects may be divided into two different regions. According to above problems, the proposed algorithm presents region selection algorithm with following steps.

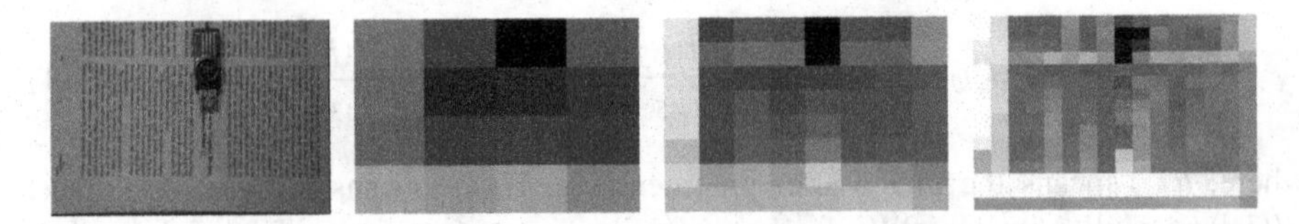

Fig. 3. Local gradient accumulation results with different region size.

Firstly, the region size is fixed. The region size should be proper to avoiding introduce the non-detailed area and noise.

Secondly, moving the accumulation region with fixed step and calculating the gradient accumulation value within the region. Moving direction is from top-left to bottom-right, then local gradient accumulation value is accumulated as follows:

$$S(i,j) = \sum_{x=i\bullet v_l}^{i\bullet v_l + l}\sum_{y=j\bullet v_h}^{j\bullet v_h + h} Sum(G_{IR}(x,y)) \tag{2}$$

where, $G_{IR}(x, y)$ means the gradient image within the chosen region, *Sum*() means accumulate the gradient value within the region; l and h means the width and height of the region respectively; v_l and v_h means moving speed (step length) in two directions, as shown in Fig. 4; i and j means step number in two directions; $S(i,j)$ means the accumulation value within the region.

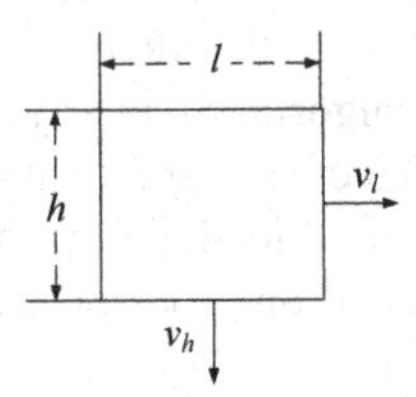

Fig. 4. The accumulation region moving direction

It can be found that if the moving step is small, it is prone to find the most detailed region of the image, but real-time performance decreases. Combing with real-time performance and accuracy, the moving step is set as the half length of the width and height of the accumulation region, as follows:

$$\begin{cases} v_h = (1/2) \bullet h \\ v_l = (1/2) \bullet l \end{cases} \tag{3}$$

Finally, choosing the region with maximum gradient accumulation value as the focusing region.

3 Experiment and Analysis

To verify the feasibility and validity of the proposed algorithm. Testing has been made to fix the parameter and compare the performance. The image data for testing is captured by camera or from the websites. Images with different background are chosen (see Fig. 5).

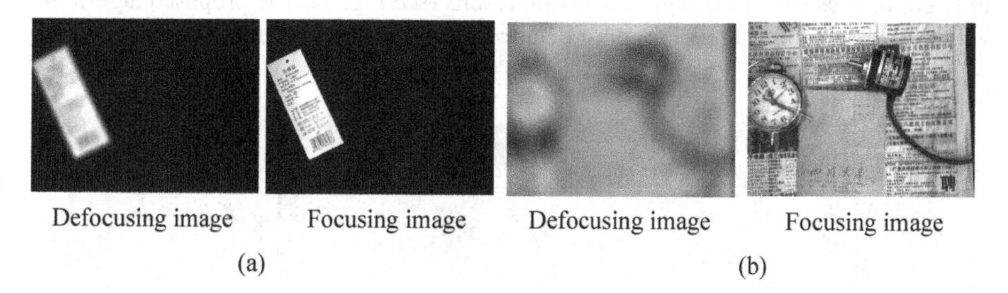

Fig. 5. Test images with different background. (a) Images with simple background, (b) Images with complicate background.

The experiments are implemented in Matlab (R2015a) on a 2.50 GHz Intel(R) Core (TM) i7-6500 machine with 8.00 GB of RAM under Windows 10 of 64 bit.

3.1 Fixation of Region Size

In the proposed algorithm, the region size needs to be fixed beforehand. In the experiments, we introduced different sizes to obtain focusing curves (see Fig. 6).

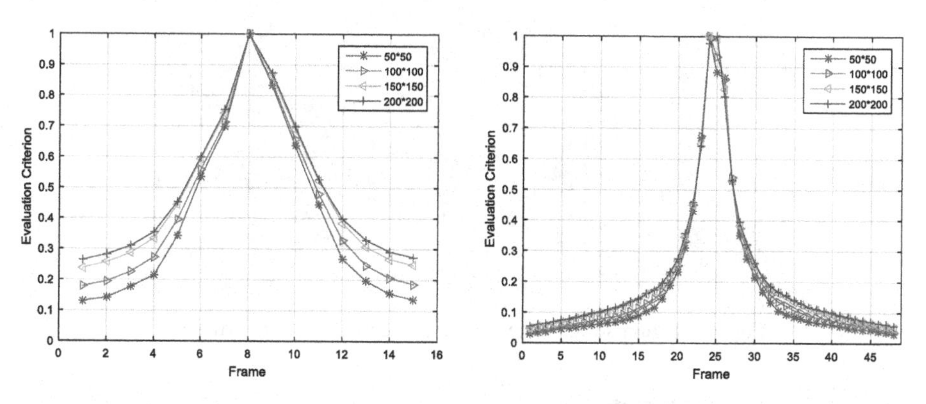

Fig. 6. Auto-focusing curve with Roberts operator for different size of region

In the experiments, we choose the most intensive area as focusing region by our region selection algorithm, and with different region sizes (50×50, 100×100, 150×150 and 200×200). It can be concluded that region with different sizes nearly have the same shape of curve, and region of 50 × 50 owns the best sensitiveness. Considered that the smaller region is prone to be affected by noise, we choose the size near 100 × 100 as the region size.

3.2 Performance of Global Searching with Fixed Step

To test performance of global searching with fixed step in the algorithm, we also use different images to test the region selection results (see Fig. 7). The proposed algorithm with global searching finds the more detailed region in the image.

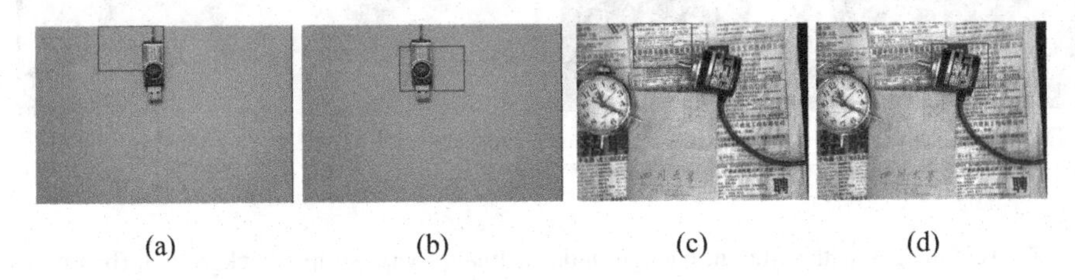

(a) (b) (c) (d)

Fig. 7. Region selection result comparisons for algorithms with and without global searching with fixed step: (a) and (c) without global searching, (b) and (d) with global searching.

We also test the stability of the proposed algorithm through locating region for focusing/defocusing images, as shown in Fig. 8. Although different images are in different levels of focusing, the proposed algorithm can locate the same region for them.

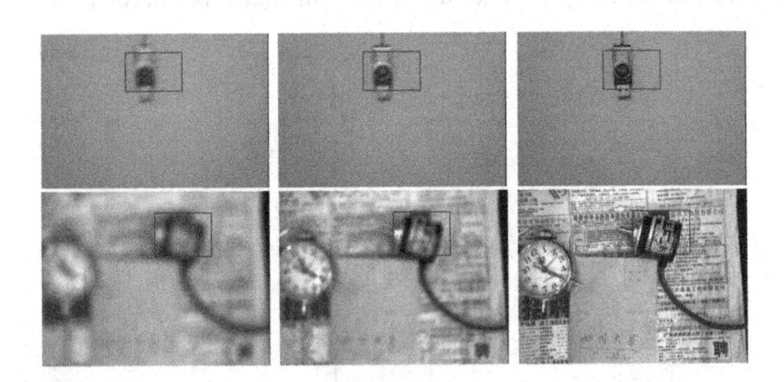

Fig. 8. Region selection result for focusing/defocusing images

3.3 Comparison with Different Algorithms

In the experiment, some other region selection algorithms are also introduced to compare the performance in accuracy and real-time performance. The introduced algorithms include the first order region selection algorithm [7] and entropy-based focusing region selection algorithm [9]. Figure 9 shows the selection results with three

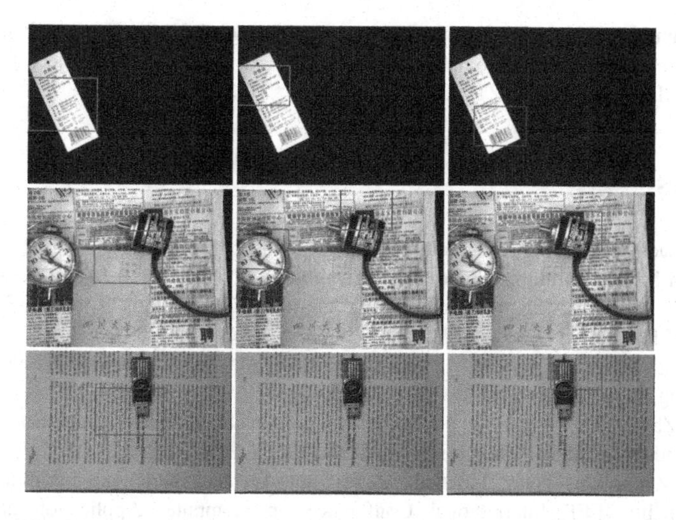

Fig. 9. Region selection results comparison for different algorithms

different algorithms. Compared with other algorithms, the proposed algorithm locates better. As the entropy-based algorithm use threshold for choosing region, the number of regions for focusing is unstable, even no region is selected when the number of the regions is small.

The real-time performance comparison with different algorithms is shown in Table 2. Although the proposed algorithm has the longest consuming time compared with two other algorithms, but the difference is acceptable.

Table 2. Real-time performance of different algorithms

Different algorithms	The first order region selection algorithm	Entropy-based focusing region selection algorithm	The proposed algorithm
Time (s)	0.0592	0.0262	0.0890

4 Conclusion

In this paper, we analysis the region selection algorithms for auto-focusing method based on depth from focus. Then we present a novel region selection algorithm based on gradient image obtained by Scharr operator with improved templates and focusing region selection by global searching maximum gradient accumulation value with fixed step. We also experiment to validate the feasibility and validity of the proposed algorithm. Through analyzing, it can be concluded that the proposed region selection algorithm is suitable for the object in the same plane. If the different subjects are in different distances, the image maybe blurred when using the proposed algorithm.

Acknowledgement. This work is partially supported by the Research Project from the Education Department of Fujian Province (JK2015031) and the Research Project from Fujian University of Technology (GY-Z160038).

References

1. Fu, G., Cao, Y., Lu, M.: A fast auto-focusing method of microscopic imaging based on an improved MCS algorithm. J. Innov. Opt. Health Sci. **8**(5), 1550020-1–1550020-10 (2015)
2. Lin, K.C.: Auto focusing under microscopic views. In: Proceedings of the IEEE Workshop on Signal Processing Systems, Shanghai, China, pp. 533–538 (2007)
3. Yuhu, Y., Tong, L., Jiawen, L.: Survey of the auto-focus methods based on image processing. Laser Infrared **43**(2), 132–136 (2013)
4. Yin, A., Zhang, Y., Yang, B., et al.: Roberts focused evaluation method and its application in multi-windows mode. J. Chongqing Univ. **34**(11), 25–30 (2011)
5. Liu, L., Zheng, Y., Feng, J., et al.: A fast auto-focusing technique for multi-objective situation. In: IEEE International Conference on Computer Application and System Modeling, Taiyuan, China, pp. 607–610 (2010)
6. Kim, Y., Lee, J.S., Morales, A.W., et al.: A video camera system with enhanced zoom tracking and auto white balance. IEEE Trans. Consum. Electron. **48**(3), 428–434 (2002)
7. Zhang, L., Jiang, W., Gao, Z.: Automatic focusing region selection algorithm based on first order of digital image. Opt. Tech. **34**(2), 163–165 (2008)
8. Wang, Y., Jiang, W., Zheng, Y.: Dynamic autofocusing region selection based on improved PSO. In: 3rd International Congress on Image and Signal Processing, Yantai, China, pp. 2760–2764 (2010)
9. Jeon, J., Yoon, I., Kim, D., et al.: Fully digital auto-focusing system with automatic focusing region selection and point spread function estimation. IEEE Trans. Consum. Electron. **56**(3), 1204–1210 (2010)
10. Liang, M., Wu, Z., Chen, T.: Auto-focusing adjustment of theodolites by largest the gradient method. Opt. Precis. Eng. **17**(12), 3016–3021 (2009)
11. Lee, S., Kumar, Y., Cho, J., et al.: Enhanced autofocus algorithm using robust focus measure and fuzzy reasoning. IEEE Trans. Circ. Syst. Video Technol. **18**(9), 1237–1246 (2008)
12. Liao, Y., Guo, L.: New no-reference image quality assessment method based on decomposition of gradient similarity. J. Comput. Appl. **33**(3), 691–694 (2013)

Graph Theory Modeling – A Petri Nets Based Approach

Shi-Jian Liu[1,2](✉), Xing-Si Xue[1,2], and Jing Zhang[1,2]

[1] College of Information Science and Engineering, Fujian University of Technology, No. 3 Xueyuan Road, University Town, Minhou, Fuzhou 350118, Fujian, China
liusj2003@fjut.edu.cn

[2] Fujian Provincial Key Laboratory of Big Data Mining and Applications, Fujian University of Technology, No. 3 Xueyuan Road, University Town, Minhou, Fuzhou 350118, Fujian, China

Abstract. There are many classic problems in graph theory which can be applied to research fields ranging from computer vision to transportation planning. Petri Nets (PNs) are mathematical objects which can be demonstrated by live graphic elements such as Place and Transition. They are competent in formalizing and solving issues in Graph theory. Take maximum flow problem as an example, though it has been well studied, its properties are rarely explored in the perspective of PNs to the best of our knowledge. In this paper, a Petri Nets based maximum flow modeling approach is proposed. Specifically, PN models of flow networks and the corresponding residual networks are firstly introduced. Based on the proposed models, the way of finding a maximum flow for a given flow network is presented. The PNs based maximum flow approach not only can solve the problem accurately, but also is intuitive and easy to understand resulting from its graphic simulation processes. Additionally, it is feasible to extent this work to other problems in graph theory as well.

Keywords: Maximum flow · Graph theory · Petri nets · Modeling

1 Introduction

There are many classic problems in graph theory, including the shortest path problem, maximum flow problem, etc. Many algorithms have been proposed to solve these problems, but it is undeniable that both the theories and solutions of

This work is supported by the Scientific Research Project in Fujian University of Technology (GY-Z160130, GY-Z160138, GY-Z160066), the Natural Science Foundation of Fujian Province of China (2017J05098), Young and Key Project of Fujian Education Department Funds (JZ160461) and Project in Fujian Provincial Education Bureau (JAT160328).

P. Krömer et al. (eds.), *Proceedings of the Fourth Euro-China Conference on Intelligent Data Analysis and Applications*, Advances in Intelligent Systems and Computing 682, DOI 10.1007/978-3-319-68527-4_12

the graph theory problems are generally based on abstract concepts and difficult to understand especially for junior students. Petri nets (PNs) are mathematic objects demonstrated by live graphic elements, which give them great advantages in formalization and modeling problems in graph theory. Many researchers have investigated methods for the combination, for example, Genrich et al. [1] use PNs to simulate and analyze the metabolic pathways. Hua and Hai [2] and Zhang et al. [3] carried out system failure analysis by solving the minimal cut sets based on PNs. Hu et al. [4] employed rough set theory and PNs to evaluate the reliability of stochastic flow network. But modeling and analysis for the properties of maximum flow problem are rarely explored in the perspective of PNs.

Maximum flow problem can be treated as a combinatorial optimization problem or special linear programming problem [5], which commonly existed in the segmentation for point clouds [6], meshes [7,8], images [9,10], and other aspects need for control and decision such as traffic network analysis [11,12]. Classic solutions of maximum flow problem include pre-flow push method, augmenting path method, etc. It can be extended to most reliable maximum flow on uncertain graph, maximum flow in directed planar networks with both node and edge capacities [13], and so on.

In order to demonstrate the PNs based graph theory modeling, this paper presents a PNs based maximum flow modeling approach as an example. The basic idea is similar to the augmenting path method. The proposed method not only can achieve accurate results, but also is intuitive and easy to understand resulting from its graphic simulation processes. The main contributions of this work include:

- A PNs based modeling method for given flow networks is proposed in the paper for maximum flow demonstration.
- A solution for maximum flow problem is proposed by simulation on proposed PN models for residual networks similar to the augmenting path method.

The rest of the paper is organized as follows. Problem statements and preliminaries are given in Sect. 2. The PNs based modeling for flow networks and residual networks are described in Sect. 3. Section 4 presents a solution for maximum flow, finally Sect. 5 concludes the paper.

2 Problem Statement and Preliminaries

2.1 Flow Networks and Flows

Before discussing the maximum flow problem, some basic notions should be given, including flow network, flow and its value.

Definition 1 (Flow networks [14]). *A flow network $G = (V, E)$ is a directed graph in which each edge $(u, v) \in E$ has a nonnegative capacity $C(u, v) \geq 0$. We further require that if E contains an edge (u, v), then there is no edge (v, u) in the* *reverse direction. If $(u,v) \notin E$, then $C(u,v) = 0$, and we disallow self-loops. We distinguish two vertices in a flow network: a source s and a sink t. We assume that each vertex lies on some path from the source to the sink.*

Definition 2 (Flow [14]**).** *Let $G = (V, E)$ be a flow network with a capacity function c. Let s be the source of the network, and let t be the sink. A flow in G is a real-valued function $f : V \times V \to R$ that satisfies the following two properties:*

(1) (Capacity constraint) For all $u, v \in V$, we require $0 \leq f(u,v) \leq c(u,v)$.
(2) (Flow conservation) For all $u \in V - \{s,t\}$, we require $\sum_{v \in V} f(v,u) = \sum_{v \in V} f(u,v)$.

When $(u,v) \notin E$, there can be no flow from u to v, and $f(u,v) = 0$. We call the nonnegative quantity $f(u,v)$ the flow from vertex u to vertex v.

Definition 3 (Value of a flow [14]**).** *The value $|f|$ of a flow f is defined as $|f| = \sum_{v \in V} f(s,v) - \sum_{v \in V} f(v,s)$.*

Generally speaking, because the total flow into the source is zero, therefore the flow in a flow network equals to total flow out of the source, which also is the total flow into the sink.

Up to now, the maximum flow problem can be defined formally, as shown in Definition 4. Sometimes, acquiring the value of a maximum flow in G is the aim of an application, but more commonly, finding out the value of flow for each edge is preferred when a maximum flow achieved [15].

Definition 4 (Maximum flow problem [14]**).** *Given a flow network G with source s and sink t, the maximum flow problem wishes to find a flow of maximum value.*

2.2 Petri Nets

The concept of Petri nets are firstly introduced by Dr. Carl Adam Petri in 1939, they are famous mathematical modeling tools for the description of distributed systems. It also known as a Place/Transition net, which consists of basic elements such as Places (e.g., conditions, represented by circles) and Transitions (e.g., events, represented by bars). Tokens (e.g., resources, represented by spots) can be transmitted from one Place to another through Transition between them following some rules. We use $|P|$ to denote the amount of tokens within a Place P (e.g., represented by numbers within the Place). A formal definition of PNs is presented in Definition 5.

Definition 5 (Petri Nets) [16]**.** *A Petri net is a triple $N = (S, T; F)$, satisfying (1) $S \bigcup T \neq \emptyset$, (2) $S \bigcap T = \emptyset$, (3) $F \subseteq (S \times T) \bigcup (T \times S)$, and (4) $dom(F) \bigcup cod(F) = S \bigcup T$, where dom means domain of F and meets $dom(F) = \{x | \exists y : (x,y) \in F\}$, cod means codomain of F and meets $cod(F) = \{y | \exists x : (x,y) \in F\}$. S and T denote the set of Place and Transition respectively, and F denotes the flow relation between them.*

Furthermore, $\sum = (N, K, W, M)$ is a PN system, where K, W and M are capacity function of Places, weight function of arcs and case (i.e., the distribution of tokens) of a system respectively.

3 Modeling of Maximum Flow Using Petri Nets

In this Section, Ford-Fulkerson (i.e., the augmenting path) method is firstly described in Algorithm 1. Then PNs based flow network and residual network modeling methods are proposed.

Algorithm 1. Ford-Fulkerson method [14]

Input: Flow network G with source s and sink t
Output: A maximum flow f

1 initialize flow f to 0;
2 **while** *there exists an augmenting path p in the residual network* G_f **do**
3 | augment flow f along p;
4 **end**
5 **return** f

3.1 Modeling for Flow Network

A primary issue when utilizing PNs for modeling is to determine the meaning of each Place and Transition. As for the modeling of flow network $G = (V, E)$ shown in Fig. 1(a), we decide to use Places stand for vertexes V, while Transitions will be associated with edges E. Our modeling method for a given flow network $G = (V, E)$ is described as follows.

(1) For each $v \in V$, we add a Place P_v into N_G corresponding to v in G, where N_G denotes the Petri net model of G.
(2) For each $(u, v) \in E$, where $u, v \in V$, we add a Transition T into N_G between P_u and P_v. Two directed arcs also will be added into N_G pointing both from P_u to T and from T to P_v with parameter m_{uv}.

Figure 1(b) shows the result of our modeling method above given $G = (V, E)$ shown in Fig. 1(a) as input. The weight m_{uv} of arcs in N_G is a variable that stands for flow $f(u, v)$ between $u, v \in V$. Therefore, in our work, such model is used for demonstration. In contrast, we solve the maximum flow by a PNs based residual network model which will be introduced in the next Section.

3.2 Modeling of Residual Networks

Given a flow network $G = (V, E)$, its corresponding residual network G_f can be constructed accordingly. Since G_f contains not only the edges in G but also some reversed edges, we propose a PNs based residual network modeling method, where a flow network $G = (V, E)$ with source s and sink t is given as the input, a Petri net model N_{G_f} of residual network G_f corresponding to G is generated as follows.

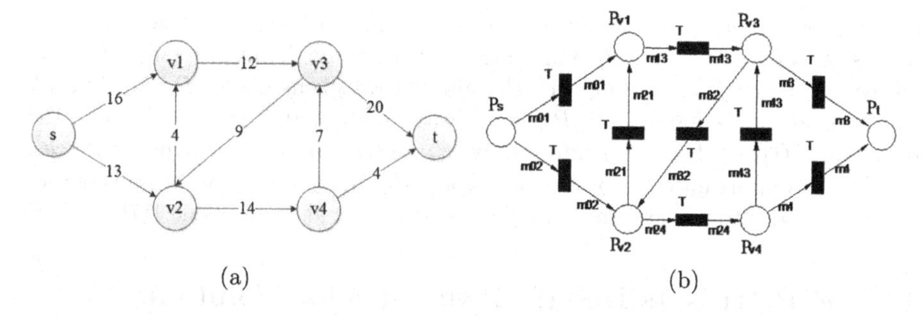

Fig. 1. A flow network (a) and its corresponding PNs model (b).

(1) For each $v \in V$, we add a Place P_v into N_{G_f} corresponding to v in G;
(2) For each $(u, v) \in E$, where $u, v \in V$, we add a Transition T into N_{G_f} between P_u and P_v;
(3) For each $(u, v) \in E$ and suppose its associated Transition is T, we add a foreword Place P_f into N_{G_f}, s.t. $P_f \in T^{\bullet}$ and $|P_f| = f(u, v)$. We also add a revise Place P_r into N_{G_f}, s.t. $P_r \in {}^{\bullet}T$ and $|P_r| = c(u, v) - f(u, v)$, where ${}^{\bullet}T$ and $T^{\bullet}$ denote the pre-set and post-set of the Transition T respectively. Furthermore, if $v \neq t$, a Transition T' will be added into N_{G_f}, s.t. ${}^{\bullet}T' = \{P_v, P_f\}$ and $T'^{\bullet} = \{P_u, P_r\}$ as well.

Figure 2(a) demonstrates an instance of method described above. Specifically, for $(u, t) \in E, f(u, t) = 0$ and $c(u, t) = 20$, then P_f and P_r are added into N_{G_f} with number of tokens equals 0 and 20 respectively. While, for $(u, v) \in E \bigwedge v \neq t$, beside P_f and P_r, T' also will be added as shown in Fig. 2(b). The reason for treating sink t differently is that adding a reverse edge for the sink in the residual network is meaningless, resources should not leave the sink after they have arrived their destination.

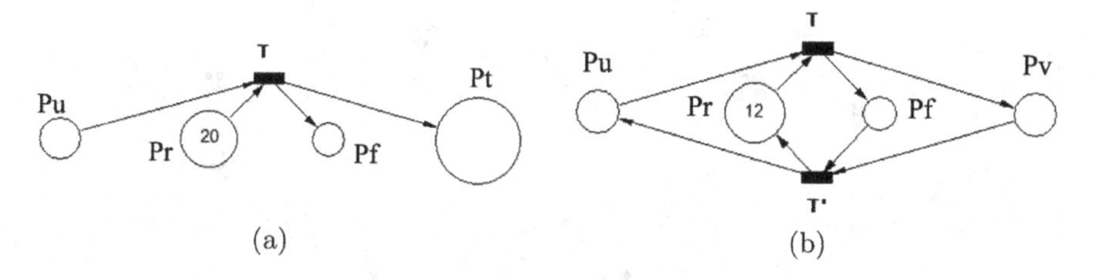

Fig. 2. Demonstration of PNs based residual network modeling method given a flow network $G = (V, E)$ as input. (a) and (b) are the results for $(u, v) \in E$ for $v = t$ and $v \neq t$ respectively, where t stands for the sink.

It is worth to point out that the P_f and P_r described in Algorithm 1. are referred to as forward Place and reverse Place in this paper, because according to PN rules, if (u, v) is a forward edge of flow network G, then the amount of tokens out of P_u and into P_f will be exactly the same. That is to say, $|P_f|$ will always equals to $f(u, v)$. Similarly, the amount of tokens out of P_v and into P_r will be exactly the same. So, $|P_r|$ will always be $c(u, v) - f(u, v)$, namely, the revise flow $f(v, u)$ if $v \neq t$. Additionally, P_f and P_r are complementary Places, which is an important concept in PNs theory [16], they make sure that no contact will be added into the system, namely, the system will remain safe [17].

4 The Petri Nets Based Maximum Flow Solution

Benefit from the simulation of PNs, we can achieve a maximum flow for a given flow network using the proposed residual network model N_{G_f} as shown in Fig. 3, which is intuitive to be observed and distinguish our method from others. The core idea of our method is that if we initialize the source Place P_s with enough tokens (e.g., 100 tokens in our experiments) and let the modeling system run under PN rules automatically until it reaches a status that there exists no path for a token flows from P_s to sink Place P_t as shown in Fig. 3, then a maximum flow is achieved. At the same time, a min-cut can be found as curve line demonstrated in Fig. 3. It separates V of flow network $G(V, E)$ into two categories, which are denoted as C_S and C_T in this paper. While C_S contains vertexes corresponding to Places in the biggest live sub-net of N_{G_f} (e.g., P_s, P_{v1}, P_{v2} and P_{v4} in Fig. 3), C_T contains the rests.

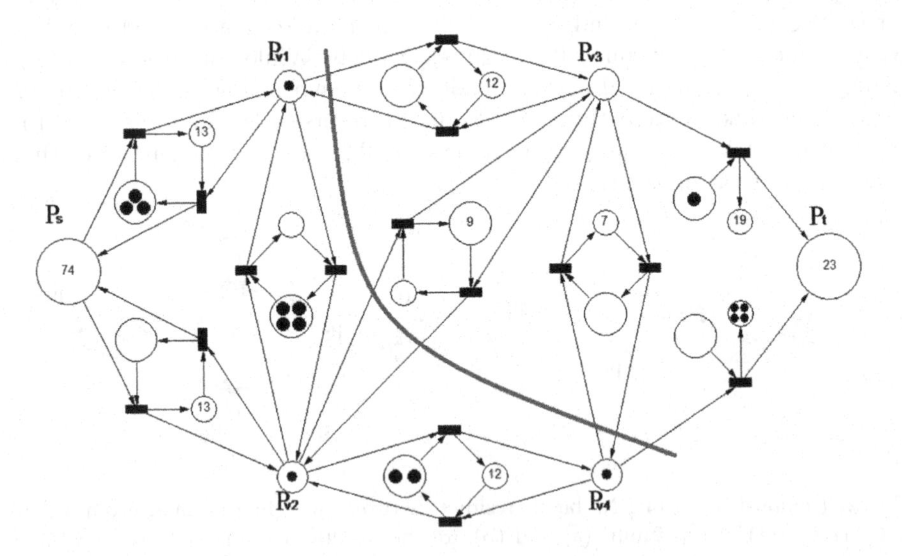

Fig. 3. Solve the maximum flow by the proposed residual network model.

Our method not only can find the maximum flow, but also has the ability to acquire flow values of every edges when maximum flow achieved. A back-flow strategy is used, specifically, we let the tokens flow back to P_s, after that, the amount of tokens reserved in the forward Places P_f shall be $f(u, v)$ of each edge as described above.

5 Conclusion and Future Works

Benefit from animated graphical notations of PNs, a PN based maximum flow modeling approach is introduced in this paper. Specifically, modeling methods for both flow networks and residual networks are firstly proposed. Based on the proposed models, a PN system for solving the maximum flow problem is presented. Namely, given a flow network, we can construct a PN system accordingly. A maximum flow is achieved when there exists no path for a token transmitted from source Place to sink Place. The entire procedures can be intuitive demonstrated by automatically executed simulation.

It is worth to point out that, though our method can be used to solve the maximum flow problem, it is meaningless to compare our method with other methods focus on efficiency, since the aim of our work is the modeling and analyzing of the problem. Furthermore, for there are many advantages for PNs based modeling and analyzing, it is feasible for us to extent this work to other problems in graph theory, which will be one of our future works.

References

1. Genrich, H., Küffner, R., Voss, K.: Executable Petri net models for the analysis of metabolic pathways. Int. J. Softw. Tools Technol. Transf. **3**(4), 394–404 (2001). doi:10.1007/s100090100058
2. Hua, H.C., Hai, C.X.: A top-to-down method getting the systems minimal cut sets by Petri net. Syst. Eng. Electron. **22**(4), 74–76 (2000)
3. Fa, Z.Y., Qi, C., Wen, Z.X.: Using Petri nets model to improve the algorithm of MCS solution. Nucl. Power Eng. **28**(5), 63–68 (2007)
4. Yan, L.L., Ping, W.X., Xin, T.S.: Assessment method of system reliability for stochastic flow network based on rough sets theory and Petri nets. Control Decis. **25**(8), 1273–1276 (2010)
5. Ying, H.X., Bin, H., Long, Z.J., Yin, L.T., Yin, J.G.: Concurrent approach to the max flow of network based on Petri net. J. WUT (Inf. Manag. Eng.) **32**(1), 27–30 (2010)
6. Pan, R., Taubin, G.: Automatic segmentation of point clouds from multi-view reconstruction using graph-cut. Vis. Comput. **32**(5), 601–609 (2016). doi:10.1007/s00371-015-1076-0
7. Fan, L., Lic, L., Liu, K.: Paint mesh cutting. Comput. Graph. Forum **30**(2), 603–612 (2011). doi:10.1111/j.1467-8659.2011.01895.x
8. Liu, L., Sheng, Y., Zhang, G., Ugail, H.: Graph cut based mesh segmentation using feature points and geodesic distance. In: 2015 International Conference on Cyberworlds (CW), pp. 115–120. IEEE Computer Society, Los Alamitos (2015). doi.ieeecomputersociety.org/10.1109/CW.2015.31
9. Liu, S.T., Yin, F.L.: The basic principle and its new advances of image segmentation methods based on graph cuts. Acta Automatica Sinica **38**(6), 912–922 (2012)

10. Peng, Y., Chen, L., Ou-Yang, F.X., Chen, W., Yong, J.H.: Jf-cut: a parallel graph cut approach for large-scale image and video. IEEE Trans. Image Process. **24**(2), 655–666 (2015). doi:10.1109/TIP.2014.2378060
11. Yan, X.H., Lin, Z., Bo, Y.: Optimization of road network structure based on maximum-flow theory. J. Southwest Jiaotong Univ. **44**(2), 284–288 (2009)
12. Hua, K.W., Ping, L.Z.: Maximum flow assignment algorithm for transshipment nodes with flow demands in transportation network. J. Southwest Jiaotong Univ. **44**(1), 118–121 (2009)
13. Chao, Z.X., He, J., Liang, C.G.: Minimum cuts and maximum flows in directed planar networks with both node and edge capacities. Chin. J. Comput. **29**(4), 544–551 (2006)
14. Cormen, T.H., Leiserson, C.E., Rivest, R.L., Stein, C.: Introduction to Algorithms, 3rd edn. The MIT Press, Cambridge (2009)
15. Sedgewick, R.: Algorithms in C, Part 5: Graph Algorithms, 3rd edn. Addison-Wesley Professional, Boston (2001)
16. Yi, Y.C.: Principle & Applications of Petri Nets. Publishing House of Electronics Industry (2005)
17. Jian, L.S., Bo, Y.X., Zheng, Z.: Additional theorem proving and analysis of s-complement in the Petri net. J. Syst. Simul. **40**, 1–5 (2009)

Sampling as a Method of Comparing Real and Generated Networks

Eliska Ochodkova, Milos Kudelka(✉), and David Ivan

FEI, VSB - Technical University of Ostrava,
17. listopadu 15, 708 33 Ostrava, Czech Republic
eliska.ochodkova@vsb.cz, milos.kudelka@vsb.cz

Abstract. In this paper, we combine the use of sampling methods and a network generator to assess the degree of similarity between real and generated networks. Generative network models provide a tool for studying essential network features. These include, for example, the average and distribution of node degree, cluster coefficient and community size. The aim of the generators based on these models is to create networks with properties close to real networks. Even with a high similarity of global properties of real and generated networks, the local structures of these networks often differ considerably. On the other hand, when the network is reduced by a sampling method, global features of networks are strongly influenced by local structures. In the paper, we compare properties of a real-world network and a generated network and also properties of their small samples. In experiments, we show how the distribution of the properties of individual networks change by using different sampling methods and how these distributions differ for both networks and their small samples.

Keywords: Network sampling, Co-authorship network · Network model

1 Introduction

The sampling and modeling of networks belong to the traditional tasks of network analysis. Both tasks are especially important for the analysis of large-scale networks. The aim of sampling [8] is to find a reduced network of a substantially smaller size to preserve the properties of the original network. The use of sampling allows studying, in particular, global properties that are otherwise difficult to examine due to the size of the original network. The aim of generative models of networks [7] is, on the contrary, to find the essence of network growth so that it is possible to generate a large-scale network with expected global properties known from real-world networks. In both areas, there are many methods providing very good results when studying global network properties. However, even in these cases, local properties of network samples or generated networks may differ

P. Krömer et al. (eds.), *Proceedings of the Fourth Euro-China Conference on Intelligent Data Analysis and Applications*, Advances in Intelligent Systems and Computing 682, DOI 10.1007/978-3-319-68527-4_13

substantially from real-world networks (e.g. community structure or the decomposition of a network into a set of connected components). In our paper, we use both approaches together. In the experiment, we use a real-world co-authorship network and a network growth model that allows generating networks with properties similar to co-authorship networks. Our experimental results show not only how the real-world network and the generated one resemble each other but also how they preserve their properties due to sampling.

The paper is organized as follows: In Sect. 2, we discuss the related work from two different sides: from the side of the sampling of large-scale networks and from the side of network models serving to represent real network data. The brief description of our model is presented in Sect. 3. In Sect. 4, we focus on the experiment and its results. Section 5 concludes the paper.

2 Related Work

2.1 Network Sampling

Data sampling, in general, is a statistical analysis technique which serves to create a representative sample from large data collections. In network analysis and network modeling, we sometimes want to or have to work with a sample of a network rather than with all its nodes and edges. This can occur when the analysis is demanding on computational performance and memory or when we do not have access to the whole network and have to instead work with what is possible to obtain (e.g. online social media). If the network is represented by a graph $G = (V, E)$, where V denotes the set of nodes ($n = |V|$) and E is the set of edges ($m = |E|$), then the network sample $G_S = (V_S, E_S)$ is a subgraph of graph G where $V_S \subset V, E_S \subset E$ and $n_S = |V_S| \ll n$, $m_S = |E_S| \ll m$.

The way we obtain a network sample can have a substantial influence on every analysis we conduct. Sampling methods can generally be divided into two groups according to whether we have or do not have access to the whole network. If we can store the whole network, the methods will include into the sample any node or edge uniformly randomly or according to a certain property, such as node degree. Otherwise, we choose a starting node or edge and explore its neighbors in the part of the network we have access to. For exploration, we use various strategies such as breadth-first search or random walk.

Some information about a network is, however, lost in sample collection and it is, therefore, crucial to understanding the change in network structure caused by sampling [12]. The success of sampling can be determined for instance by a simple comparison of distributions of certain network properties. The aim of sampling, in general, is to create the most accurate samples in terms of as many network properties as possible so that samples can represent original networks.

Subgraph sampling has a long history starting with snowball sampling used in sociology [5]. A lot of recent works have investigated sampling of large-scale networks, with a focus on recovering topological characteristics such as degree distribution, clustering coefficients and others. Leskovec and Faloutsos [12] focused on empirically observed network properties such as shrinking diameter and comparing the performance of different sampling techniques. Another comparison of sampling techniques was presented in [11]. An innovative sampling technique based on the Metropolis algorithm was proposed in [9] and was tested to assess the degree of consistency with network properties as well. One of the most popular directions [6] is the focus on sampling of online social networks.

Real-world networks commonly reveal communities that are densely connected clusters of nodes loosely connected between. Communities play important roles in real-world systems, mainly in social or collaborative networks. Sampling has not previously been applied to the problem of community detection. Maiya and Berger-Wolf show in [13] how to produce samples representative of all or most of the communities in the network. Also in [3] authors studied the presence of characteristic groups of nodes in different social and information networks and analyzed the changes in network group structure introduced by sampling.

2.2 Network Models

In the last two decades, the analysis of real-world networks received extraordinary attention and gave rise to many network models. Models of networks have been designed based on features observed in real networks, such as the small-world effect, the power-law degree distribution or community structure. Underlying processes that take place during the evolution of real-world networks have also been examined. Some network models are based on analyzing these processes using the formally described underlying process as a generative mechanism. Such a mechanism can generate networks possessing one or more known properties. Co-authorship and collaboration networks, in general, are long-investigated sources in this area. A common feature of this type of network is that underlying processes proceed in cliques, which then become fundamental building blocks of the network.

Barabasi et al. [1] presented and analyzed in detail a network model inspired by the evolution of co-authorship networks. Their measurements of real-world networks revealed that the distribution of degrees in collaboration networks has a power law and that they are small-world networks with an increasing average degree. Newman confirmed the above observations in [16] and extended the knowledge about the properties of collaboration networks by assessing that they have a high clustering coefficient and a positive assortativity coefficient. The model presented by Ramasco et al. [18] combines preferential edge attachment with a bipartite structure and depends on the act of collaboration. They also demonstrated that assortativity in collaboration networks depends on the aging of nodes. Among more recent work, is the model of a multilayer network proposed by Battiston et al. [2]; their model captures the multifaceted character of actors in collaborative networks. Layers here represent different areas of interest and interactions can take place both within the same layer and between nodes in different layers.

3 3-Lambda Model

3-lambda model briefly described below is based, similarly to Milojević [14], on the Poisson distribution of a number of nodes participating in network growth interactions. The model is based on the following premises: (1) one step in the growth of the network is the interaction of one proactive node with nodes in three different roles; (2) the variables describing the numbers of co-authors in different roles are independent random variables that follow Poisson distribution; (3) the proactive node, its neighbors (old connections), and new connection nodes are selected at random; (4) the expected values of λ_1 (old connections), λ_2 (newbies), and λ_3 (new connections) of the Poisson distributions are preselected.

λ_1, λ_2, and λ_3 significantly affect the density of the network. If we assume that a randomly selected interaction has, on the basis of the corresponding distributions, b neighbors of a proactive node, n new nodes, and e nodes unconnected to the proactive node, then the number of nodes involved in this interaction (interaction size) is as shown in Eq. 1.

$$s = 1 + b + n + e \tag{1}$$

and the following applies:

- n new nodes which must connect to a proactive node and with each other are created.
- There are b nodes adjacent to the proactive node which must first become connected with each other, then with e nodes that are not adjacent to the proactive node (these edges may already exist prior to the interaction). Finally, they must connect with the new n nodes.
- There are e nodes not adjacent to the proactive node, which must become connected with it. Next, they must become connected with each other (these edges may already exist prior to the interaction) and n new nodes.

The network generator uses a simple algorithm which comes directly from the model description. The only extra step is setting up the initial network state. The model is memory-less, which allows working with an arbitrary initial state. For the generator, a complete graph with a number of nodes equal to the round of $(1 + \lambda_1 + \lambda_2 + \lambda_3)$ was chosen as the default state. The time complexity of the algorithm is $O(s^2 \cdot \frac{N}{\lambda_2})$ (for details see [10]).

4 Experiments

In our experiments, we have adopted methodology published in [12]. We use a large static undirected real-world network (based on DBLP dataset) and undirected network generated by the 3-lambda model. As can be seen in Table 3, these networks do not differ much in most global properties. The purpose of the experiments described below was to show how these properties change due to sampling, even in small samples, where differences in local properties necessarily occur. Our goal was to show that if the model fits the real-world network well, then the global features of compared networks should be stable even with the use of very small samples.

4.1 Data

The first data collection was created from the real-world co-authorship network DBLP.[1] The data contained 3,688,962 publications with 1,870,930 authors at the time of preparing this paper (April 2017). For the purpose of sampling and experiments, we extracted from the source XML format only publications from 2011. The resulting undirected network consists of only the largest component of the generated network. Network nodes represent authors, and an undirected edge between two authors exists in case these authors are co-authors of at least one publication. More co-authored publications between two authors are neglected, and edge weights are equal to 1. The generated network contains 158,632 nodes and 398,521 edges.

The second data collection represents the 3-lambda model and was obtained by using its generator, see Sect. 3, with parameters $\lambda_1 = 1, \lambda_2 = 0.67, \lambda_3 = 0.03$. This network consists of 150,001 nodes and 386,630 edges.

4.2 Measured Properties

Researchers have suggested several techniques to evaluate individual methods. One of the strategies is to calculate the distribution of the original network and its sample in order to show their similarity. The following properties and their distributions are compared:

- Degree distribution and average degree <k>.
- The distribution of the clustering coefficient CC and average <CC>. The distribution of the average clustering coefficient <CC> for all nodes with degree k was generated.
- The average shortest path <l> and diameter l.
- The distribution of sizes of connected components and number of connected components $\#_{comp}$.
- Hop-plot ($Hops$): the number $P(h)$ of reachable pairs of nodes at distance h or less where h is the number of hops [17].
- The distribution of the first left singular vector (S_{vec}) of the graph adjacency matrix versus the rank.
- The distribution of singular values (S_{val}) of the graph adjacency matrix versus the rank.
- Assortativity r [15].
- Distribution of community sizes IM and modularity Q_{IM} (according to Infomap algorithm [20]).

[1] http://dblp.uni-trier.de/.

4.3 Sampling Methods

Network sampling methods can be, in general, divided into two groups (see Sect. 2.1). The first group includes methods based on random selection. *Uniform node sampling (RN)* is the simplest way to obtain a sample network. It selects a set of nodes V_S uniformly randomly with a probability p. The resulting sample G_S is an induced subgraph with V_S. *The Hybrid approach (HYB)* combines Random Node-Edge sampling and Random Edge sampling with parameter $p = 0.8$ [12].

The second group includes methods based on exploration. In *the Random Walk (RW)* method, the starting node from which is we simulate a random walk in the network is (uniformly randomly) selected.

The method *Random Jump (RJ)* is similar to RW. The difference is that we randomly jump to any node in the network with the probability $c = 0.15$.

In the *Forest Fire (FF)* method, the initial node is selected randomly, and the burning of the links and corresponding nodes starts from it. If the link is burned, the second end node gets the chance to burn its links on its own, etc. The model has two parameters: forward (p_f) and back (p_b) (for directed networks) probabilities of burning. We set $p_f = 0.7$ [12].

Previous methods based on random walks favor nodes with high degrees, and the distribution of the original network is not maintained. The solution is the Metropolis-Hastings algorithm in combination with a random walk (*Metropolis-Hastings Random Walk (MHRW)*) [9]. The *Topologically Divided Stratums (DS)* method [4] takes topological structure into account. The topological structure can reveal the real topology relation and social relation of networks. The DS works with the diameter of the original network. In the beginning, one of the two most distant nodes of the original network is selected. Subsequently, the nodes of the original network are divided into subsets according to the distance from the initial node. From each subset, p nodes are selected where p represents the desired sample size. The goal is to select nodes and edges evenly across the original network. Parameter k was experimentally tested and set $k = 0.7$.

The main problem with the RW method is the threat of being stuck in a small isolated component or a locally dense area. The success of the method depends on the initial node selection, and the results can be very different when selecting another starting node. For this reason, the m-Multi-dimensional Random Walk method, also called *Frontier Sampling (FS)* [19], which performs m random walks, has been proposed.

The above-mentioned algorithms, from both groups, were used for the experiments. For each method, sampling was performed 5 times. The main criterion determining the success of the method is how similar the two property distributions for the sample and original network are. The indicator of this criterion is the D-value representing the maximum difference between cumulative distributions. Typically, the D-statistic is used as a part of the Kolmogorov-Smirnov test to reject the null hypothesis. Here we use it to measure agreement between the two distributions. The D-statistic does not address the scale problem but compares the shape of (normalized) distributions. The smaller the test value, the larger the probability that two samples obey the same distribution.

For our experiments, the sample size was set to 15% of the original network size [12]. We supported this size by our tests when a total of 10 samples of different sizes from 85% to 5% of the original network size were created. We determined the quality of the sample based on the average D-value of all distributions.

4.4 Results

In the experiments, we used an ensemble approach and worked with eight methods in total, the results of which for individual properties in 15% samples were averaged for the evaluation. For both networks, the RW and FF methods were the most accurate (see Tables 1 and 2, for each column we bold the best test value). However, the means of all these methods indicate that 15% sampling is more accurate for the 3-lambda generated network, especially in the CC, IM and S_{val} parameters. All these parameters are associated with strongly interconnected local network structures (community structures).

Table 1. Average D-values for 3-lambda

	<k>	$\#_{comp}$	CC	IM	$Hops$	S_{vec}	S_{val}	Avg
RN	0.420	1.000	0.088	0.651	0.178	0.363	0.108	0.401
HYB	0.824	1.000	1.000	0.71	0.414	0.860	0.280	0.726
RW	0.274	**0.000**	0.066	**0.111**	0.067	**0.043**	0.072	**0.090**
RJ	**0.131**	1.000	0.053	0.203	**0.064**	0.828	0.128	0.344
FF	0.161	**0.000**	0.077	**0.024**	0.107	0.365	0.136	**0.124**
MHRW	**0.097**	**0.000**	0.088	0.167	0.139	0.593	0.088	0.167
DS	**0.107**	1.000	**0.047**	0.403	**0.014**	0.205	**0.006**	0.254
FS	0.169	1.000	**0.036**	0.148	0.081	0.804	0.052	0.327
Avg	0.317	0.7	0.158	0.361	0.150	0.494	0.123	0.329

In Figs. 1, 2 and 3, the cumulative distributions of three global properties (degree, clustering coefficient vs. degree, community size) are shown for the original networks and the 15% samples created by the most successful methods. The degree and community size distributions have similar and consistent behavior for both networks. For the clustering coefficient vs. degree distribution, it is obvious for the DBLP network that most methods do not provide good results for 15% of samples. Higher consistency between the original network and samples for the 3-lambda model is probably related to the fact that, unlike the DBLP network, this network has a stronger and more regular interconnected local (community) structure.

Table 2. Average D-values DBPL

	<k>	$\#_{comp}$	CC	IM	$Hops$	S_{vec}	S_{val}	Avg
RN	0.409	1.000	**0.124**	0.867	0.124	0.777	**0.114**	0.487
HYB	0.802	1.000	0.63	0.657	0.395	0.875	0.338	0.671
RW	0.255	**0.000**	0.25	**0.057**	**0.032**	0.387	0.264	**0.177**
RJ	**0.158**	1.000	0.391	0.276	0.147	0.662	0.276	0.415
FF	0.209	**0.000**	0.318	**0.034**	0.073	**0.072**	0.158	**0.123**
MHRW	**0.099**	**0.000**	0.313	0.303	0.185	0.505	0.14	0.220
DS	**0.072**	1.000	0.22	0.784	**0.02**	0.402	**0.136**	0.376
FS	**0.121**	1.000	**0.188**	**0.121**	0.065	0.47	**0.07**	0.290
Avg	0.309	0.7	0.280	0.466	0.132	0.473	0.191	0.364

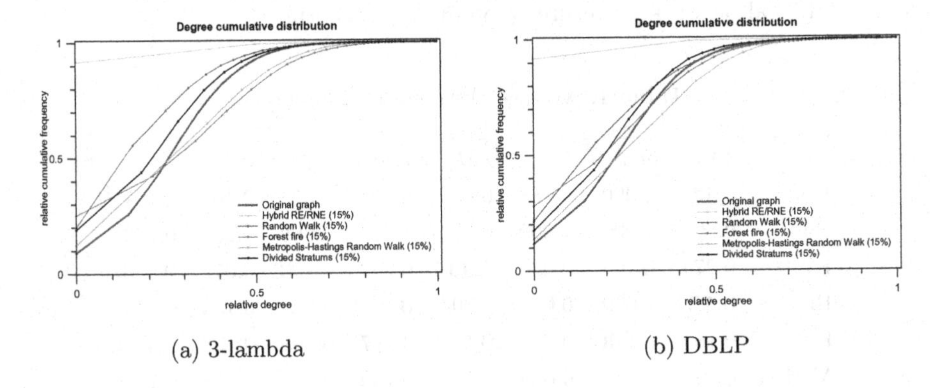

(a) 3-lambda (b) DBLP

Fig. 1. Degree distribution

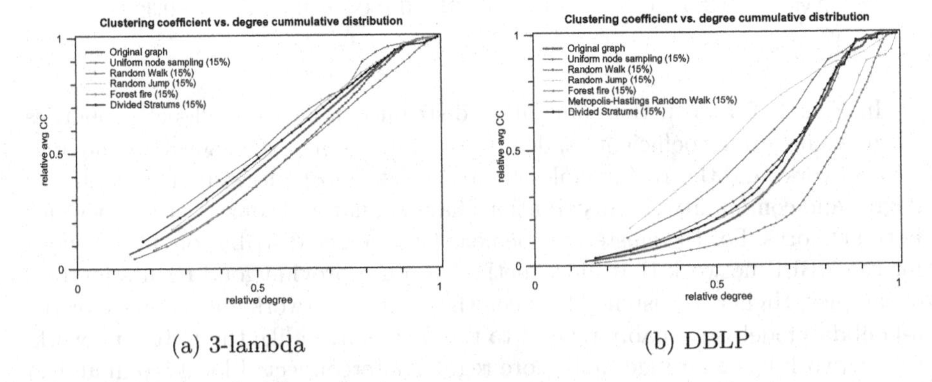

(a) 3-lambda (b) DBLP

Fig. 2. Clustering coefficient distribution

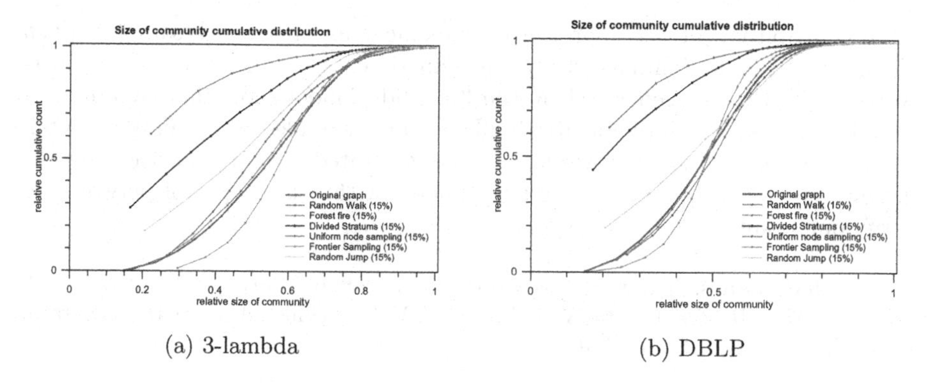

(a) 3-lambda (b) DBLP

Fig. 3. Distribution of size of communities

Table 3 shows that in the main parameters neither the original networks nor the samples differ substantially. The only exception is assortativity (r). For the DBLP network, it is unnaturally high, which again may be due to weaker local network connections. It can be caused by the fact that only the year 2011 was used for the analysis, in which may lack links from previous years.

Table 3. Global properties

	n	m	<k>	<l>	$\#_{comp}$	<CC>	r	Q_{IM}
3-lambda (100%)	150,001	386,630	5.15	8.93	1	0.66	0.155	0.846
FF (15%)	22,500	41,927	3.72	13.42	1	0.37	0.072	0.878
RW (15%)	22,500	38,640	3.43	9.96	1	0.328	0.146	0.836
MHRW (15%)	22,500	36,426	3.25	30.63	1	0.377	0.134	0.908
Avg	22,500	38,997.6	3.466	18.003	1	0.358	0.117	0.874
DBLP (100%)	158,632	398,521	5.02	10.18	1	0.562	0.451	0.85
FF (15%)	23,795	51,328	4.31	10.39	1	0.313	0.668	0.86
RW (15%)	23,795	37,731	3.17	8.48	1	0.234	0.129	0.78
MHRW (15%)	23,795	37,523	3.15	26.59	1	0.344	0.459	0.90
avg	23,795	42,194	3.543	15.153	1	0.297	0.4187	0.846

5 Conclusion

The main principle of the quality assessment of a network model is a measurement of selected properties of the networks generated by this model. The goal is to set the parameters of the generator so that the generated network fits well with the real network. Traditional approaches only use the resulting generated network when comparing properties of real-world and generated networks. The main contribution of our approach is assessment using network samples. Even though this is a preliminary study, our experiments with a co-authorship network and a network generated by the 3-lambda model show that by applying sampling methods we are able to, thanks to small samples, assess in more detail the match between the real-world and the generated network. Further investigation is needed to provide more experiments with other types of networks to verify our results.

Acknowledgments. This work was supported by grant of Ministry of Health of Czech Republic (MZ CR VES16-31852A) and by SGS, VSB-Technical University of Ostrava, under the grant no. SP2017/100.

References

1. Barabâsi, A.L., Jeong, H., Néda, Z., Ravasz, E., Schubert, A., Vicsek, T.: Evolution of the social network of scientific collaborations. Phys. A: Stat. Mech. Appl. **311**(3), 590–614 (2002)
2. Battiston, F., Iacovacci, J., Nicosia, V., Bianconi, G., Latora, V.: Emergence of multiplex communities in collaboration networks. PloS ONE **11**(1), e0147451 (2016)
3. Blagus, N., Šubelj, L., Weiss, G., Bajec, M.: Sampling promotes community structure in social and information networks. Phys. A: Stat. Mech. Appl. **432**, 206–215 (2015)
4. Du, X., Ye, Y., Li, Y., Song, G.: A new relational networks sampling algorithm using topologically divided stratums (2014)
5. Frank, O.: Sampling and estimation in large social networks. Soc. Netw. **1**(1), 91–101 (1978)
6. Gjoka, M., Kurant, M., Butts, C.T., Markopoulou, A.: A walk in Facebook: uniform sampling of users in online social networks. arXiv preprint arXiv:0906.0060 (2009)
7. Goldenberg, A., Zheng, A.X., Fienberg, S.E., Airoldi, E.M., et al.: A survey of statistical network models. Found. Trends® Mach. Learn. **2**(2), 129–233 (2010)
8. Hu, P., Lau, W.C.: A survey and taxonomy of graph sampling. arXiv preprint arXiv:1308.5865 (2013)
9. Hübler, C., Kriegel, H.P., Borgwardt, K., Ghahramani, Z.: Metropolis algorithms for representative subgraph sampling. In: Eighth IEEE International Conference on Data Mining, ICDM 2008, pp. 283–292. IEEE (2008)
10. Kudelka, M., Ochodkova, E., Zehnalova, S.: Around average behavior: 3-lambda network mode. arXiv preprint arXiv:1701.01274 (2017)
11. Lee, S.H., Kim, P.J., Jeong, H.: Statistical properties of sampled networks. Phys. Rev. E **73**(1), 016102 (2006)
12. Leskovec, J., Faloutsos, C.: Sampling from large graphs. In: Proceedings of the 12th ACM SIGKDD International Conference on Knowledge Discovery and Data Mining, pp. 631–636. ACM (2006)
13. Maiya, A.S., Berger-Wolf, T.Y.: Sampling community structure. In: Proceedings of the 19th International Conference on World Wide Web, pp. 701–710. ACM (2010)
14. Milojević, S.: Principles of scientific research team formation and evolution. Proc. Natl. Acad. Sci. **111**(11), 3984–3989 (2014)
15. Newman, M.E.: Assortative mixing in networks. Phys. Rev. Lett. **89**(20), 208701 (2002)

16. Newman, M.E.: Coauthorship networks and patterns of scientific collaboration. Proc. Natl. Acad. Sci. **101**(suppl 1), 5200–5205 (2004)
17. Palmer, C.R., Gibbons, P.B., Faloutsos, C.: ANF: a fast and scalable tool for data mining in massive graphs. In: Proceedings of the 8th ACM SIGKDD International Conference on Knowledge Discovery and Data Mining, pp. 81–90. ACM (2002)
18. Ramasco, J.J., Dorogovtsev, S.N., Pastor-Satorras, R.: Self-organization of collaboration networks. Phys. Rev. E **70**(3), 036106 (2004)
19. Ribeiro, B., Towsley, D.: Estimating and sampling graphs with multidimensional random walks. In: Proceedings of the 10th ACM SIGCOMM Conference on Internet Measurement, pp. 390–403. ACM (2010)
20. Rosvall, M., Bergstrom, C.T.: Maps of random walks on complex networks reveal community structure. Proc. Natl. Acad. Sci. **105**(4), 1118–1123 (2008)

The Relative Homogeneity Between-class Thresholding Method Based on Shape Measure

Hong Zhang[1(✉)], Wenyu Hu[2], and Fan Yang[1]

[1] School of Automation, Xi'an University of Posts and Telecommunications, Xi'an, Shaanxi, China
zhmlsa@xupt.edu.cn, 703905407@qq.com

[2] Fujian Provincial Key Laboratory of Big Data Mining and Applications, Fuzhou, Fujian, China
huwenyu@fjut.edu.cn

Abstract. For images with unequal distribution variances and heterogeneous foreground or background, a relative homogeneity between-class thresholding criterion is deduced. For images with distorted gray distribution, as geometric characteristic, shape measure function is introduced. To segment the images with unequal, heterogeneous and distorted distribution, two criterions are fused, a relative homogeneity thresholding method based on shape measure is proposed. The segmentation results show that the accuracy of proposed criterion is improved greatly.

Keywords: Thresholding method · Focusing on objects · Relative homogeneity · Shape measure

1 Introduction

Otsu's measure is regarded as one of the classic image thresholding techniques and clustering criterion [1, 2]. Nonetheless, when gray level distribution have unequal variances, Otsu's method will provide a biased threshold value.

In order to overcome the inherent defect of Otsu's method, Hou [3] proposed a generalized criterion, minimum class variance thresholding (MCVT). Chen [4] analyzed the limitations of Otsu's criterion and developed a new binarization method, obtained correct threshold value for unequal distribution variances images. In addition, extensive research [5–7] has been already conducted to introduce new and more robust thresholding techniques based on class variance information.

Based on Chen's [4] method, we have a more detailed research [8], deduced a relative homogeneity between-class thresholding criterion. In real images, background distribution is complexity, and image is affected by light, environment and other factors, the gray distribution does not correctly represent correct distribution information of object and background. For this case, we introduced shape measure function as a characteristic information of object geometric features, which is less affected by light and other factors, can represent the actual image information. By fusing the two criterions, we constructed a new thresholding function, in which, the problem of gray distribution deviation can be modified, the completeness and accuracy of the object extraction is greatly improved.

P. Krömer et al. (eds.), *Proceedings of the Fourth Euro-China Conference on Intelligent Data Analysis and Applications*, Advances in Intelligent Systems and Computing 682, DOI 10.1007/978-3-319-68527-4_14

2 Thresholding Method Based on Relative Homogeneity Between-class

2.1 Binarization Focusing on Objects

Otsu's method views both object and background as having uniform or homogeneous gray distribution. However, for some images, object pixels may have more distribution uniformity or homogeneity than background pixels, that is, background possesses more likely heterogeneity and non-uniformity. Therefore, a biased threshold estimate will possibly be resulted by adopting a single mean to represent background.

For remedying such shortcomings of Otsu's method, Chen [4] defined an alternative discriminant criterion, which assumes object has gray level homogeneity, an optimal gray level $t^* \in \{0, 1, 2, \cdots, L-1\}$ is selected, which makes the following criterion $J_{LC}(t)$ minimized:

$$J_{LC}(t) = \left(\frac{P_1(t)}{P_2(t)}\right)^{\alpha} \frac{\sum\limits_{(x,y)\in o} \left[\lambda(g(x,y)-m)^2 + (1-\lambda)(\bar{g}(x,y)-m)^2\right]}{\sum\limits_{(x,y)\notin o} \left[\lambda(g(x,y)-m)^2 + (1-\lambda)(\bar{g}(x,y)-m)^2\right]} \tag{1}$$

Then, the optimal threshold t^* is:

$$t^* = Arg \min_{0 \le t \le L-1} J_{LC}(t) \tag{2}$$

Here, (x, y) is gray level of pixel (x, y) , $\bar{g}(x, y)$ is neighboring average gray level, O is the set of pixels belonging to object, $m = \frac{1}{|O|} \sum\limits_{(x,y)\in O} g(x, y)$, $P_1(t) = \frac{1}{|O|}$, $P_2(t) = \frac{1}{N-|O|}$, W is a window centered at (x, y) , N is the total number of pixels of image, $\alpha(\alpha \ge 0)$ is an exponent and adjusts $(P_1(t)/P_2(t))^{\alpha}$ to achieve some trade-off and $\lambda(0 \le \lambda \le 1)$ also is an parameter to trade off the proportion between gray levels and its local average gray levels.

In formula (1), the numerator measures the object-class similarity or scatter degree, the more similar (compact) the pixels in object class, the smaller the scatter and thus the smaller the numerator value is. And the denominator measures the background-class dissimilarity to the object class, a larger value implies that the two classes are better separated. This criterion more focuses on both the similarity of object class itself and the dissimilarity of background to object, better avoids the problem probably incurred by the heterogeneity of background. For images with both heterogeneous background and uniform foreground (object), a better segmentation effect can be obtained.

2.2 Thresholding Based on Relative Homogeneity Between-class

In the discriminant of Sect. 2, it is implied that m can represent all, and the threshold t^* can also represent all. Further, if we take into account the difference of gray level histogram and neighborhood average histogram, a more detailed description can be given.

The thresholding criterion based on histogram information is obtained, i.e.

$$J_O(t) = \left(\frac{P_1(t)}{P_2(t)}\right)^{\alpha} \frac{\sum_{i \in O} \left[\lambda \cdot p(i)(i - \mu(t))^2 + (1 - \lambda) \cdot \bar{p}(i)(i - \bar{\mu}(t))^2\right]}{\sum_{i \notin O} \left[\lambda \cdot p(i)(i - \mu(t))^2 + (1 - \lambda) \cdot \bar{p}(i)(i - \bar{\mu}(t))^2\right]} \quad (3)$$

where, for an image with $M \times N$ size, the gray value of pixel (x, y) is i, $f(i)$ is the total pixels number of gray level i, $p(i) = \frac{f(i)}{M \times N}$, $i = 0, 1, \cdots, L-1$. $P_1(t) = 1/\sum_{i \in O} p(i)$, $P_2(t) = 1/\sum_{i \notin O} p(i)$ are the prior probability of object and background, $\mu(t) = \sum_{i \in O} ip(i)/\sum_{i \in O} p(i)$ denotes the object area mean value of original image, $\bar{\mu}(t)$ denotes the object area mean value of neighborhood average image.

In research, we found that the function $J_O(t)$ only includes the degree of consistency of object (background) with respect to the other pixels, the less the pixels, the better the internal relative uniformity. That is, with the increase of pixels number, the relative uniformity may be more and more poor, then $J_O(t)$ show monotonicity with the increase of t, the optimal threshold value can't be obtained according to the criterion $J_O(t)$.

In images, the pixels with good homogeneity may be located in low value region, also be located in high value region of gray level, therefore, the discriminant function $J_O(t)$ can also be described as follows:

$$J_{O1}(t) = \left(\frac{P_1(t)}{P_2(t)}\right)^{\alpha} \frac{\sum_{i=0}^{t-1} \left[\lambda \cdot p(i) \cdot (i - \mu_1(t))^2 + (1 - \lambda) \cdot \bar{p}(i) \cdot (i - \overline{\mu_1}(t))^2\right]}{\sum_{j=t}^{L-1} \left[\lambda \cdot p(j) \cdot (j - \mu_1(t))^2 + (1 - \lambda) \cdot \bar{p}(j) \cdot (j - \overline{\mu_1}(t))^2\right]} \quad (4)$$

$$J_{O2}(t) = \left(\frac{P_2(t)}{P_1(t)}\right)^{\alpha} \frac{\sum_{i=t}^{L-1} \left[\lambda \cdot p(i) \cdot (i - \mu_2(t))^2 + (1 - \lambda) \cdot \bar{p}(i) \cdot (i - \overline{\mu_2}(t))^2\right]}{\sum_{j=0}^{t} \left[\lambda \cdot p(j) \cdot (j - \mu_2(t))^2 + (1 - \lambda) \cdot \bar{p}(j) \cdot (j - \overline{\mu_2}(t))^2\right]} \quad (5)$$

Where, $P_1(t) = \frac{1}{\sum_{i=0}^{t-1} p(i)}$, $P_2(t) = \frac{1}{\sum_{i=t}^{L-1} p(i)}$, then $\frac{P_1(t)}{P_2(t)} = \frac{\sum_{i=t}^{L-1} p(i)}{\sum_{i=0}^{t-1} p(i)}$,

$$\frac{P_2(t)}{P_1(t)} = \frac{\sum_{i=0}^{t-1} p(i)}{\sum_{i=t}^{L-1} p(i)} \circ \mu_1(t) = \frac{\sum_{i=0}^{t-1} ip(i)}{\sum_{i=0}^{t-1} p(i)}, \ \mu_2(t) = \frac{\sum_{i=t}^{L-1} ip(i)}{\sum_{i=t}^{L-1} p(i)}.$$

$J_{O1}(t)$ denotes the criterion in which homogeneity is better in low gray level region, $J_{O2}(t)$ denotes the function that homogeneity is better in high gray level region. In addition to the possible monotonicity, for an image, the information will be lost if only using the criterion $J_{O1}(t)$ or $J_{O2}(t)$ to indicate the homogeneity. For taking into account the regional internal uniform information, we can combine the two criterions and construct a new thresholding function as follows:

$$J_{OB}(t) = J_{O1}(t) + J_{O2}(t) \tag{6}$$

For simplified, we choose the parameter $\alpha = 0$, $\lambda = 1$, then the criterion function is:

$$J_{OB}(t) = \frac{\sum_{i=0}^{t-1}\left[\lambda \cdot p(i) \cdot (i-\mu_1(t))^2\right]}{\sum_{j=t}^{L-1}\left[\lambda \cdot p(j) \cdot (j-\mu_1(t))^2\right]} + \frac{\sum_{i=t}^{L-1}\left[p(i) \cdot (i-\mu_2(t))^2\right]}{\sum_{j=0}^{t}\left[p(j) \cdot (j-\mu_2(t))^2\right]} \tag{7}$$

The optimal threshold t^* is:

$$t^* = \mathrm{Arg} \min_{0 \le t \le L-1} J_{OB}(t) \tag{8}$$

For any images, whether object is in low or high gray level region, this criterion function (7) is applicable, the uniformity and integrity of object or background can be taken better.

Comparing formula (7) with within-class Otsu's thresholding criterion [1], we can find that the numerators of them are same. However, it is important that, in (7), the denominator is the measure of relative uniformity degree between classes. In this way, the distribution difference of classes is considered, segmentation deviation is reduced. This method is defined as thresholding technique based on relative homogeneity between-class.

Comparing formula (7) with formula (3), for each class segmented, although there is greater uniformity difference, but compared to the other class, there is a better homogeneity characteristics. It is more reasonable for real images using $J_{OB}(t)$ than $J_O(t)$.

3 Relative Homogeneity Measure Method Based on Shape Measure

In Sect. 2, using the relative homogeneity method between object and background, the accuracy of threshold selection can be improved, and more reasonable segmentation effect can be obtained. However, for some images with complex background and specific or distorted gray distributions, the applicability is still limited, there is large threshold selection bias. For such images, the thresholding criterion need to be adjusted combining with other image feature, such as spatial or shape information.

Usually, spatial feature may be increase information dimension and computation complexity, but shape feature can avoid that, moreover, can reduce the effect from specific or deformed gray distribution.

3.1 Shape Measure Function

As a measure of image segmentation performance, shape measure function is used to measure the geometric features of objects in an image, to obtain clear edge contour information, it is calculated as [9]:

$$J_{SM}(t) = \frac{\sum_{(x,y)\in F} sign(f(x,y) - \bar{f}(x,y))\Delta(x,y)sign(f(x,y) - t)}{C_F} \tag{9}$$

Here,

$$\Delta(x,y) = \left[\sum_{k=1}^{4} D_k^2 + \sqrt{2}D_1(D_3 + D_4) - \sqrt{2}D_2(D_3 - D_4)\right]^{\frac{1}{2}} \tag{10}$$

In which,

$$\begin{aligned} D_1 &= f(x+1,y) - f(x-1,y), D_2 = f(x,y-1) - f(x,y+1), \\ D_3 &= f(x+1,y+1) - f(x-1,y-1), \\ D_4 &= f(x+1,y-1) - f(x-1,y+1). \end{aligned}$$

$$\bar{f}(x,y) = \frac{1}{8}\left[\sum_{i=x-1}^{x+1}\sum_{j=y-1}^{y+1} f(i,j) - f(x,y)\right] \tag{11}$$

$$sign(x) = \begin{cases} +1; & if\, x \geq 0 \\ -1; & if\, x < 0 \end{cases} \tag{12}$$

$$C_F = \max_t \left\{ \sum_{(x,y)\in F} sign(f(x,y) - \bar{f}(x,y))\Delta(x,y)sign(f(x,y) - t) \right\} \tag{13}$$

According to the shape measure function, the optimal thresholding value t^* can be obtained:

$$t^* = Arg \max_{0 \leq t \leq L-1} [J_{SM}(t)] \tag{14}$$

3.2 Relative Homogeneity Thresholding Method Based on Shape Measure

As described in Sect. 2, the method based on relative homogeneity between-class can better avoid the problem probably incurred by the heterogeneity of one class. No matter the images with both heterogeneous background or foreground, it is compatible. However, it is limited by specific or distorted.

Shape measure function is geometric feature measure of object, can avoid the effect of gray distribution. So, the geometric feature function can be selected as a component to build criterion function. According to the extreme value direction of two criterions, a synthetic threshold formula is constructed as:

$$J_{SM\_OB}(t) = \frac{1}{J_{SM}(t)} * [J_{OB}(t)] \tag{15}$$

the optimal thresholding value t^* of criterion function is:

$$t^* = \mathrm{Arg} \min_{0 \le t \le L-1} [J_{SM\_OB}(t)] \tag{16}$$

The method applying with the formula (15) is defined as relative homogeneity thresholding method based on shape measure.

4 Results and Analysis

In order to verify the correctness and efficiency, the proposed method (denoted as SM_OB) is compared with Otsu's method (denoted as 1d_Otsu) [1], relative homogeneity method between object and background(denoted as 1d_OB), and shape measure method (denoted as SM). The two aspects of real images segmentation effect and segmentation performance assessment are analyzed below.

4.1 Experimental Analysis

Experimental results of four representative images are shown in Figs. 1, 2, 3 and 4: Color, SAR, Cube and Lymp with sizes 158 × 159, 340 × 304, 486 × 500 and 130 × 130, respectively. Figures 1, 2, 3 and 4 (a) are original images, Figs. 1, 2, 3 and 4 (b), (c), (d), and (e) show the results of four methods. Table 1 lists the segmentation thresholds of four methods.

Comparing with the results of four methods, for Color and SAR images, background is complexity and with large heterogeneous, 1d_Otsu, 1d_OB and SM methods cannot extract the object correctly, but the proposed SM_OB method can obtain complete object results. For Cube and Lymp images, foreground is complexity and with large heterogeneous, applying with the proposed SM_OB method, the extraction integrity of object is the best.

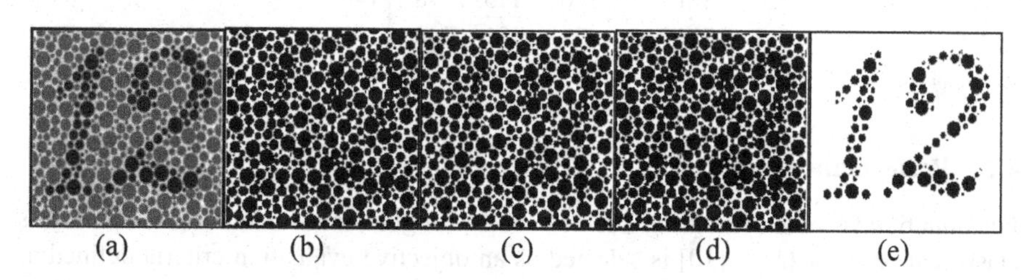

Fig. 1. Color image: (a) original image, (b) 1d_Otsu, (c) 1d_OB, (d) SM, (e) SM_OB.

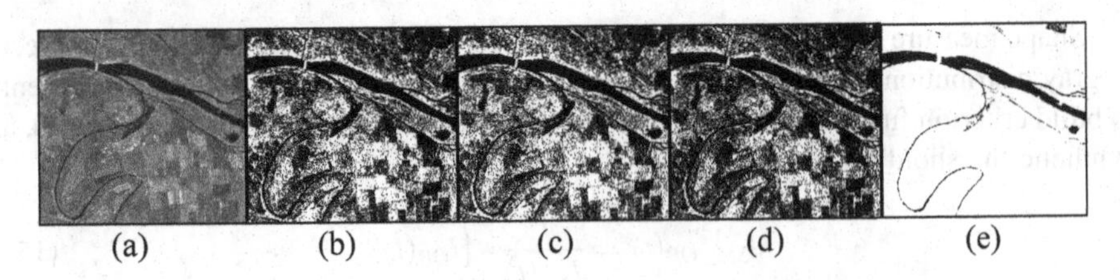

Fig. 2. SAR image: (a) original image, (b) 1d_Otsu, (c) 1d_OB, (d) SM, (e) SM_OB.

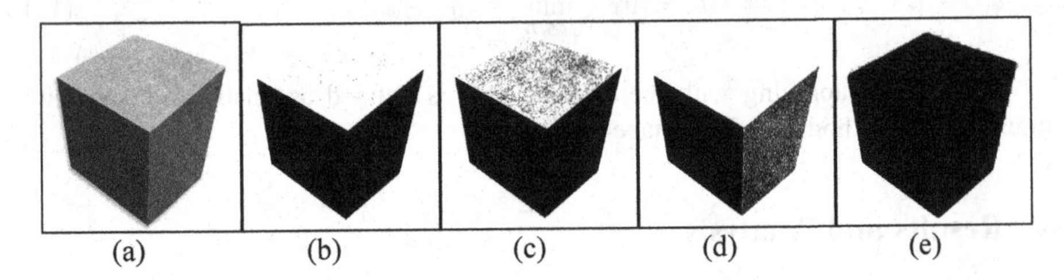

Fig. 3. Cube image: (a) original image, (b) 1d_Otsu, (c) 1d_OB, (d) SM, (e) SM_OB.

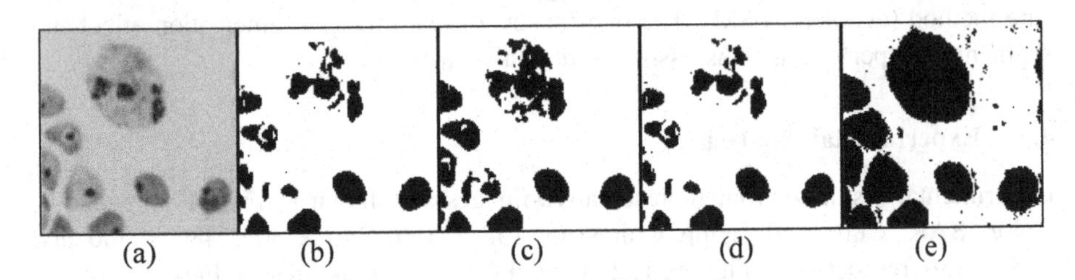

Fig. 4. Lymp image: (a) original image, (b) 1d_Otsu, (c) 1d_OB, (d) SM, (e) SM_OB.

Table 1. The results of four methods

Method	Image			
	Color	SAR	Cube	Lymp
1d _Otsu	134	105	141	150
1d_OB	103	102	165	164
SM	120	113	78	148
SM_OB	33	191	246	195

4.2 Performance Evaluation

For quantitatively analyzing the performance of segmentation method, common misclassification error (*ME*) [10] is selected as an objective evaluation criteria defined as:

$$ME = 1 - \frac{|B_O \cap B_T| + |F_O \cap F_T|}{|B_O| + |F_O|}, \tag{17}$$

Where, B_O and F_O denote the background and object in the ground-truth image, respectively, B_T and F_T denote the background and object in the test image, respectively, and | • | is the operator to get the cardinality of a set.

The original images and ground-truth images downloaded from site Mehmet Sezgin [11] are used to test the performance, two representative images which sizes are 93 × 81 and 166 × 60, are shown in Figs. 5 and 6. Where, Figs. 5, 6 (a) and (b) are original and ground-truth images. The segmentation threshold value and *ME* value of two methods are given in Table 2, *ME* value of our proposed method is less than the other three methods.

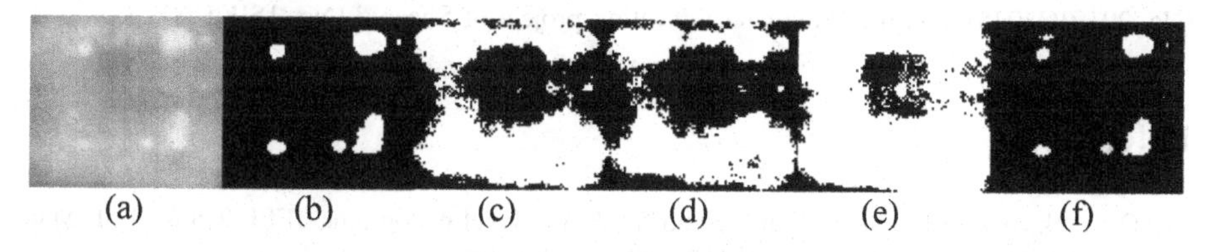

Fig. 5. Test image1 of 93 × 81: (a) original image, (b) ground-truth, (c)1d_Otsu, (d)1d_OB, (e) SM, (f)SM_OB.

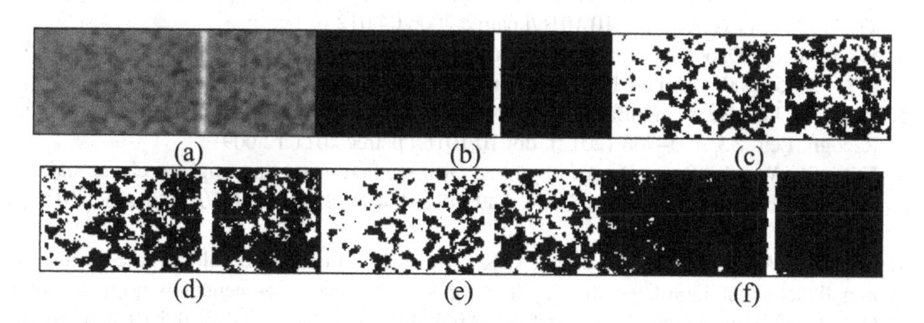

Fig. 6. Test image2 of 166 × 60: (a) original image, (b) ground-truth, (c)1d_Otsu, (d)1d_OB, (e) SM, (f)SM_OB.

Table 2. The results of four methods

Image	Method			
	1d_Otsu (t^*, ME)	1d_OB (t^*, ME)	SM (t^*, ME)	SM_OB (t^*, ME)
Test1	(174,0.4747)	(177,0.3924)	(159,0.7585)	(205,0.0163)
Test2	(89,0.5642)	(97,0.3789)	(85,0.6310)	(117,0.0168)

5 Conclusion

In this paper, we analyzed the limitation of method focusing on object, deduced relative homogeneity between-class segmentation criterion. This method is suitable for images with unequal gray level distribution variances, especially, for images with heterogeneous foreground or background. Taking into account the changes in gray distribution resulted by light, environment and other factors, as geometric characteristic of images, shape measure function is introduced. By fusing the relative homogeneity and shape measure function, a new thresholding criterion function is constructed. The experimental results of real images show that the proposed criterion can deduced the classification error and improve the completeness of object segmented.

Acknowledgements. This work is supported in part by the National Science Foundation of China (No. 61671377), the Provincial Natural Science Foundation research project of Shanxi (No. 2012JQ8045), and the Provincial Education project of Shaanxi (No. 15JK1682).

References

1. Otsu, N.: A threshold selection method from gray-level histograms. IEEE Trans. Syst. Man Cybern. **9**, 62–66 (1979). doi:10.1109/TSMC.1979.4310076
2. Yukun, L., Paul, L.R.: Efficient circular thresholding. IEEE Trans. Image Proc. **23**, 992–1001 (2014). doi:10.1109/TIP.2013.2297014
3. Hou, Z., Hu, Q., Nowinski, W.: On minimum variance thresholding. Pattern Recogn. Lett. **27**, 1732–1743 (2006). doi:10.1016/j.patrec.2006.04.012
4. Songcan, C., Daohong, L.: Image binarization focusing on objects. Neurocomputing **69**, 2411–2415 (2006). doi:10.1016/j.neucom.2006.02.014
5. Fan, J.L., Lei, B.: A modified valley-emphasis method for automatic thresholding. Pattern Recogn. Lett. **33**, 703–708 (2012). doi:10.1016/j.patrec.2011.12.009
6. Yuan, X., Wu, L., Peng, Q.: An improved Otsu method using the weighted object variance for defect detection. Appl. Surf. Sci. **349**, 472–484 (2015). doi:10.1016/j.apsusc.2015.05.033
7. Bhandari, A.K., Kumar, A., Singh, G.K.: Modified artificial bee colony based computationally efficient multilevel thresholding for satellite image segmentation using Kapur's, Otsu and Tsallis functions. Expert Syst. Appl. **42**, 1573–1601 (2015). doi:10.1016/j.eswa.2014.09.049
8. Hong, Z., Wenyu, H.: Thresholding method based on the relative homogeneity between the classes. Third Euro-China Conf. Intell. Data Anal. Appl. **11**, 622–626 (2016). doi:10.1007/978-3-319-48499-0_14
9. Chang, C.I., Du, Y., Wang, J., et al.: Survey and comparative analysis of entropy and relative entropy thresholding techniques. IEE Proc. Vis. Image Sig. Process **153**, 837–850 (2006)
10. Sezgin, M., Sankur, B.: Survey over image thresholding techniques and quantitative performance evaluation. J. Electron. Imaging **13**, 146–168 (2004)
11. Mehmet Sezgin: blt_image_references. http://mehmetsezgin.net

Mining of Multiple Fuzzy Frequent Itemsets with Transaction Insertion

Tsu-Yang Wu[1,2], Jerry Chun-Wei Lin[3(✉)], and Yuyu Zhang[3]

[1] Fujian Provincial Key Laboratory of Big Data Mining and Applications, Fujian University of Technology, Fuzhou 350118, China
wutsuyang@gmail.com

[2] National Demonstration Center for Experimental Electronic Information and Electrical Technology Education, Fujian University of Technology, Fuzhou, China

[3] School of Computer Science and Technology, Harbin Institute of Technology Shenzhen Graduate School, Shenzhen, China
jerrylin@ieee.org, yuyuzhang.hit@gmail.com

Abstract. In this paper, we thus present an algorithm to efficiently update the multiple fuzzy frequent itemsets from the quantitative dataset with transaction insertion. The designed approach is based on the Fast UPdated (FUP) concept to divide the transformed linguistic terms into four cases, and each case is performed by the designed approach for updating the discovered information. Also, the fuzzy-list (FL) structure is adopted to reduce the generation of candidates without multiple database scans. Experiments are conducted to show that the proposed algorithm outperforms the state-of-the-art approach.

Keywords: Fuzzy data mining · FL-strcutrue · Incremetal · Dynamic database · Insertion

1 Introduction

Data mining can efficiently discover useful and implicit information from a very large database, which can be used to aid manages or retailers for making efficient-decision. Association-rule mining (ARM) is the fundamental approach to find the implicit information among items, and Apriori algorithm [1] was the first one to discover association rules in the level-wise manner. Since Apriori algorithm takes costly computation to find the association rules, FP-tree structure and its mining approach called FP-growth [5] were designed to mine the frequent itemsets based on minimum support threshold.

When the size of the database is changed, for example, some transactions are inserted into the original database, most algorithms running on the batch mode need to process the whole updated database to obtain the up-to-date information. Thus, the already discovered information become useless, as well as the previous computation time. To solve this problem, Cheung et al. proposed the

P. Krömer et al. (eds.), *Proceedings of the Fourth Euro-China Conference on Intelligent Data Analysis and Applications*, Advances in Intelligent Systems and Computing 682, DOI 10.1007/978-3-319-68527-4_15

Fast-UPdated (FUP) [2] concept to incrementally maintain and update the discovered information in ARM. It divides the original databases and new inserted transactions into four cases. For each case, the process is respectively designed to update the discovered information. Form the experimental results, it showed better performance than that of the traditional batch mode. Lin et al. [7] then presented a Fast Updated Frequent Pattern (FUFP)-tree for incrementally updating the discovered frequent itemsets, which showed better performance than the Apirori-like [2] FUP concept.

Fuzzy-set theory [18] was used to represent the information as the linguistic terms, which can efficiently handle the quantitative database. The fuzzy-set theory is based on the pre-defined membership functions to transform the quantitative value into the representation of linguistic terms. Several algorithms [3,6,8,10,17] were presented to mine the set of fuzzy frequent itemsets (FFIs). Lin et al. respectively presented the fuzzy frequent pattern (FFP)-tree [11], compressed fuzzy frequent pattern (CFFP)-tree [12], and upper-bound fuzzy frequent pattern (UBFFP)-tree structure [13] to mine FFIs. Since the previous algorithms mine the FFIs with maximal cardinality, Hong et al. [9] then designed a multiple fuzzy frequent pattern (MFFP)-tree algorithm to mine multiple FFIs (MFFIs) from the database, which can provide more complete information for decision-making. To speed up mining process of the MFFIs, Lin et al. then respectively proposed CMFFP-tree [14] and UBFFP-tree algorithms [15] to mine MFFIs. In order to speed up mining performance, Lin et al. proposed an algorithm called MFFI-Miner [16] to mine the multiple fuzzy frequent itemsets (MFFIs) without candidate generation. This algorithm adopts the fuzzy-list (FL)-structure to efficiently reduce the computation for mining MFFIs.

In real-life situations, when some transactions are inserted into the original database, some information may arise and some rules may become invalid. In this paper, we present an incremental algorithm to efficiently update the discovered MFFIs with transaction insertion.

2 Preliminaries and Problem Statement

Let $I = \{i_1, i_2, \ldots, i_m\}$ be a finite set of m distinct items (attributes) in a quantitative database $D = \{T_1, T_2, \ldots, T_n\}$, in which each transaction $T_q \in D$ and (1) is a subset of I; (2) contains several items with its purchase quantities v_{iq}; (3) has an unique identifier, called *TID*. An itemset X is a set of k distinct items $\{i_1, i_2, \ldots, i_k\}$, where k is the length of an itemset called k-itemset. X is said to be contained in a transaction T_q if $X \subseteq T_q$. A minimum support threshold is defined as δ. The user-specified membership functions is set as μ. An example is shown in Table 1 for the original quantitative database.

The membership functions used to transform the quantitative value into the linguistic terms with their fuzzy degrees (values). For example, the membership functions are presented in Fig. 1.

Table 1. A quantitative database.

TID	Items with their quantities
1	$(A{:}3)$, $(B{:}1)$, $(E{:}1)$
2	$(B{:}3)$, $(D{:}1)$, $(E{:}3)$
3	$(A{:}4)$, $(D{:}5)$, $(E{:}2)$
4	$(C{:}3)$, $(D{:}3)$, $(E{:}1)$
5	$(A{:}3)$, $(C{:}2)$, $(E{:}1)$
6	$(B{:}1)$, $(D{:}5)$

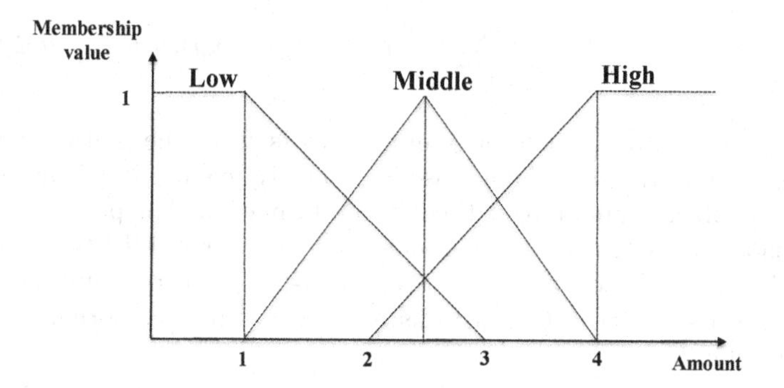

Fig. 1. The membership functions of fuzzy linguistic 3-terms.

Definition 1. The linguistic variable R_i is an attribute of a quantitative database whose value is the set of fuzzy linguistic terms represented in natural language as $(R_{i1}, R_{i2}, \ldots, R_{ih})$; this variable can be defined in the membership functions μ.

Definition 2. The quantitative value of i denoted as v_{iq}, is the quantitative of the item i in transaction T_q.

Definition 3. The fuzzy set, denoted as f_{iq}, is the set of fuzzy linguistic terms with their membership degrees (fuzzy values) transformed from the quantitative value v_{iq} of the linguistic variable i by the membership functions μ as:

$$f_{iq} = \mu_i(v_{iq})(= \frac{fv_{iq1}}{R_{i1}} + \frac{fv_{iq2}}{R_{i2}} + \cdots + \frac{fv_{iqh}}{R_{ih}}), \tag{1}$$

where h is the number of fuzzy linguistic terms of i transformed by μ, R_{il} is the l-th fuzzy linguistic terms of i, fv_{iql} is the membership degree (fuzzy value) of v_{iq} of i in the l-th fuzzy linguistic terms R_{il} and $fv_{iql} \subseteq [0, 1]$.

Definition 4. The transformed fuzzy linguistic term R_{il} is represented and denoted as the fuzzy itemset in the field of fuzzy data mining.

Definition 5. The support of the fuzzy itemset, denoted $supp(R_{il})$, is the sum of its transformed fuzzy values, which can be defined as:

$$supp(R_{il}) = \sum_{R_{il} \subseteq T_q \wedge T_q \in D'} fv_{iql}, \quad (2)$$

where D' is the quantitative database D transformed by membership functions $(= \mu)$, and the size of D' is the same as the original D.

Definition 6. The support of fuzzy k-itemset ($k \geq 2$), denoted as $sup(X)$, is the sum of the minimum fuzzy values among k items of X, which can be defined as:

$$sup(X) = \{X \in R_{il} | \sum_{X \subseteq T_q \wedge T_q \in D'} min(fv_{aql}, fv_{bql}), a, b \in X, a \notin b\} \quad (3)$$

Most previous work focus on mining MFFIs from the static database. In real-life situation, the size of database is frequently changed; it is an important issue to dynamically update the discovered information. The problem statement of this paper aims to efficiently update the discovered MFFIs based on the Fast UPdated (FUP) concept [2] and the fuzzy-list (FL) structure [16]. We thus assume that a set of new transactions are inserted into the original database, shown in Table 2.

Table 2. A quantitative database.

TID	Items with their quantities
7	(A:3), (C:2), (E:1)
8	(A:1), (D:1)

Thus, the designed incremental multiple fuzzy frequent itemset mining (IMFFIM) is to efficiently update the set of the discovered MFFIs. Thus, an itemset X is a MFFI in the updated database ($D + d$) if its support count of each fuzzy itemset X is no less than the pre-defined minimum support count. The set of MFFIs is thus formally defined as:

$$MFFIs \leftarrow \{X | sup(X) \geq \delta \times (|D| + |d|)\}, \quad (4)$$

where δ is the minimum support threshold, $|D|$ is the size of original database, and $|d|$ is the size of inserted transactions.

3 Proposed Incremental Algorithm

In the past, Lin et al. proposed a MFFI-Miner algorithm [16] to mine the multiple fuzzy frequent itemsets (MFFIs) without candidate generation. The algorithm designed the fuzzy-list (FL) structure, which could efficiently reduce the computation of multiple database scans for mining MFFIs. In this paper, the FL

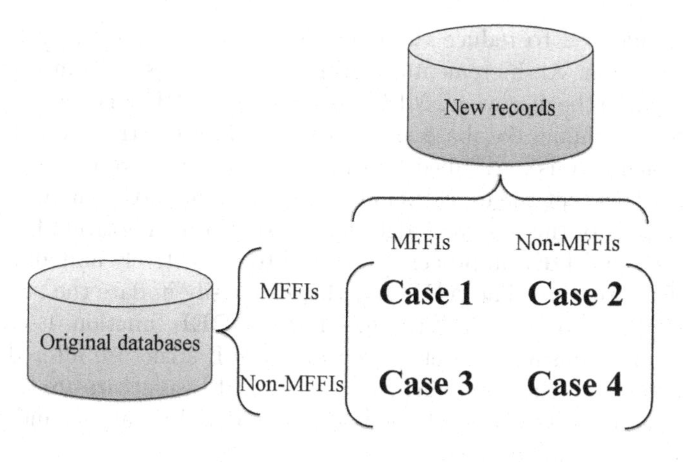

Fig. 2. Four cases of the designed approach.

Algorithm 1. Proposed algorithm

Input: D, a quantitative database; δ, the minimum support threshold; μ, the user-specified membership functions; FLs, the built FL structure from D; d, a set of inserted transactions.

Output: the sets of MFFIs.

```
1  set MFFIs.U := null;
2  for each transformed R_il ∈ d do
3      calculate sup(R_il)^d;
4      if sup(R_il)^d ≥ (|D| + |d|) × δ then
5          1-MFFIs.d := 1-MFFIs.d∪R_il;
6  for each R_il ∈ T_q ⊆ d do
7      if R_il ∈ 1-MFFIs.D∈ FLs then
8          sup(R_il)^U := sup(R_il)^D + sup(R_il)^d;
9          if sup(R_il)^U ≥ (|D| + |d|) × δ then
10             ADD(FLs);
11             MFFIs.U := MFFIs.U ∪ R_il;
12         else
13             DEL(FLs) ;
14     else
15         scan_set := scan_set∪R_il;
16 for R_il ∈ scan_set do
17     calculate sup(R_il)^D;
18     calculate sup(R_il)^U := sup(R_il)^D + sup(R_il)^d;
19     if sup(R_il)^U ≥ (|D| + |d|) × δ then
20         ADD(FLs);
21     if FLs ≠ null then
22         Construct(FLs);
23         update MFFIs.U;
24 return MFFIs.U;
```

structure is adopted to reduce the multiple database scans and keep the necessary information. An incremental algorithm is thus designed in this paper to efficiently update the discovered MFFIs based on the FUP concept. It divides the itemsets in the original database and in the new inserted transactions into four cases, and each case is performed by the designed procedure to update the discovered MFFIs except the itemsets in Case 4. The four cases can be summarized in Fig. 2. Details of the designed algorithm is shown in Algorithm 1.

The **ADD** and **DEL** approaches are used to respectively add and delete the R_{il} in the FL structure. The **ADD** function can easily update the fuzzy value of the itemsets based on the FL structure. For the **DEL** function, it can directly remove the unpromising itemsets based on the FL structure after database is updated. After that, the remaining itemsets in the FL structure are then checked against to the minimum support count in the updated database, and the actual MFFIs can thus be maintained.

4 Experimental Results

In this section, the performance of the proposed algorithm is compared with the state-of-the-art MFFI-Miner algorithm [16]. All algorithms were implemented in Java and experiments were carried on a computer having an Intel(R) Core(TM) i7-6700 3.41 GHz processor with 8 GB of main memory, running the 64 bit Microsoft Windows 10 operating system. A mushroom dataset [4] is used in the experiments to evaluate the performance of two compared algorithms. The quantity of items is randomly assigned in the range of [1, 11] interval by adopting normal distribution. The membership functions for 2-terms and 3-terms were used in the experiments. Besides, the insertion ratio of the new transactions is denoted as IR, which is used to set the size of the newly inserted transactions compared to the original database.

4.1 Runtime

In this experiments, the runtimes of two algorithms are then compared under varied minimum support thresholds with a fixed insertion ratio. The results are shown in Fig. 3.

From Fig. 3, the experiments are carried out with 2-terms and 3-terms membership functions. The runtime decreases along with the increasing of minimum support thresholds. The reason is that the algorithms generate fewer MFFIs with higher minimum support threshold, thus the runtime to mine or update the discovered information can be decreased. It is clear to see that the designed algorithm outperforms the state-of-the-art MFFI-Miner algorithm since the MFFI-Miner needs to process the updated database in the batch manner. In summary, the proposed algorithm performs better than the MFFI-Miner algorithm in terms of varied linguistic terms.

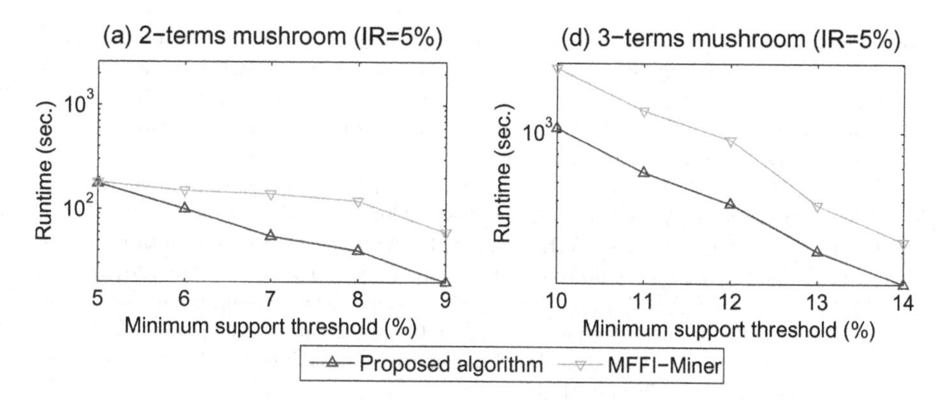

Fig. 3. Runtimes under varied minimum support thresholds.

5 Conclusion

In this paper, we present an incremental algorithm to fast update the discovered multiple fuzzy frequent itemsets with transaction insertion. The designed algorithm is based on the Fast UPdate (FUP) concept and the fuzzy-list (FL) structure to maintain the discovered information without candidate generation. The designed algorithm can easily maintain the FL structure and generate the required multiple fuzzy frequent itemsets (MFFIs) without multiple database scans. Experiments showed that the designed algorithm outperforms the state-of-the-art approach in terms of varied linguistic terms.

Acknowledgment. This research was partially supported by the National Natural Science Foundation of China (NSFC) under grant No. 61503092 and by the Research on the Technical Platform of Rural Cultural Tourism Planning Basing on Digital Media under grant 2017A020220011.

References

1. Agrawal, R., Srikant, R.: Fast algorithms for mining association rules in large databases. In: International Conference on Very Large Data Bases, pp. 487–499 (1994)
2. Cheung, D.W., Wong, C.Y., Han, J., Ng, V.T.: Maintenance of discovered association rules in large databases: an incremental updating techniques. In: The International Conference on Data Engineering, pp. 106–114 (1996)
3. Delgado, M., Marin, N., Sanchez, D., Vila, M.A.: Fuzzy association rules: general model and applications. IEEE Trans. Fuzzy Syst. **11**, 214–225 (2003)
4. Fournier-Viger, P., Lin, J.C.W., Gomariz, A., Gueniche, T., Soltani, A., Deng, Z., Lam, H.T.: The SPMF open-source data mining library version 2 and beyond. In: The European Conference on Machine Learning and Principles and Practice of Knowledge Discovery, pp. 36–40 (2016)

5. Han, J., Jian, P., Yin, Y., Mao, R.: Mining frequent patterns without candidate generation: a frequent-pattern tree approach. Data Min. Knowl. Disc. **8**(1), 53–87 (2004)
6. Hong, T.P., Kuo, C.S., Chi, S.C.: Mining association rules from quantitative data. Intell. Data Anal. **3**(5), 363–376 (1999)
7. Hong, T.P., Lin, C.W., Wu, Y.L.: Incrementally fast updated frequent pattern trees. Expert Syst. Appl. **34**(4), 2424–2435 (2008)
8. Hong, T.P., Lan, G.C., Lin, Y.H., Pan, S.T.: An effective gradual data-reduction strategy for fuzzy itemset mining. Int. J. Fuzzy Syst. **15**(2), 170–181 (2013)
9. Hong, T.P., Lin, C.W., Lin, T.C.: The MFFP-tree fuzzy mining algorithm to discover complete linguistic frequent itemsets. Comput. Intell. **30**(1), 145–166 (2014)
10. Kuok, C.M., Fu, A., Wong, M.H.: Mining fuzzy association rules in databases. ACM SIGMOD Rec. **27**(1), 41–46 (1998)
11. Lin, C.W., Hong, T.P., Lu, W.H.: Linguistic data mining with fuzzy fp-trees. Expert Syst. Appl. **37**(6), 4560–4567 (2010)
12. Lin, C.W., Hong, T.P., Lin, T.C.: An efficient tree-based fuzzy data mining approach. Int. J. Fuzzy Syst. **12**(2), 150–157 (2010)
13. Lin, C.W., Hong, T.P., Lu, W.H.: Mining fuzzy frequent itemsets based on UBFFP trees. J. Intell. Fuzzy Syst. **27**(1), 535–548 (2014)
14. Lin, J.C.W., Hong, T.P., Lin, T.C.: A CMFFP-tree algorithm to mine complete multiple fuzzy frequent itemsets. Appl. Soft Comput. **28**(C), 431–439 (2015)
15. Lin, J.C.W., Hong, T.P., Lin, T.C., Pan, S.T.: An UBMFFP tree for mining multiple fuzzy frequent itemsets. Int. J. Uncertain. Fuzziness Knowl. Based Syst. **23**(6), 861–879 (2015)
16. Lin, J.C.W., Li, T., Fournier-Viger, P., Hong, T.P., Wu, J.M.T., Zhan, J.: Efficient mining of multiple fuzzy frequent itemsets. Int. J. Fuzzy Syst. **19**(4), 1032–1040 (2017)
17. Shitong, W., Chung, K.F.L., Hongbin, S.: Fuzzy taxonomy, quantitative database and mining generalized association rules. Intell. Data Anal. **9**(2), 207–217 (2005)
18. Zadeh, L.A.: Fuzzy sets. Inf. Control **8**, 338–353 (1965)

Applications of Intelligent Data Analysis

Efficient Mobile Middleware for Seamless Communication of Prehospital Emergency Medicine

Bor-Shing Lin[1(✉)], Po-Hsun Cheng[2], Bor-Shyh Lin[3], Ren-Hao Wu[4], and Sao-Jie Chen[4]

[1] Department of Computer Science and Information Engineering, National Taipei University, New Taipei City 23741, Taiwan
bslin@mail.ntpu.edu.tw
[2] Department of Software Engineering and Management, National Kaohsiung Normal University, Kaohsiung 82444, Taiwan
[3] Institute of Imaging and Biomedical Photonics, National Chiao Tung University, Tainan 71150, Taiwan
[4] Graduate Institute of Electronics Engineering, National Taiwan University, Taipei 10617, Taiwan

Abstract. In the field of the pre-hospital emergency, immediate and medical processes must be provided to patients to increase their chance of survival. When a patient is in transit to a hospital, real-time or uninterrupted patient data shall be transmitted to the hospital so that physicians could realize the present status and make decisions of appropriate instructions and assessments. For moving ambulances, obtaining uninterrupted network services to wirelessly transmit data is the most important issue. In the proposed work, a seamless and ubiquitous stream control transmission protocol (SCTP) tunnel was developed to improve wireless communication connections. Cognitive radios can optimize spectrum usage by detecting their operating environment; therefore, it is necessary to integrate existing wireless communication services in an efficient mobile middleware (MM) and achieve a seamless transmission in a heterogeneous network. The developed efficient MM for seamless transmission, which is migrated to an embedded device, is consistent of existent network environments without requiring modification of the original system architecture. Moreover, the handoff when switching between wireless networks is shorter than that obtained in related studies. Thus, medical personnel can seamlessly transmit video and vital data to hospitals through the proposed MM while walking or in a moving vehicle.

Keywords: Cognitive radio · Mobile middleware · Prehospital emergency medicine · Uninterrupted communication

1 Introduction and Related Work

When a person suffers a severe injury, stabilizing them as soon as possible is critical before transporting them to a hospital for further medical assistance. In life-threatening cases, the time interval in which these procedures must occur is called the golden

P. Krömer et al. (eds.), *Proceedings of the Fourth Euro-China Conference on Intelligent Data Analysis and Applications*, Advances in Intelligent Systems and Computing 682, DOI 10.1007/978-3-319-68527-4_16

hour [1]. To maximize the benefits of prehospital time and enhance survival rates, medical personnel must thoroughly understand the implications of prehospital emergency medical services.

Before an ambulance transporting a patient arrives at a hospital, the real-time transmission of vital figures regarding a patient's health to the hospital through video and pictures is necessary. A doctor can then assess the patient in advance and provide professional advice to the emergency medical service (EMS) personnel. EMS workflow is extremely time critical, and successful implementation depends on well-trained cooperation between EMS personnel [2]. In addition to enabling doctors to assess the state of patients, sending real-time information to hospitals can enable medical staff to prepare the necessary medical apparatus and operating rooms during the transition period. Maximizing such preparations during the golden hour can increase survival rates. The main challenge to achieving this objective is in maintaining a stable and uninterrupted communication between moving ambulances and hospitals.

In recent years, the development of prehospital EMS systems has been extensively studied [2–4]. These EMS systems have been combined with various technologies such as wireless sensor networks and application development. However, although these previous research works addressed on the integration of wireless transmissions and EMS, the quality of wireless transmission has not been analyzed. Wireless communication has evolved considerably in recent years. The widespread availability of uncharged Wi-Fi hotspots is a great discover for pre-hospital EMS because of the enhancement of the communication, mobility and being feasible of real-time data. However, the tough environments and real-time access of medical processes has necessitated the introduction of quality-of-service (QoS) provisions in medical wireless networks [5]. Moreover, according to the existence of numerous wireless network services, another major consideration for real-time EMS transmission is the selection on optimal wireless channels for ambulances and the maintenance of reliable transmission [6–10].

One solution for robust prehospital EMS transmission is seamless communication. At present, using a fixed spectrum allocation scheme in traditional wireless communication systems results in low spectrum utilization. Because of the rapid development of wireless communication techniques, the spectrum resources available for allocation have become scarce. Thus, improving spectrum allocation and rendering unused spectra temporarily available for use is necessary. The concept of cognitive radio was proposed in 1999 [11]. Cognitive radios can analyze operating environments and optimize spectrum usage. Therefore, an efficient mobile middleware (MM) based on cognitive radio techniques for use in a prehospital EMS is proposed in this paper; the MM can not only select optimal transmission paths when presented with multiple links but can also provide a reliable transmission in order to prevent data loss during network switching.

2 Methods

2.1 Channel Decision Algorithm

In an intricate concept of the seamless communication presented in [12], the decision algorithm represents the determinant duty of eliminating handoff delays for wireless medical applicants. An adaptive reasoning and learning framework (ARALF) has been previously exploited as cognitive engines to manage cognition tasks [13]. A cognitive cycle has three tasks. First, the radio-scene analysis wherein respective environmental circumstances are sensed and variant configurations are probed. Second, the channel-state information is evaluated by the channel identification and the capabilities of the performance under different configurations are also forecasted. Third, radio-configuration selection determines what configuration is used to send the signal. An ARALF differs from typical cognitive engines because of not only the information with respect to its environment being received but also user-specific information; thus, an ARALF is able to arrange radio parameters according to users' preferences. Moreover, the proposed framework seamlessly integrates the adaptivity together with the mobility so that all users of cognitive radio are not negatively influenced on problems resulted by environmental alterations.

For integrating an ARALF inside the proposed MM, a simplified version of an ARALF, ARALF Lite, was designed (Fig. 1). The main characteristic of the ARALF Lite is: collecting data, consisting of the strength of signals and quality of links from USB wireless network adapters. Subsequently, wireless network connections can be adjusted and the optimal solutions are applied to the proposed MM by ARALF Lite.

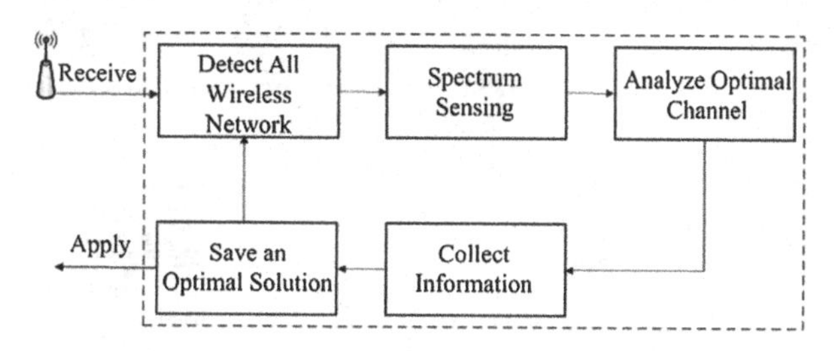

Fig. 1. Design flow of adaptive reasoning and learning framework (ARALF) lite.

2.2 Protocol

Stream control transmission protocol, abbreviated as SCTP, has been designed to run on transport layer and request for comments (RFC) 4960 [14]. Multihoming is a critical feature of the SCTP. A multihoming host has more than one network interface and can be bound to multiple Internet protocol (IP) addresses when the endpoint initializes an association. The SCTP is capable of switching among various networks without interrupting ongoing data transfer. The multihoming feature of the SCTP acts a great important duty in seamless communication, which is why the idea was adopted in this study to replace among multiple classes of wireless network. As designated in RFC 4960, in the process of an association, one path will be configured as the primary path and others are backups, indicating that one of the designated IP addresses to receive data must be selected as the primary address.

Routing is one of the core topics that must be addressed for IP address switching. Because wireless connections can be unstable and traffic load often varies, establishing a routing protocol that is based on traffic status is imperative. A routing protocol is a set of rules used to decide how routes exchange information and communicate with each other; such protocols also perform activities such as discovering main networks and updating routing tables. Among these activities, network discovery involves actively searching for the next suitable connecting node. Because reducing the search time during network discovery is critical, ARALF Lite was used in the proposed system to periodically choose a connecting path with the optimal quality in a heterogeneous network. Adopting ARALF Lite not only minimizes the time required to search for the next appropriate network but also eliminates transmitting errors after accessing the most suitable node to connect.

2.3 Overall Architecture of the System

To attain the goal of seamless communication for wireless medical applicants, the proposed equipment combines a third-generation mobile network, a Wi-Fi network, and a mobile middleware for transmission data among a pair of SCTP client and server (Fig. 2).

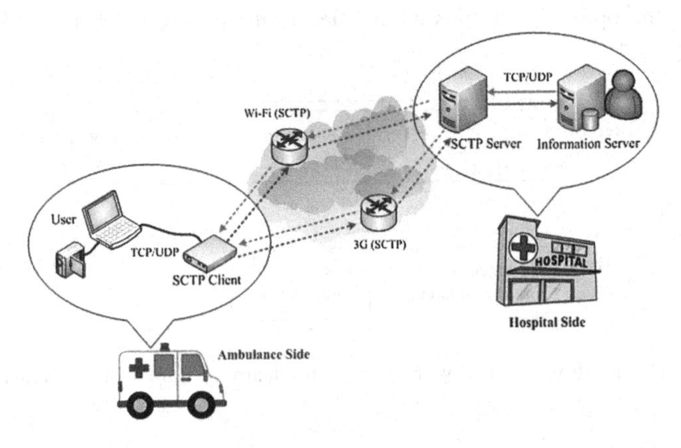

Fig. 2. System overview.

In Taiwan, most hospitals adopt transmission control protocol (TCP)-based information systems and connection-oriented communication systems. Each endpoint is bound to a single interface by TCP protocol; thus, when it is necessary to switch connectivity among multiple wireless networks, implementing a TCP for a communication protocol is unsuitable because a TCP connection must be rebuilt. Furthermore, although user datagram protocol (UDP) is a more agile mechanism of reducing handoff than TCP, it cannot guarantee that transmitted data is actually arrived under a UDP transmission because a UDP communication lacks a reliable mechanism. Therefore, the SCTP was used for multi-homing to conquer present challenges of peer-to-peer connection.

The developed MM consists of two components: a pair of SCTP server and client. A SCTP server was set up in a hospital and runs on Linux. The SCTP client, which was portable because it has to connect to a mobile device, was designed in an embedded equip. Target board of a SCTP client is BeagleBoard-xM (BeagleBoard.org, United States). Before a patient's data is going to be sent by a first aid officer to the information system of a hospital, a connection under SCTP among a pair of SCTP client and server will be created. Next, after receiving data from a mobile device via a TCP/UDP connection, TCP/UDP data packets shall be encapsulated into SCTP data packets by the SCTP client and then send those SCTP packets to the SCTP server. Eventually, the received SCTP packets embedded into the TCP/UDP packets shall be unpacked by the SCTP server and send these TCP/UDP packets to the data center of the hospital. Equally, a database is able to transmit packets to a mobile device by being packed and unpacked from a pair of SCTP server and client, respectively.

Figure 3 illustrates the mechanism of the seamless switching. Phase 1 is about creating a connection under SCTP to achieve multiple-paths transmission among a database and a mobile equip, and the data is also obtained with respect to accessible network interfaces. Afterward, an ARALF Lite is used to calibrate radio parameters according to environmental alterations and users' preferences. In Phase 2, the current optimal connecting path is selected; this path is designated by ARALF Lite to be the primary path of a SCTP communication, following which replacing to that connecting path is possible by altering the contents of a routing table in the kernel of the system. In Phase 3, the present selected path is set to be the primary path and previous primary paths are set to be backup paths. Phases 2 and 3 are repeated when a new signal of switching produced by an ARALF Lite is obtained.

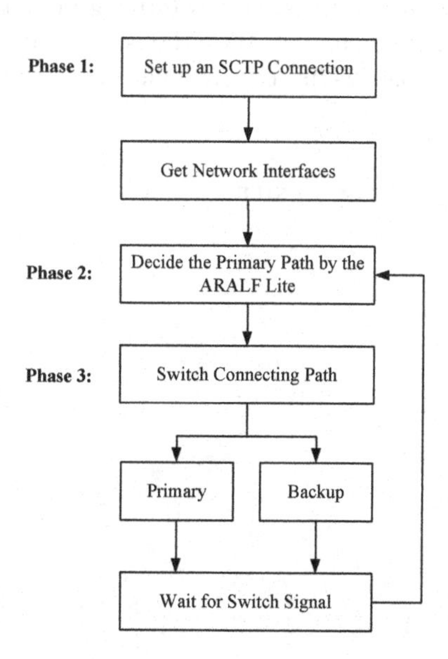

Fig. 3. Seamless switching mechanism.

3 Field Experiment Results

This section demonstrates the experimental results of the developed middleware of seamless communication for mobile cognitive radio devices. To attain seamless mobility when connected to heterogeneous wireless networks, the handoff latency during network switching was required to be less than 100 ms to prevent a noticeable delay that would impair interactivity [15]. In this study, the proposed MM was tested in walking and driving scenarios in Taipei City, Taiwan.

In these scenarios, the developed MM was tested under the transferring of medical information such as electrocardiography (ECG) and saturation of peripheral oxygen (SpO_2) records and real-time video frames. In this study, the ECG records comprised open data from MIT-BIH database [16], and the SpO_2 records comprised open data from the MIMIC II database [17]. Furthermore, WebRTC—according to a resolution of 640 × 360 pixels at 30 frames per second—was used for browser-based, and real-time video communication. WebRTC [18] is an application programming interface definition to enable browser-to-browser video communication; therefore, it can be easily integrated into medical service systems. The content of the web page served by our embedded system was established using PHP programs, and each web page should be updated every 0.2 s to capture updated data. Furthermore, the use of the information system for sending data from an ambulance to a hospital was tested using a real-life experiment, and the experimental route selected was from National Taiwan University (NTU) to National Taiwan University Hospital (NTUH), as illustrated in Fig. 4.

Each of the two aforementioned scenarios (driving or walking) has three configurations. In the first configuration, only Wi-Fi hotpots are available. Though a Wi-Fi service can support a high-bandwidth service, limited coverage interrupts connections.

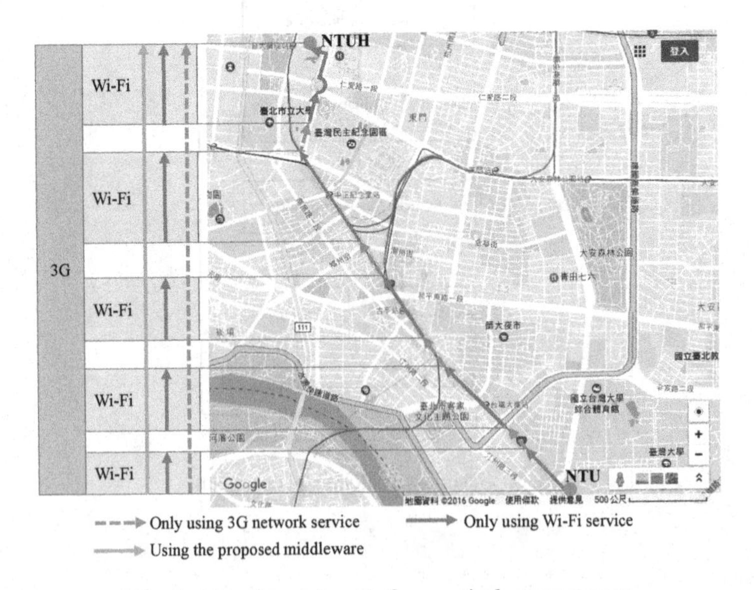

Fig. 4. Experimental route for practical measurement.

In the second configuration, only 3G networks are used, which does not involve disconnection problems reasoned in restricted coverage, yet the speed of a 3G connection is lower than that obtained with a Wi-Fi service. The third configuration works as follows. The proposed MM is used for accessing Wi-Fi, accompanying with a high-bandwidth service and connecting to 3G mobile networks in a complementary manner to keep obtaining an uninterrupted network serving whenever gaps of Wi-Fi coverage were encountered. Thus, three types of measurement were obtained: those measured using the proposed MM, using Wi-Fi alone, and using 3G networks alone. During the measurement process, four data categories were individually sent to the server: only SpO_2 data, only ECG data, only real-time video, or all three types of data; an average data-transmission rate of 10 s was recorded in each experiment. The experimental distance in both scenarios was 3.5 km. The walking and driving scenario experiments involved holding a 6 km/h-walking speed and 40 km/h-average-driving speed, respectively. The experimental results related to entire data (including real-time frames of video, SpO_2 records, and ECG records) transmissions via only Wi-Fi services, only 3G networks, or the proposed MM are presented in Fig. 5 for the walking scenario and Fig. 6 for the driving scenario.

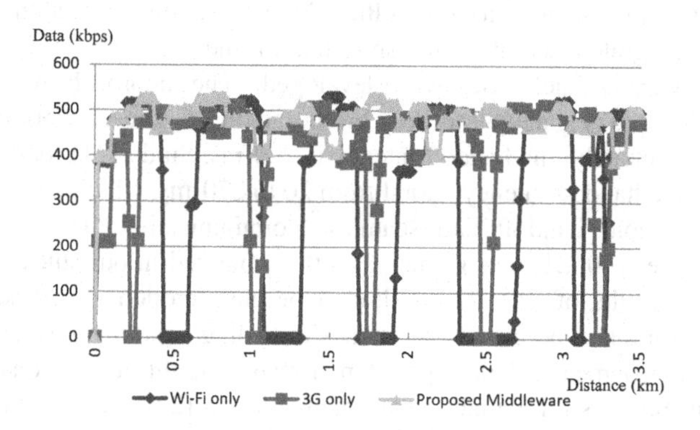

Fig. 5. Results of transmitting all data during walking via only Wi-Fi services, only 3G networks, or the proposed MM.

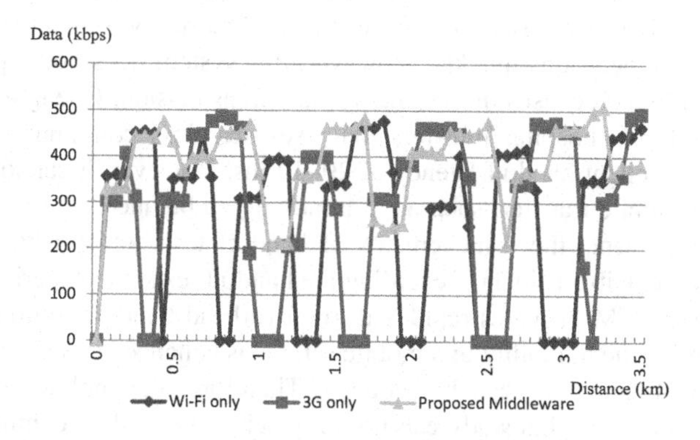

Fig. 6. Results of transmitting all data during driving via only Wi-Fi services, only 3G networks, or the proposed MM.

4 Discussion

Because this study presents a new middleware for providing seamless communication to mobile clients, this section compares the handoffs observed in this study with those achieved in other studies. When people are away from the coverage of one access point (AP) and need to connect to a closer AP, connectivity switching must occur. The connectivity switch involves a transition called a handoff. A handoff can be further divided into two different types: intracell handoff and intercell handoff. Both handoffs can occur simultaneously. An intracell handoff is required to switch connectivity between APs and can be considered a form of micromobility; the intercell handoff, however, can be considered a form of macromobility. An intercell handoff is required to switch connectivity when the user moves between regions connected to different network domains.

In [19], the mobile-IP (MIP) method of handoff was implemented. Approximately, the handoff latency is 760 ms because of the delay occurring during MIP registrations and AP re-associations. According to the above two factors, which lead to severe delays, exist in actual environments of wireless networks; therefore, they should be overcome using the proposed system in which seamless communication with the proposed MM is implemented for controlling AP association and enabling handoffs. In [20, 21], an integrated seamless transmission of the third generation mobile and 802.11-based wireless networks was developed. The approach reported in [20] involved using a proxy gateway to be a bridge among a core network and an access network. Approximately, the handoff latency is 200 ms. Study [21] focused on routing protocols, and the handoff latency varied from 20 to 270 ms, which nearly achieves the objective of a rapid handoff and seamless communication. The handoff latency obtained in studies [20, 21] was greater than that obtained in our study. However, the validity of the results of [20, 21] is limited because motion experiments were not conducted. In summary, because most studies regarding handoffs have conducted tests in simulated environments, their experimental results might not demonstrate the suitability of their methods for actual wireless networks. In [22], Neyem *et al.* proposed a cloud-based mobile system for prehospital emergency services to support team collaboration and decision-making. The system is used to transmit patients' vital information when patients are being transferred to a hospital and calculates the rapid emergency medicine score, used to evaluate patients' mortality so that physicians in the hospital can make decisions quickly. However, this system does not exploit seamless communication, which causes discontinuous data transmission. Cheng *et al.* proposed an application for in-hospital transfers that uses seamless communication [9]. The system has an excellent handoff latency of 20–30 ms. However, it cannot be deployed for high-speed mobile transfer, such as in a moving ambulance.

This paper presents the achievements of handoff tests and practical evaluations exploited in Taipei City, Taiwan. Regarding the handoff tests, the experiments revealed that the proposed MM provides rapid and low overhead handoff. With multi-homing wireless networks, the reliability of a mobile service is enhanced and costs of accessible network during evaluations can be reduced. Therefore, the mobile cognitive radio middleware proposed in this study can not only achieve seamless communication but also enhance the quality of the current mobile transmission of medical services in Taipei City, Taiwan.

5 Conclusion

In this study, an efficient MM of performing seamless and ubiquitous SCTP tunnel for use by EMSs was proposed. The architecture and a protocol for seamless transmission over a heterogeneous wireless network to improve the quality of mobile transmission in Taipei City, Taiwan were presented. In Taipei City, the critical challenge to mobile medical applicants is the selection of sufficient network links for maintaining an uninterrupted wireless communication. The proposed system was designed to prioritize connection through Wi-Fi services, which are widely available and free in Taipei City, and to use 3G networks whenever Wi-Fi coverage gaps are encountered. Because this approach is integrated with cognitive radio techniques, it can not only select optimal transmission paths using multiple wireless links but can also support reliable transmission to prevent data loss during network switching. Moreover, to enable the implementation of the proposed system in current EMS systems without performing system modifications, the proposed MM was also ported into an embedded system.

The performance of the proposed MM was demonstrated through practical experiments. The proposed design achieved the goal of seamless communication, thereby to enable mobile clients in the Taipei City to obtain wireless networks out of interruptions. Henceforth, as the wireless network coverage in Taipei City increases, the performance of the proposed MM will be further enhanced. With the developed architecture, new mobile communication service (e.g., 4G communication network) could be subjoined in the proposed system to heighten the solid of the seamless transmissions among a heterogeneous network.

References

1. Voice of America (VOA): Injectable Foam Could Prove Life Saver http://www.voanews.com/content/new_technology_stop_internal_bleeding_special_forces/1569177.html. Accessed 10 May 2017
2. Beul, S., Mennicken, S., Ziefle, M., Jakobs, E.: What happens after calling the ambulance: information, communication, and acceptance issues in a telemedical workflow. In: Proceedings of International Conference on Information Society (i-Society), London, UK, pp. 98–103 (2010)
3. Venkatesh, V., Vaithyanathan, V., Manikandan, B., Raj, P.: A smart ambulance for the synchronized health care – a service oriented device architecture-based. In: Proceedings of International Conference on Computer Communication and Informatics (ICCCI), Coimbatore, India, pp. 1–6 (2012)
4. Lin, B.S.: A seamless ubiquitous emergency medical service for crisis situations. Comput. Methods Programs Biomed. **126**, 89–97 (2016)
5. Zvikhachevskaya, A., Markarian, G., Mihaylova, L.: Quality of service consideration for the wireless telemedicine and e-health services. In: Proceedings of IEEE International

Conference on Wireless Communications and Networking Conference (WCNC), Budapest, Hungary, pp. 1–6 (2009)
6. Wang, J., Ghosh, M., Challapali, K.: Emerging cognitive radio applications: a survey. IEEE Commun. Mag. **49**, 74–81 (2011)
7. Kitano, T., Juzoji, H., Nakajima, I.: Elevation angle of quasi-zenith satellite to exceed limit of satellite visibility of space diversity which consisted of two geostationary satellites. IEEE Trans. Aerosp. Electron. Syst. **48**, 1779–1785 (2012)
8. Gallego, J.R., Hernandez-Solana, A., Canales, M., Lafuente, J., Valdovinos, A., Fernandez-Navajas, J.: Performance analysis of multiplexed medical data transmission for mobile emergency care over the UMTS channel. IEEE Trans. Inf. Technol. Biomed. **9**, 13–22 (2005)
9. Cheng, P.H., Lin, B.S., Yu, C., Hu, S.H., Chen, S.J.: A seamless ubiquitous telehealthcare tunnel. Int. J. Environ. Res. Public Health **10**, 3246–3262 (2013)
10. Thelen, S., Czaplik, M., Meisen, P., Schilberg, D., Jeschke, S.: Using off-the-shelf medical devices for biomedical signal monitoring in a telemedicine system for emergency medical services. IEEE J. Biomed. Health Inform. **19**, 117–123 (2015)
11. Mitola, J., Maguire, G.Q.: Cognitive radio: making software radios more personal. IEEE Pers. Commun. **6**, 13–18 (1999)
12. Isaksson, L.: Seamless communications: seamless handover between wireless and cellular networks with focus on always best connected. Ph.D. dissertation, Department of Telecommunication Systems, Blekinge Institute of Technology, Sweden (2007)
13. Chen, S.J., Hsiung, P.A., Yu, C., Yen, M.H., Sezer, S., Schulte, M., Hu, Y.H.: ARAL-CR: an adaptive reasoning and learning cognitive radio platform. In: Proceedings of International Conference on Embedded Computer Systems (SAMOS), Samos, Greek, pp. 324–331 (2010)
14. Internet Engineering Task Force (IETF): Stream Control Transmission Protocol, https://tools.ietf.org/html/rfc4960. Accessed 10 May 2017
15. Amir, Y., Danilov, C., Hilsdale, M., Musaloiu-Elefteri, R., Rivera, N.: Fast handoff for seamless wireless mesh networks. In: Proceedings of the 4th International Conference on Mobile Systems, Applications and Services (Mobisys), Uppsala, Sweden, pp. 83–95 (2006)
16. PhysioNet: MIT-BIH Arrhythmia Database. http://www.physionet.org/physiobank/database/mitdb. Accessed 10 May 2017
17. PhysioNet: MIMIC II Database. https://mimic.physionet.org. Accessed 10 May 2017
18. WebRTC: WebRTC, http://www.webrtc.org. Accessed 10 May 2017
19. Ramachandran, K.N., Buddhikot, M.M., Chandranmenon, G., Miller, S., Belding-royer, E. M., Almeroth, K.C.: On the design and implementation of infrastructure mesh networks. In: Proceedings of IEEE Workshop on Wireless Mesh Networks (WiMesh), Santa Clara, CA, USA, pp. 1–12 (2005)
20. Yang, P., Deng, H., Ma, Y.: Seamless integration of 3G and 802.11 wireless network. In: Proceedings of the 5th ACM International Workshop on Mobility Management and Wireless Access (MobiWac), Chania, Crete Island, Greece, pp. 60–65 (2007)
21. Cao, J., Xie, K., Wu, W., Liu, C., Yao, G., Feng, W., Zou, Y., Wen, J., Zhang, C., Xiao, X., Liu, X., Yan, Y.: HAWK: real-world implementation of high-performance heterogeneous wireless network for Internet access. In: Proceedings of 29th IEEE International Conference on Distributed Computing Systems Workshops (ICDCS Workshops), Montreal, QC, USA, pp. 214–220 (2009)
22. Neyem, A., Carrillo, M.J., Jerez, C., Valenzuela, G., Risso, N., Benedetto, J.I., Rojas-Riethmuller, J.S.: Improving healthcare team collaboration in hospital transfers through cloud-based mobile systems. Mob. Inf. Syst. (2016). Article ID 2097158

A Method to Evaluate the Research Direction of University

Xian-le Cao[1] and Xiao-qiang Xi[2](✉)

[1] School of Science, Xi'an University of Posts and Telecommunications, Xi'an, China
1960700694@qq.com
[2] Institute of Internet of Things and IT-Based Industrialization, Xi'an University of Posts and Telecommunications, Xi'an, China
xxq@xupt.edu.cn

Abstract. The era is changing from information technology to data technology, big data is used very well in the field of financial, medical, e-commerce and so on, but not very well in the field of education. The idea of "Data - driven schools, analysis of change education" make the need for the educational data mining more and more prominent. Data mining in education can help us to connect the relevant areas of education and find the key educational variables, which can make the education and teaching decision simple and accurate. In this paper, by using the Chinese word segmentation algorithm, association rule and RStudio tool, we analyse the title of master's thesis in four universities that have same discipline structure. The title data is obtained from http://www.cnki.net, which is an authority database in China. The results show that the research directions of the four university tend to be wireless network, mobile communication and algorithms.

Keywords: Education data mining · RStudio · Association rules · Chinese word segmentation algorithm

1 Introduction

As the high-speed development of science and technology, interpersonal interaction become very strong and life is more and more convenient, big data is the product of this high-tech era. With the help of big data, people can view the complex world from a more comprehensive and finer perspective. For education, big data contains great potential to promote personalized learning, improve teaching materials and method, and ultimately improve student's achievement, and these effects will also be reflected in higher education, especially in the situation of Internet + education. Massive data is produced and recorded in universities' daily work. It is very important to grasp the construction of those data and turn it into a strong support for educational decision-making and comprehensive information for the development of university. The results of data analysis will effectively promote the cultivation of high-quality and innovative talents. How to dig out useful information from big data to improve educational management and learning performance? The emergence of this paper

P. Krömer et al. (eds.), *Proceedings of the Fourth Euro-China Conference on Intelligent Data Analysis and Applications*, Advances in Intelligent Systems and Computing 682, DOI 10.1007/978-3-319-68527-4_17

prompted the appear of educational data mining (EDM). Since 2005, artificial intelligence education applications, intelligent tutor system and other international conferences carried out a number of "educational data mining" theme seminar, the first education data mining international academic Conference held in Canada in 2008, now this seminar has hold 9 times [1]. In 2009, the fifth advanced data mining and application of the International Conference [2] hold in Beijing Normal University, in which the "data mining in the application of education" theme is joined for the first time. Guiyang International Big Data Industry Expo hold in 2015, in which Alibaba founder Ma Yun said that the future internet is the era of big data, Ma strongly believe that the data is the first productive force, who owns the data who will have a huge wealth. Baidu founder Li Yanhong said that only the data is not feasible, we also need data mining technology to dig out the hidden value of the data [3]. The Platform for Action to Promote big data Development is issued by the State Council on August 31, 2015 [4], which promote the education of basic data to accompany the collection and sharing of the country, explore the use of big data to change the way of education and promote the quality of education. US 2016 National Education Technology Program "Future Study Preparation: Remodeling the Role of Technology in Education" [5], as the part of the study, they propose to meet the individualized learning through data collection and analysis. In the evaluation part, they propose to use different types of evaluation data to improve the study better. EDM is the embodiment of digital education research, it is the inevitable requirement of the development of educational information, and it is an emerging and interesting research field. Researchers have already started to use different data mining methods to explore the problems and laws in education [6], such as Baker's four EDM key applications [7], Castro presented EDM main applications. Some new EDM application trends continue to emerge, such as association rules most commonly used in feedback to teachers to help them make decisions [8], the use of neural network technology to assess student performance in order to achieve the purpose of predicting the performance of students [9, 10], the use of clustering technology to students personalized grouping to improve learning efficiency and so on [11, 12].

In this paper, we try to mine some useful information about higher education from public database. The data means the title of master thesis, from which we try to find the research direction of subject, which is very useful to evaluate the subject characteristic.

2 Data Sources and Research Methods

2.1 Data Collection and Cleaning

First of all, we need to determine the target data and mining object, which can help to determine the key aspects of data mining technology. The target data were collected from http://www.cnki.net. According to the needs of the study, A, B, C, D four universities that have similar disciplined structure are chosen, the title of their master thesis is collected, 15364 items during 2010–2016. Some inconsistency data will be eliminated in the cleaning processing. Then the title data is divided and stored in the database to be called. As shown in Table 1:

Table 1. Details of sample data

	A	B	C	D
Number of thesis	6000	6845	1742	777
Time span	2013–2015	2010–2016	2012–2016	2010–2016

2.2 Research Methods and Tools

The Chinese word segmentation algorithm and association rules are used to quantitatively analyze the content of the literature. RStudio is used to process data, R is a free, open source software belonging to the GNU system and an excellent tool for statistical calculation and statistical mapping, it is a complete set of data processing, computing and mapping software systems. Rwordseg package is used to realize the keyword frequency statistics, and wordcloud package is used to realize visual analysis. Rwordseg package is realized by rJava Chinese word segmentation tool Ansj, which is based on the Chinese Academy of Sciences ictlas Chinese word segmentation algorithm open source tools. The third party package arules is used to explore the association rules of the keyword matrix, and the R expansion package arulesViz is used to visually display the associated results.

2.3 Research Process

The research process includes four aspects: (1) extract the title of master's thesis from http://www.cnki.net of A, B, C, D universities, clean the titles manually and store them in the database for calling; (2) use segmentCN function in Rwordseg package to realize word segmentation, and delete the commonly stop words, extract the key words and its frequency, and display them by word cloud; (3) use the Apriori function, which is one of the most influential algorithms for mining the frequent itemsets of Boolean association rules, to analyse the association rule of the keywords, then plot the support histogram, set the scatter plot of support and confidence thresholds, and the correlation plot between the keywords; (4) explore the research hotspots and future research trends of the four universities by using the analysing results.

3 Analysis of the Results

Key words can reveal the core information of the thesis, which can reflect the main research field and direction of the paper from one aspect, which can predict the development direction of the school. Csv file is read by the segmentCN function in the Rwordseg package, and word segmentation of text is realized. A corpus will be created by using Corpus function. The stop word will be read by read.table function. And then use the tm_map function in the tm package to delete the stop word and blank, use the subset function to select the number of words greater than 1. The number of words selected more than 50 times for A, B university, more than 10 times for C, D universities, then use overcloud function to produce a word cloud for visual analysis. As shown in Fig. 1, the size of the keyword in the figure represents the frequency of the keyword, the larger the size, the higher the frequency. The result shows that network has the highest frequency in A, B, D university. The next following are communication, mobile, algorithms and wireless, which are all have strong connection with network. Algorithm is the highest frequency in C university, the following are network, wireless, positioning, images and so on. So the network, algorithms, communications and other aspects are the most studied in the four universities, which also meet our expectations.

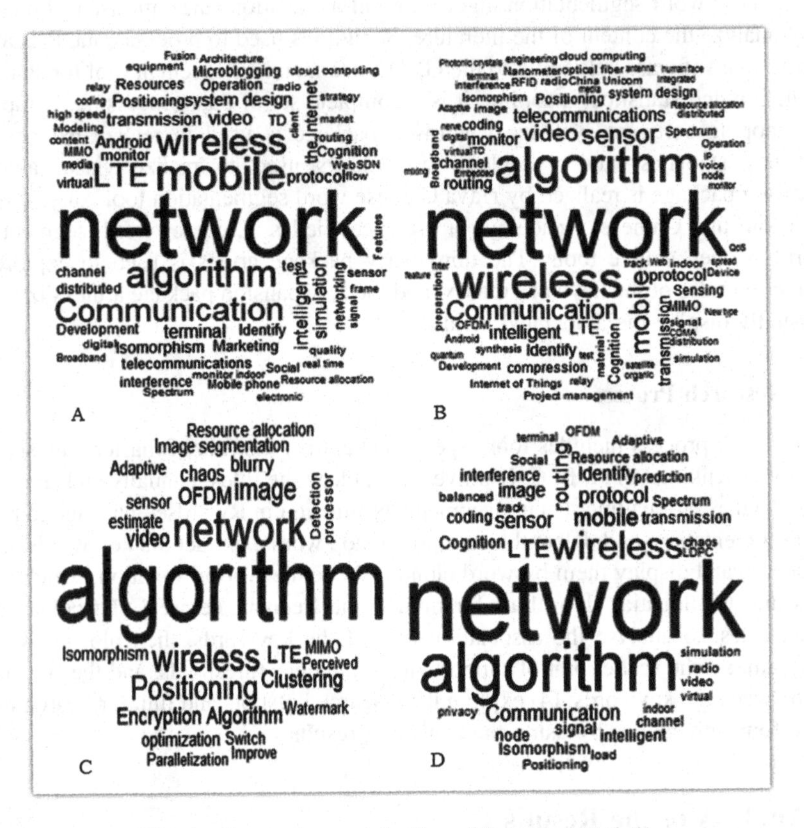

Fig. 1. Comparison graph of word frequency statistics

The frequency of the keyword can tell us the research direction of the four universities, then how about the relation between the keywords? To find the relation between the keywords we need to use association rules. Support, confidence and lift should be used to quantify the correlation between two things. The read.transactions function is used to read the data and create a sparse matrix, the summary function is used to look at the data set and the relative statistics summary information. The support can be obtained by using the frequency function, and the inspect function can be used to see the information. Here we intercept the top ten keywords in terms of support, as shown in Fig. 2, we can find that the support of the network in A, B, C, D university is

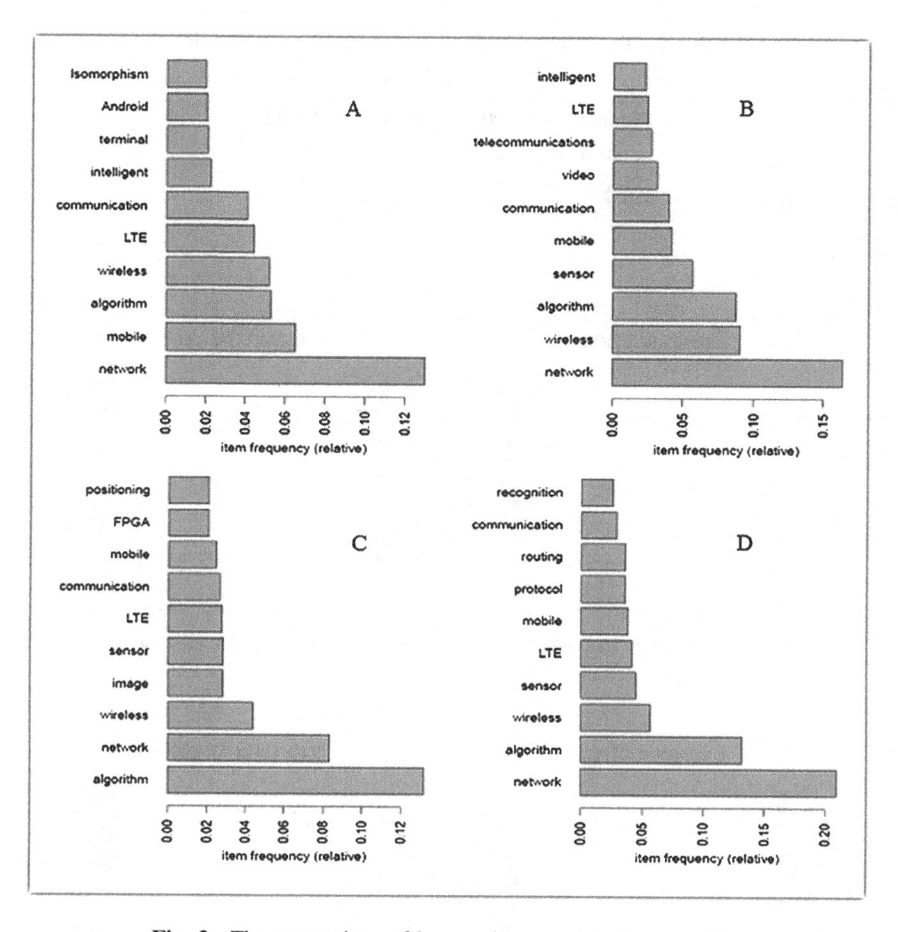

Fig. 2. The comparison of keywords support in 4 universities

about 0.13, 0.16, 0.08, 0.22 respectively, the algorithm is about 0.05, 0.09, 0.13, 0.13 respectively, the wireless is about 0.05, 0.09, 0.04, 0.06 respectively.

We care about the whole distribution of the keywords, the support and confidence distribution range of each keyword is very important for us to understand the correlation between the keywords. Using the apriori function in arules package can help us to find the keywords correlation. The scatter plot can be drawn by plot function. If introducing interactive parameters (interactive = TRUE) to the plot function, the interaction scatter plot can be drawn. The association rules below lift will be filtered out. Because the keyword matrix sparseness is relatively large, and the number of keywords is also very large, so we set the support threshold of the four universities as small as possible, the values are shown in Table 2. The results are shown in Fig. 3, The horizontal axis is the support and the vertical axis is the confidence. Each small square in the figure represents a keyword greater than the support threshold and the confidence threshold. The depth of the square color represents the bigger lift degree, the greater the degree, the better the data quality; the smaller the degree, the more uneven the data.

Table 2. The threshold and the number of generated rules

	A	B	C	D
Support	0.001	0.001	0.002	0.002
Confidence	0.2	0.5	0.01	0.1
Rules	525	642	623	1612

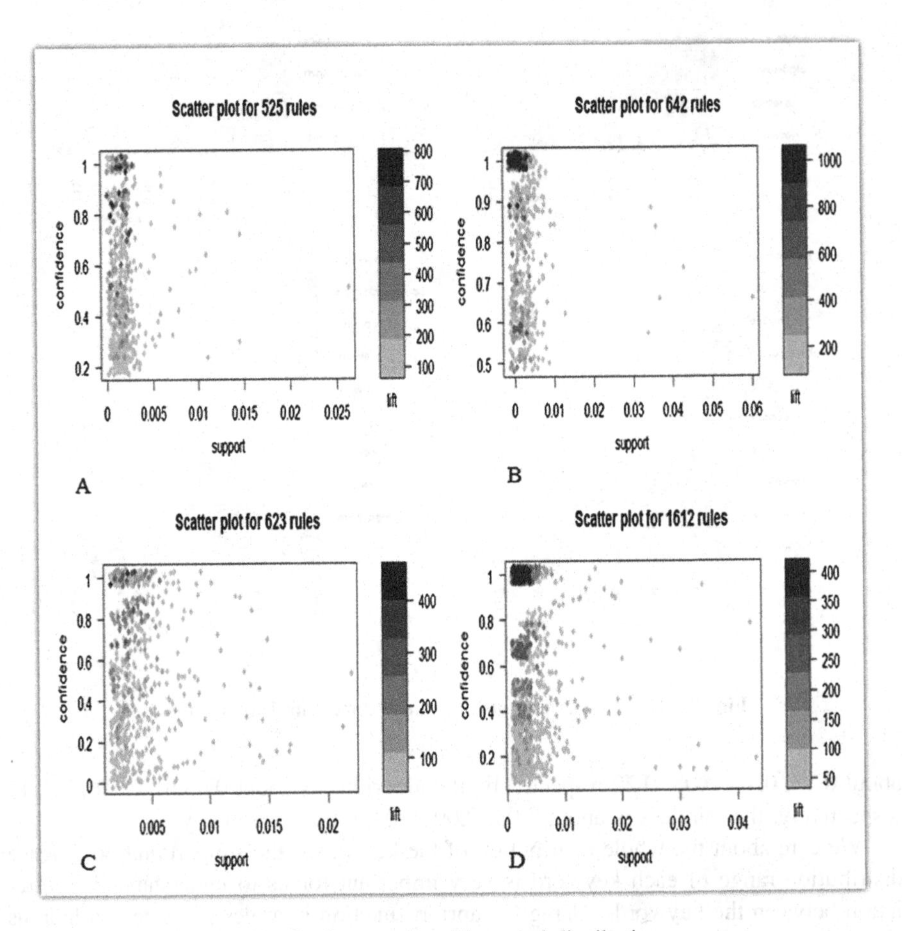

Fig. 3. Scatter plot of keyword distribution

From the figure we can see that the keywords are mostly distributed in areas with a support degree of less than 0.005, and the degree of improvement of the keyword increases with the confidence of the whole area. The support of the keyword is relatively low, most of the keywords are not associated with the degree or correlation, and strong correlation between keywords is insufficient.

Association rule analysis consists of two steps: discover frequent itemsets and generate their association rules. The frequent itemsets can be found by using the eclat

Table 3. The threshold and the number of generated rules

	A	B	C	D
Support	0.003	0.007	0.006	0.008
Confidence	0.6	0.6	0.8	0.8
Rules	12	12	19	17

function in the arules package. A certain support and confidence threshold range can determine a valid rule. In actual process a lot of data will appear, there will be a lot of rules for a simple search and then have a considerable part of the invalid rules. So one must test different threshold of support and confidence, until find the expected rules. For the data of A, B, C, D four universities, a proper support and confidence threshold is shown in Table 3. The correlation between the keywords is shown in Fig. 4, which is obtained by using the plot function, the circle's size indicates the support's size, the color's depth indicates the lift's size. One can see that the network as the center of the

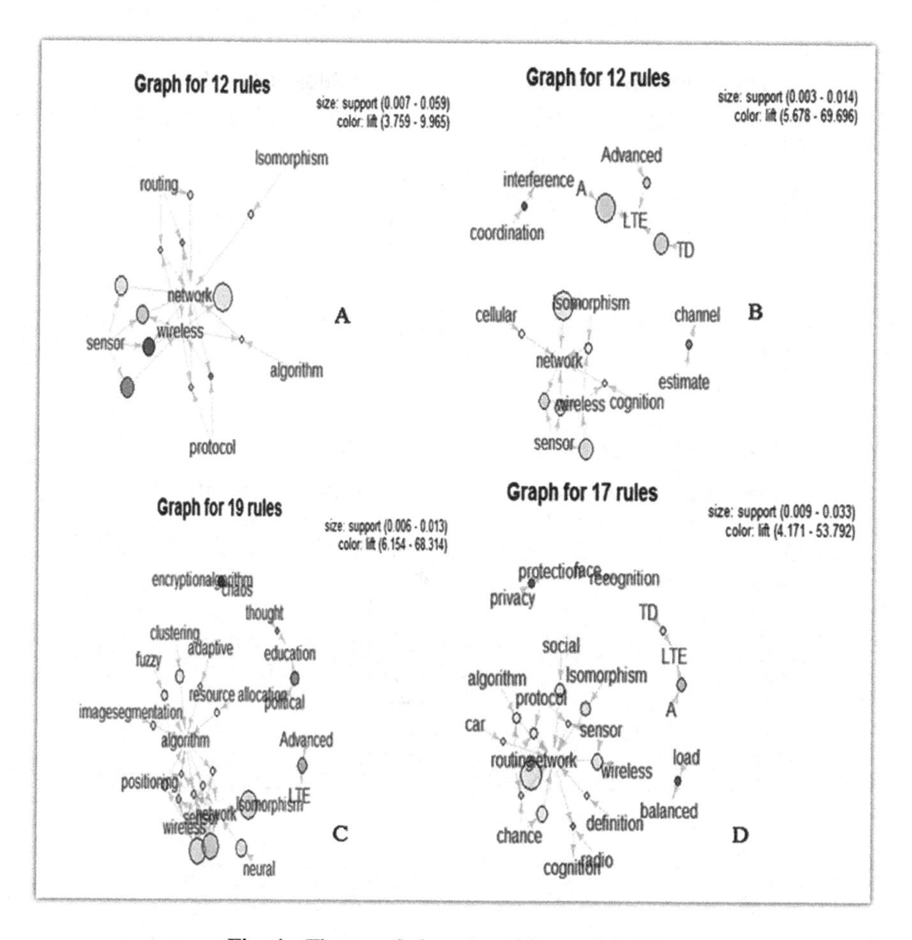

Fig. 4. The correlation plot of four universities

Table 4. The threshold and the number of generated rules

	A	B	C	D
Support	0.002	0.006	0.006	0.01
Confidence	0.6	0.5	0.6	0.7
Rules	10	11	9	9

formation of divergent outward radiation for A, B, D university, the relatively higher correlation pair are wireless and network, heterogeneous and network, mobile and network, mobile and communications, network and communications, wireless and communications, wireless and sensors. These keywords always appear in pairs that has a certain relevance. It is easy to infer that mobile communications, wireless networks and other aspects have a depth research in A, B, D university. The algorithm is the center of radiation in C university, the higher correlation pair are images and algorithms, clustering and algorithms, neural and network.

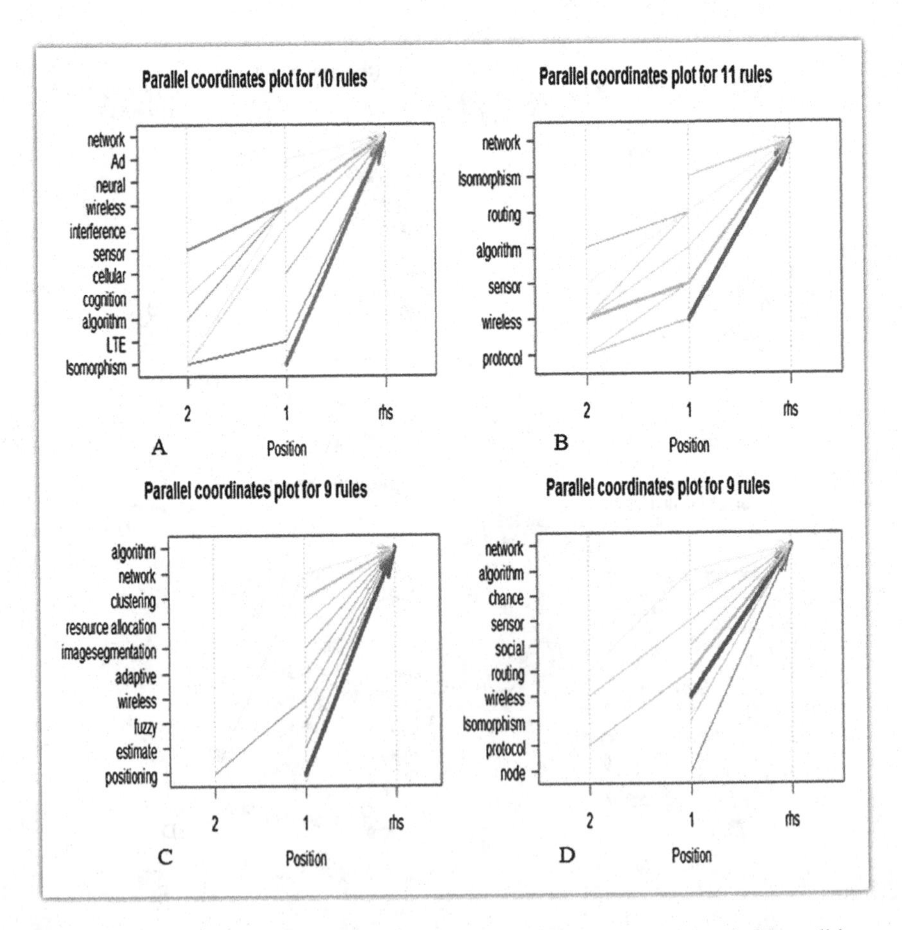

Fig. 5. The correlation plot between the keywords that satisfied the threshold conditions

The network keyword has the highest frequency in A, B, D university, so we analysis the association rules about it. Set rhs = c ("network") in the apriori function, and the threshold of support and confidence in Table 4. The corresponding results are shown in Fig. 5, the thickness of the straight line indicates the degree of correlation between two keywords. The thicker the line, the stronger the degree of association. The results show that heterogeneous networks, wireless networks, mobile networks, social networks, routing networks are the main direction of research in A, B, D university. The algorithm keyword has the highest frequency in C university, similarly process as the network in A, B, D university, it can be seen that wireless location algorithm, image segmentation algorithm, clustering algorithm, fuzzy algorithm are the hot topic in C university.

4 Conclusion and Discussion

In this paper, we analysis the title of master's thesis in A, B, C and D four universities that have same discipline structure. By using Chinese word segmentation algorithm and association rules analysis, we find that A, B and D university are focus on network research. A university is focus on heterogeneous network and sensor network research, B and D university are focus on wireless networks, mobile networks, routing networks and so on. C university is focus on the network algorithm, wireless positioning algorithm and so on. For university's administrator, these results can help them to grasp their main research direction, find the potential research direction and make a reasonable decision. For education management department, the method in this paper can help them to evaluate the level of discipline.

The research content in this paper is one part of Educational Data Mining (EDM). Compared with the analysis of student test scores, students' campus consumption, book borrowing and college enrollment et al., our content is from a different way to use the education data. Combined with artificial intelligence technology, wearable technology, virtual reality technology, depth learning technology et al., our research is a more fundamental research and need a deep development.

Acknowledgements. This research was supported in part by grants from the National Natural Science Foundation of China (No. 11475135).

References

1. International Educational Data Mining Society. http://www.educationaldatamining.org/
2. 5th International Conference, Advanced Data Mining and Applications (ADMA) 2009, Beijing, China, 17–19 August 2009
3. Guiyang International Big Data Industry Exposition and Global Big Data Age Guiyang Summit. http://www.chinadaily.com.cn/m/guizhou/2015-05/25/content_20808880.htm
4. The State Council. Circular of the State Council on Printing and Distributing the Platform for Action for the Development of big data. http://www.gov.cn/zhengce/content/2015-09/05/content-10137.htm

5. U.S. Department of Education, Office of Educational Technology. Future Ready Learning: Reimagining the Role of Technology in Education. U.S. Department of Education, Washington, D.C. (2016). https://tech.ed.gov/files/2015/12/NETP16.pdf
6. Romero, C., Ventura, S.: Educational data mining: a survey from 1995 to 2005. Expert Syst. Appl. **33**(1), 135–146 (2007). doi:10.1016/j.eswa.2006.04.005
7. Beck, J.E., Mostow, J.: How who should practice: using learning decomposition to evaluate the efficacy of different types of practice for different types of students. In: Proceedings of the 9th International Conference on Intelligent Tutoring Systems, Montreal, Canada, pp. 353–362. Springer, Berlin (2008). doi:10.1007/978-3-540-69132-7_39
8. Abdous, M., He, W., Yen, C.J.: Using data mining for predicting relationships between online question theme and final grade. Educ. Technol. Soc. **15**(3), 77–88 (2012). http://www.jstor.org/stable/jeductechsoci.15.3.77
9. Goksu, I., Idirs, B.: Learners' evaluation based on data mining in a web based learning environment. J. Comput. Educ. Res. **3**(5), 78–95 (2015). doi:10.18009/jcer.03016
10. Linan, L.C., Perez, A.A.J.: Educational data mining and learning analytics: differences, similarities, and time evolutional. J. Educ. Technol. High. Educ. **12**(3), 98–112 (2015). doi:10.7238/rusc.v12i3.2515
11. Mostow, J., Gates, D., Ellison, R., Goutam, R.: Automatic identification of nutritious contexts for learning vocabulary words. In: Proceedings of the 8th International Conference on Educational Data Mining, Madrid, Spain, pp. 266–273. Worcester Polytechnic Institute, Worcester (2015)
12. Johnson, L., Adams, B.S., Cummins, M., et al.: Horizon Report: 2016 Higher Education Edition. The New Media Consortium, Austin (2016)

Design and Implementation of a Course Selection System that Based on Node.js

Jun-ran Wang[1], Cheng-peng Xu[2], and Xiao-qiang Xi[2(✉)]

[1] School of Science, Xi'an University of Posts and Telecommunications, Xi'an, China
uwy547130739@126.com

[2] Institute of Internet of Things and IT-Based Industrialization, Xi'an University of Posts and Telecommunications, Xi'an, China
1675657@qq.com, xxq@xupt.edu.cn

Abstract. In this paper, a course selection system using single thread system and unified language is designed and realized, the system is easy to maintenance and stable in highly concurrent. The content include designing idea, system module and implementation methods. Node.js is used to improve the efficiency and reduce maintenance difficulty. Mongodb database is used to store large files, pictures, curriculum information, student information and other unstructured data, it can improve the efficiency of the system in a large degree. The result of test show that this course selection system is stable and efficiency. If change the course into person or things, the system will be changed into a personnel or an item management system for enterprise.

Keywords: Course selection system · Node.js · Single thread · Mongodb

1 Introduction

As the number of undergraduate increasing, course type and number are also increasing, manual processing of the course selection information become nearly impossible [1–3]. Most of universities choose to use course selection system, which can increase the management efficiency and save time for the students. The traditional course selection system has two weak-points, the first is that it use different language to realize, for example, background development using C and C++ languages, logic analysis using java and front page using php [4–7]. It is not very convenient for maintaining and upgrade. The second is that it is serviced by multi-threading, newly added threads would occupy more memory and increase the complexity of system, and lower the speed of server or make server collapse.

In this paper, we will combine Node.js language and Mongodb database to design a course selection system that can overcome the shortcoming of the previous system.

P. Krömer et al. (eds.), *Proceedings of the Fourth Euro-China Conference on Intelligent Data Analysis and Applications*, Advances in Intelligent Systems and Computing 682, DOI 10.1007/978-3-319-68527-4_18

2 System Requirements

Although this system is based on the need of the university that the authors belong to, it still has universality. The number of internal students is from thousands to tens of thousands, the system should satisfy to handle multiple requests, to realize the storage and maintenance of the course selection information in a short period, i.e. the system must has real-time response and high throughput concurrent connection, the database must satisfy high concurrent and has high efficiency read-write performance. Fortunately, the combination of Node.js language and Mongodb database satisfy those requirements [8].

The users of the system are students and administrators. The student module must has the ability to realize the fundamental function of course selection with fewer clicks, the pages should be simple and friendly. The management module is used to check the students' course selection information, and generate course information and the corresponding curriculum schedule.

3 System Module Design

The system includes the student and management module [9]. The student module represents the system's foreground, its functions are course selection, displaying the corresponding information, such as course lists and introduction, teacher's introduction and so on. The selection results can be submitted at last. The management module represents the system's backstage, its function are maintaining and managing the information of students and courses.

The main structures of this system is shown in Fig. 1:

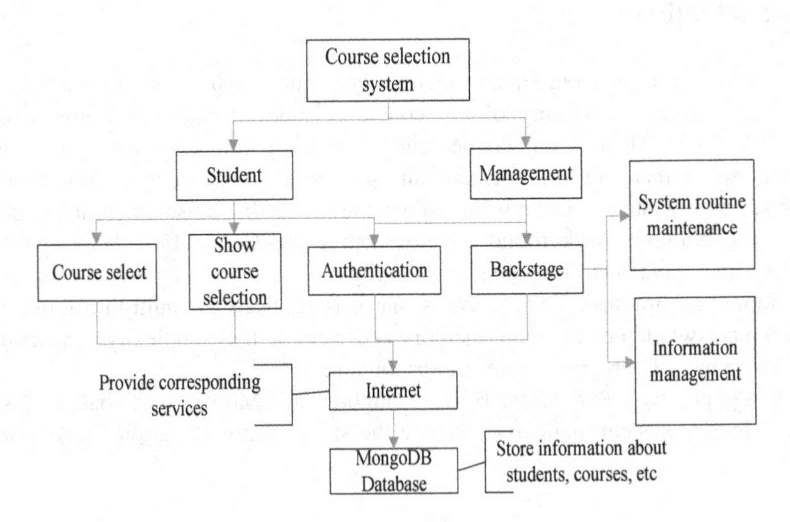

Fig. 1. Main structure of the course selection system

3.1 The Function Design of Role Module

The course selection system is divided into foreground and backstage. The foreground is for students to select courses, and process data interaction between the student and course, it can display student's information and the selecting courses. The backstage is for administrator to manage the users and courses' information, it can add, delete or modify the student and course.

The flowchart of role module design is shown in Fig. 2:

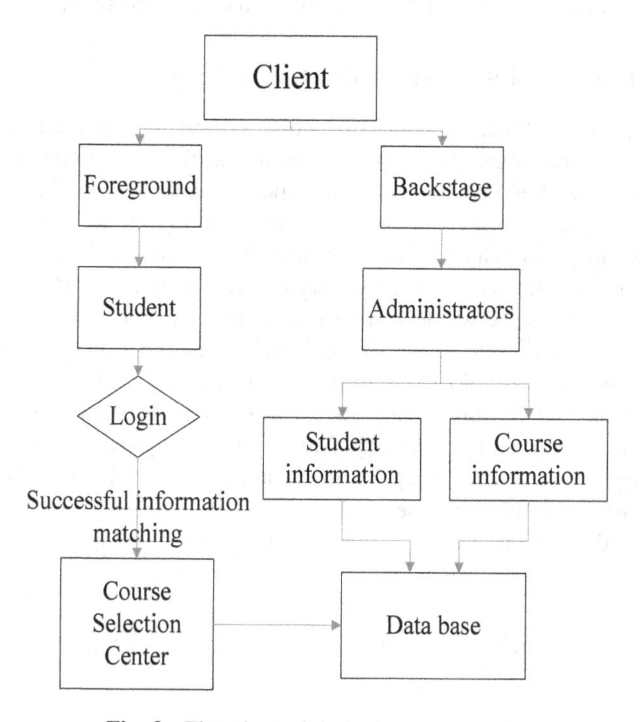

Fig. 2. Flowchart of designing role module

The administrator, who has the highest authority, first sets up the students' and courses' information, then the students can visit the course selection system. The course can be submitted only when it match the courses the database in the system.

3.2 Database Design

Database is the key part of the course selection system, every operation should be visit with database, the information need to read-write and extract from it. The efficiency of the whole system is depended on the fluent exchange between database and the background and foreground. In the logic process of this system, the students correspond to courses and administrator correspond to students and courses [10]. The students' information include 9 items, such as name, ID number, school, sex, class, phone number, e-mail, birthday, mentor and so on. The administrator' s information also include 7 items, such as name, school, sex, address, phone number, ID number and e-mail. The course information include 6 items, they are name, classroom, teacher, credit, time and classroom.

4 System Design and Implementation

Different from others course selection system, our system is written by Node.js from foreground to background [11, 12], it is easy to maintain and upgrade.

4.1 The Utilization of Single-Threaded Concurrency

The traditional mufti-threading course selection system will be hard to increase the performance by continuous optimization if its thread reach a certain magnitude. For example, a Windows Server's stack is set to 1 megabyte, its working limit will be about 1600 threads, if the stack set to 2 megabyte, the working thread can not reach 1600 [13]. The overall performance increase is limited by increase the number of thread, while, more thread will occupy more memory and consume more CPU. If one thread is collapsed, the stable of the whole system will be effected [14, 15].

The system in this paper adopts single-thread high concurrency to process data of course selection and non-blocking asynchronous I/O is used to invoke [16]. All requests are managed by this thread, and it allows tens of thousands of concurrent connections and occupies less memory. Moreover, all requests are routed through a single-threaded event, which means that users do not need to deliberately avoid traffic peak in the processing of course selection.

The single-threaded flowchart of this system is shown in Fig. 3:

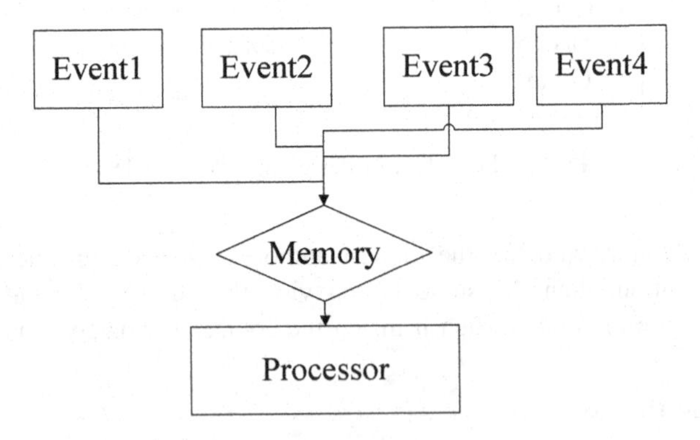

Fig. 3. Single-threaded flowchart

4.2 System Implementation

After login, the system would visit the node-based Server through the request from foreground page. After receiving the response, the system would judge the users' identity by invoking relevant information in database, and finally send the extracted information to users to manipulate through foreground page. In order to avoid conflict among the various parameters, "sid, type, result, title and target" are used to define student's information, curriculum, credit and instructor respectively.

(1) System routing configuration

Express is a development framework based on Node, in which http is used to respond requests through the ways of app.get (getting data from database) and app.post (sending information back to database). The key code are as follows:

```
app.get('/userinfo', checkLogin);

app.get('/userinfo', user.userinfo);

app.get('/user-imglist', checkLogin);

app.get('/user-imglist', user.imglist);

app.post('/user/setface', checkLogin);

app.post('/user/setface', user.dosetface);

app.get('/user/password', checkLogin);

app.get('/user/password', user.setpassword);


app.post('/user/password', checkLogin);

app.post('/user/password', user.dosetpassword);

app.post('/student-user', checkLogin);

app.post('/student-user', student.dostudentuser);

app.get('/student-info', checkLogin);

app.get('/student-info', student.studentinfo);
```

Generally, request and response are adopted to handle information in Express. Through client requests, the route resolves the URL path first and then retrieves the HTML document. Suppose the URL is http://course-select.xupt.edu.cn/, the browser will transit into the http site which address is course-select.xupt.edu.cn in a network sever, and complete corresponding data processing. The first parameter of the "get" method is the access path, where "/" represents the root path, and the second parameter is the callback function, "req" represents the request from the client, and "res" represents a response to the request. The send method in the callback function refers to sending a character string to the client. The role of "/, routes. Index" is to invoke index to get the designed page and finally present it.

Its key code are as follows:

```
var express = require('express');

var ejs = require('ejs');

var flash = require('connect-flash');

var settings = require('../settings');

//var routes = require('./routes');

var routes = require('./routes/index');

var path = require('path')

var user = require('./user');

var teacher = require('./teacher');

var student = require('./student');

app.use(function(req, res, next) {

        res.locals.error = req.flash('error').toString();

        res.locals.success = req.flash('success').toString();

        res.locals.user = req.session ? req.session.user : null;

        res.locals.userid = req.session ? req.session.userid :
null;
```

```
        res.locals.role = req.session ? req.session.role : null;

        res.locals.face = req.session ? req.session.face : null;

        res.locals.studentlist = req.session ? req.session.studentlist : null;

        next();

    });

    routes(app);

    app.use(express.static(__dirname + '/web'));
```

(2) Database usage

Before using the database, one need to create a data directory and a db sub-directory, then using -dbpath to call the database. The code for calling the database is as follows:

```
//var settings = require('./settings');

//var mongodb = require('mongodb');

//var Db = mongodb.Db;

//var Connection = mongodb.Connection;

//var Server = mongodb.Server;

//module.exports = new Db(settings.db, new Server(settings.host, Connection.DEFAULT_PORT, {}));

//var util = require('util');

//var mongoose = require('mongoose');

//var Schema = mongoose.Schema;
```

```
//var dburl = require('../config').db; // Database address

var settings = require('./settings');

//exports.connect = function (callback) {

 //   mongoose.connect(settings.db);

//}

//exports.disconnect = function (callback) {

//    mongoose.disconnect(callback);

//}

//exports.setup = function (callback) { callback(null); }

var mongoose = require('./LMS/node_modules/mongoose');

mongoose.connect(settings.db);

exports.mongoose = mongoose;
```

4.3 System Deployment

In the deployment, the coordinated operation among user terminal, Web server and database should be considered, several crucial works should be done in order to ensure the stable performance of the system. The network deployment diagram is shown in Fig. 4:

(1) Implementation of high concurrency

Web server is not stable when a large number of users try to visit the course selection system, in order to overcome this problem, the system adopts link aggregation technology to integrate server memory that multiple threads need to occupy into a single logical circuit, so as to eliminate memory constraints and instability factors in data transmission channel, at last improve the server response speed and increase the processing capacity.

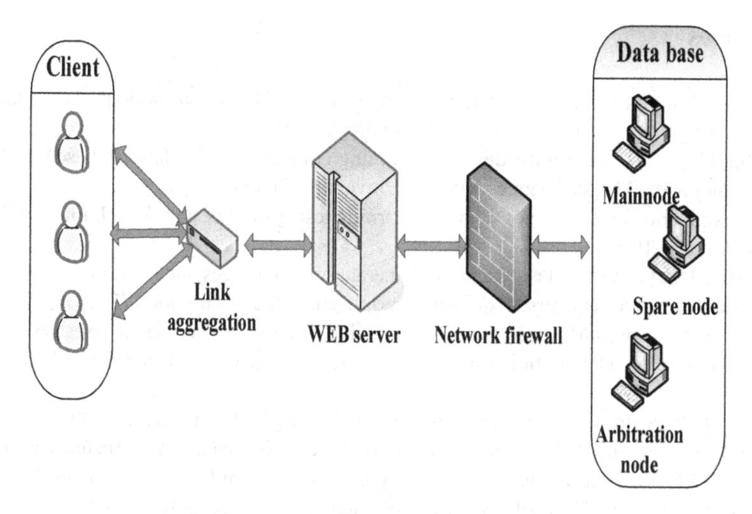

Fig. 4. The network deployment diagram

(2) Database management

The cluster mode of the database adopts a replica set cluster, in which include a main node, a standby node and one mediation node. The main node is the most important part of database, because the invocation of relevant information in client and background must pass through it. The standby node acts as a temporary database, its function include to divide the visiting flow of client requests in traffic peak and make preprocessing in information modification so as to correct errors before data through the main node. The mediation node is used to judge the visiting amount of database, if the visiting amount exceed the threshold, the standby node will be upgraded to the main node for data processing.

5 Conclusion

In this paper, the model of single-threaded online course selection system is presented, and the basic functions of course selection, administrator management and curriculum maintenance are also realized. The system can overcome the shortcoming of the traditional course selection system, such as easy to collapse and not easy to maintain. The test show that this system support multiple responses and has a better processing performance for higher concurrency.

Acknowledgements. This research was supported in part by grants from the National Natural Science Foundation of China (No. 11475135).

References

1. Na, P.: Design and implementation of Blog System based on Node.js. Dissertation for Master Degree, Dalian University of Technology (2013)
2. Zhang, L.X.: Research on the design of learning communication platform based on Node.js. Dissertation for Master Degree, Yunnan University (2015)
3. Shen, X.: Node.js and MongoDB online learning test system design. Wirel. Internet Technol. **4**(2), 30–32 (2015)
4. Shi, H.: Elective system design and implementation of colleges and universities. Dissertation for Master Degree, University of Electronic Science and Technology (2013)
5. Luo, Z.M.: Design and implementation of university public elective courses online course selection system. Dissertation for Master Degree, South China University of Technology (2013)
6. Shao, Q.: Design and implementation of the college students course-selection system. Dissertation for Master Degree, University of Electronic Science and Technology (2015)
7. Wang, Y.: Design and implementation of micro blog system based on Node.as. Dissertation for Master Degree, University of Electronic Science and Technology (2014)
8. Li, Y.Y., Fu, S., Li, L.B.: Campus online course selection system. Comput. Syst. Appl. **1**, 62–69 (2013)
9. Chaniotis, I.K., Kyriakou, K.I.D., Tselikas, N.D.: Is Node.js a viable option for building modern web applications? A performance evaluation study. Computing **97**(10), 1023–1044 (2015). doi:10.1007/s00607-014-0394-9
10. Gackenheimer, C.: Understanding Node.js, pp. 1–26. Apress (2013). doi:10.1007/978-1-4302-6059-2_1
11. Krill, P.: Node.js welcomes Microsoft's Chakra JavaScript Engine (2016). InfoWorld.com
12. Zhao, S.M., Xia, X.L., Le, J.J.: A real-time web application solution based on Node.js and WebSocket. Adv. Mater. Res. **816–817**, 1111–1115 (2013). doi:10.4028/www.scientific.net/AMR.816-817.1111
13. Tilkov, S., Vinoski, S.: Node.js: using JavaScript to build high-performance network programs. IEEE Internet Comput. **14**(6), 80–83 (2010). doi:10.1109/MIC.2010.145
14. Yu, L.L.: Design and realization of higher vocational students' course selection system. Inf. Technol. Informatiz. **3**, 76–77 (2017)
15. Luo, H.Y.: Design and implementation of college student information management system. Electron. Technol. Softw. Eng. (05), 63–63 (2017)
16. Li, S., Huang, K.M., Yang, Y., Zhang, T.R., Wang, Q.: The study and design of college students' course selection system. Comput. Knowl. Technol. **13**(9), 99–100 (2017)

A Prototype System for Three-Dimensional Cardiac Modeling and Printing with Clinical Computed Tomography Angiography

Shao-jie Tang[1(✉)], Hong Zhang[1], Wei Peng[1], Hao-yang Shi[1], Cong Guo[1], Guang-yuan Zhao[1], Wei-guo Bian[2], and Yang Chen[3,4]

[1] School of Automation, Xi'an University of Posts and Telecommunications, Xi'an, China
s.j.tang@hotmail.com

[2] Orthopaedics Department, The First Affiliated Hospital of Xi'an Jiaotong University, Xi'an, China

[3] Laboratory of Image Science and Technology, Southeast University, Nanjing, China

[4] Key Laboratory of Computer Network and Information Integration (Southeast University), Ministry of Education, Nanjing, China

Abstract. To help physicians and their assistants plan, rehearse and follow up a surgical therapy with a computer device, people may carry out a three-dimensional (3-D) cardiac modeling by digital image processing algorithms and a 3-D cardiac printing by appropriate materials, according to the DICOM data of patient's heart acquired by an advanced computed tomography angiography (CTA) device in a hospital. In this work, we introduce the design and implementation of a prototype system of 3-D cardiac modeling and printing. Firstly, we converted the CTA data of the heart from the original DICOM format to the . mat format within the Matlab programming environment. Secondly, we carried out the 3-D image segmentation with the region growing to semi-automatically determine the region of interest (ROI), i.e., the cardiac tissue, by sequentially processing the CT image slices. Thirdly, we performed the 3-D modeling and volume rendering for the binary volume image of the cardiac tissue. Lastly, we converted the 3-D modeling into STL format and sent it to the table-top 3-D printer mounted in our lab, generated the corresponding G codes by Flashprint software to manipulate the 3-D printer. It is noted that the software part of the prototype system was developed within Matlab and performed well for the expected task.

Keywords: CTA · DICOM · 3-D modeling · 3-D printing · Volume rendering

1 Introduction

In recent years, three-dimensional (3-D) printing is one of the hot topics in both academic and industrial fields. To help physicians and their assistants plan, rehearse and follow up a surgical therapy with a computer device, we carried out a 3-D cardiac modeling by digital image processing algorithms and a 3-D cardiac printing by

P. Krömer et al. (eds.), *Proceedings of the Fourth Euro-China Conference on Intelligent Data Analysis and Applications*, Advances in Intelligent Systems and Computing 682, DOI 10.1007/978-3-319-68527-4_19

appropriate materials, according to the DICOM data of a patient's heart acquired by the advanced computed tomography angiography (CTA) in a hospital. In this work, we introduced the design and implementation of a prototype system of 3-D cardiac modeling and printing with clinical CTA in details. Most of the system was developed within the Matlab programming environment. Experimental results showed that the prototype system performs well for the expected task with the CTA data of a patient's heart as an input.

2 Working Flow of the Prototype System

The working flow chart of the prototype system of the 3-D cardiac modeling and printing is shown in Fig. 1. The details will be explained below.

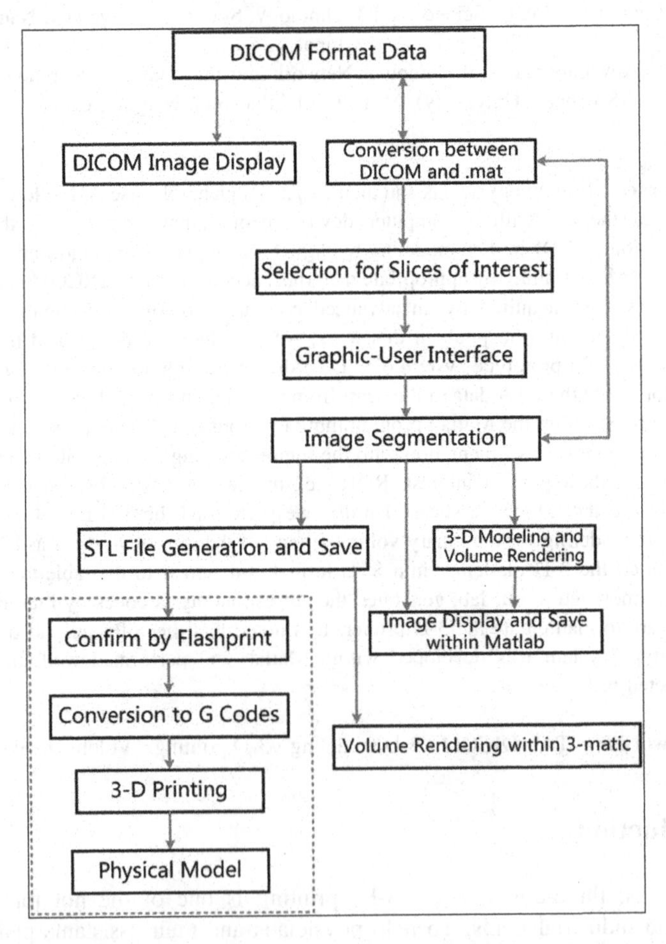

Fig. 1. The working flow chart of the developed prototype system of 3-D cardiac modeling and printing. Note that the part inside the dashed rectangle is carried out within the table-top 3-D printer mounted in our lab.

3 Format and Segmentation of CTA Data

3.1 The Structure of DICOM File

Nowadays, CT images acquired in the hospitals are almost always saved as DICOM files, which are convenient to be archived and transferred. A DICOM file consists of a number of data elements [4], which shall not appear multiple times in the original data set, but can appear repeatedly in the nested data set. All the information of file operations is available by calling the data set information. The schematic on the composition of a DICOM file is depicted in Fig. 2.

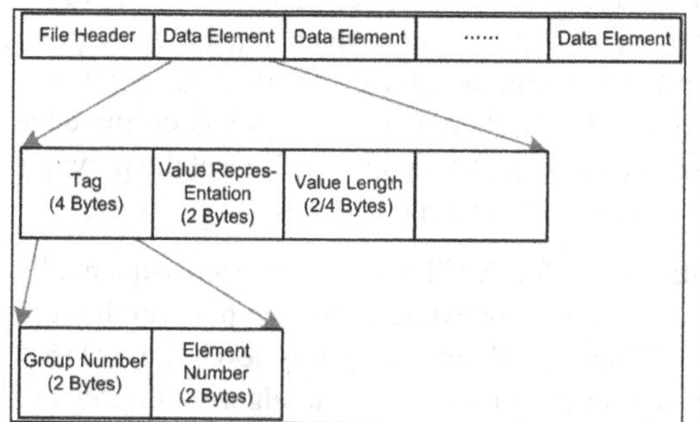

Fig. 2. Composition of a DICOM file.

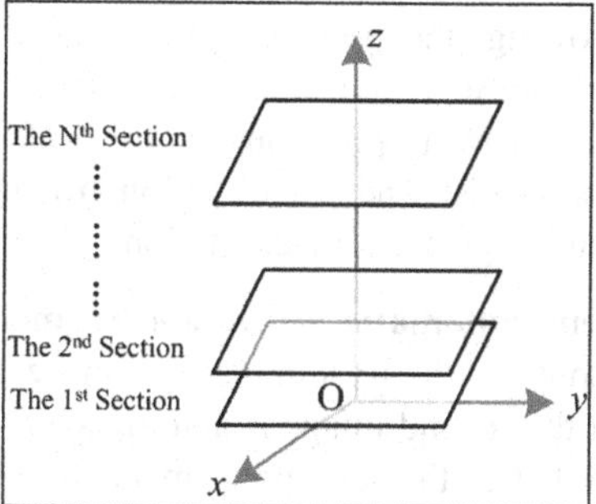

Fig. 3. Schematic of composing volume dataset

3.2 Read DICOM File Within Matlab

Within Matlab, a DICOM file can be read by the dicomread function as follows,

$$\mathrm{I} = \mathrm{dicomread}(\mathit{filename}), \tag{1}$$

which reads from a compatible DICOM file named *filename*. As a gray-level image, I is in a form of $M \times N$ array, while as a true-color image, I is in a form of $M \times N \times 3$ array.

$$[\mathrm{I}, \mathrm{map}] = \mathrm{dicomread}(\mathit{filename}), \tag{2}$$

returns the colormap of image I too. As I is a gray-level or true-color image, the map will be empty.

3.3 Conversion of Data Format

Input CT data into an array V slice by slice within Matlab. Here, V is a 3-D matrix with a size of $L \times M \times N$, wherein L, M and N are the numbers of pixels along x-, y- and

z-directions, respectively. The schematic of composing a volume dataset is shown in Fig. 3.

3.4 Image Segmentation with the Region Growing

As one of critical digital image processing techniques, the region growing is very useful to the image segmentation needed in this work [3].

Algorithmic Steps of Region Growing. a. Manually select the initial growing point (i.e., the seed point) needed for the task of image segmentation and set the coordinate of the selected pixel as (x_0, y_0); b. Set the seed point (x_0, y_0) as the growing point, and inspect its eight-neighbor pixels (x, y). If the pixel (x, y) meets the rule of the region growing, the pixel (x, y) will be merged into the region which the growing point belongs to, and is labeled; c. Select another point on the edge of the growing region as a new growing point and return back to the step b; d. When all the points on the edge don't satisfy the rule of region growing, return back to the step a; e. Repeat the previous four steps until all the regions to be segmented have been processed.

Semi-automatic 3-D Image Segmentation. After N CT slices have been sequentially input and the ROI in the first image was successfully extracted, a seed point in the ROI of the second image is automatically determined, which is of a gray-level value closest to that of the seed point manually chosen in the first image. The algorithmic steps of region growing as described in the subsection D.1 is further implemented for the second image. This process repeats for all the CT slices sequentially. Due to the inherent characteristic of the CT scanning, the difference between arbitrary two

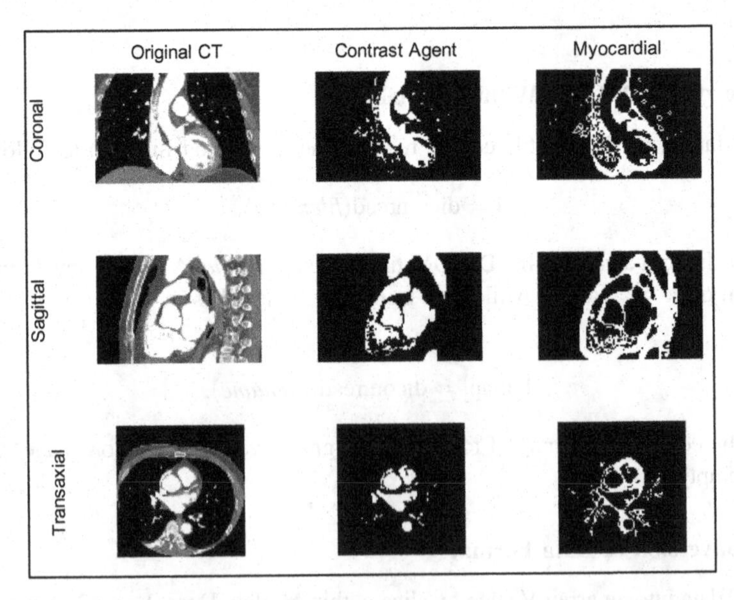

Fig. 4. The 1st, 2nd and 3rd rows correspond to the three orthogonal slices of the patient's thorax, while the left, middle and right columns correspond to the original CTA image, the contrast agent image in cardiac cavity and the cardiac tissue image, respectively.

sequential CT images is much slight, which makes the 3-D image segmentation feasible in practice. Essentially, the proposed 3-D image segmentation is a sort of semi-automatic approach.

Display of Segmented Images. From the top to the bottom in Fig. 4 are the three orthogonal slices of the patient's thorax, while from the left to the right are the images of original CTA, the contrast agent in cardiac cavity and the cardiac tissue, respectively. It can be noted that almost no extra bone and muscle tissues are included in the cardiac tissue image as shown in the right column of Fig. 4.

4 3-D Modeling and Volume Rendering

4.1 3-D Modeling of Segmented Volume Images

a. **Data Preprocessing and Smoothing.** The 3-D matrix V is usually of a huge amount of elements. In order to reduce the computational consumption and thereby increase the computational speed while maintaining the computational accuracy, the reducevolume function can be used to reduce the amount of elements in V.

$$[\mathrm{x}, \mathrm{y}, \mathrm{z}, \mathrm{V}] = \mathrm{reducevolume}(\mathrm{V}, [\mathrm{a}, \mathrm{b}, \mathrm{c}]), \tag{3}$$

$$\mathrm{V} = \mathrm{smooth}\,3(\mathrm{V}), \tag{4}$$

The principle is to sample V at coordinates x, y and z, which are determined by the ratio parameters a, b and c.

b. **Volume Rendering.** Using isosurface function to draw the contours of volume image is an indirect method for rendering 3-D implicit function within Matlab,

$$\mathrm{fv} = \mathrm{isosurface}(\mathrm{x}, \mathrm{y}, \mathrm{z}, \mathrm{D}, \mathrm{isovalue}) \tag{5}$$

where the first four parameters have been obtained and the isovalue is generally taken 1.

c. **Construction of Structural Image Patches.** Using patch function to define the light, color and other information of the sub-regions in image,

$$\mathrm{p1} = \mathrm{patch}(\mathrm{fv}, '\mathrm{FaceColor}', \mathrm{yourscol}, '\mathrm{EdgeColor}', '\mathrm{none}'), \tag{6}$$

where fv is the output of the isosurface function and yourscol is used by user to define the colors of object.

d. **Determine Normal Direction of Image Patches.** The isonormals function is used to create a new 3-D view with appropriate lighting effects,

$$\mathrm{Isonormals}(\mathrm{x}, \mathrm{y}, \mathrm{z}, \mathrm{D}, \mathrm{p1}). \tag{7}$$

e. **Determine Geometric Boundaries of Volume Image.** Using isocaps function to calculate geometric boundaries for volume image,

$$fvc = isocaps(x, y, z, D, isovalue), \tag{8}$$

where the parameters have been defined in Step b.

f. **Replace fv with the results obtained in Step e and repeat Step c.**

4.2 3-D Volume Rendering of Segmented Volume Images

Shown in Fig. 5 is the volume rendering of the segmented (a) contrast agent and

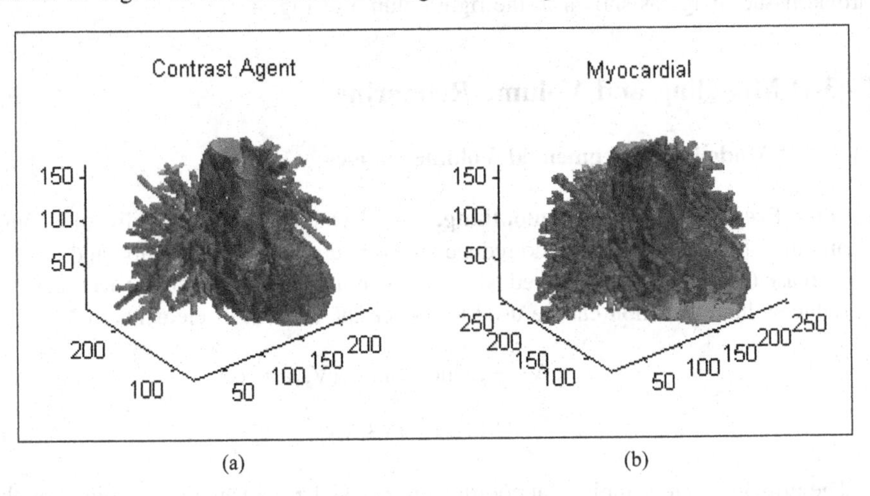

Fig. 5. The volume rendering of the segmented (a) contrast agent and (b) cardiac tissue within Matlab.

(b) cardiac tissue within Matlab [1, 7]. From Fig. 5, it can be noted that the image segmentation used in our system performs well.

5 Generation of STL File

5.1 Introduction of STL File

An STL file is a closed entity consisting of a large number of triangular patches. The triangular facets are the bridge between the physical layers, which are used to approximate the surface features of the object and finally generate a STL file. Each triangular facet stores the coordinates of the three vertices and the normal vector component data.

5.2 The Rules Obeyed by STL File

A STL file needs to follow certain rules of writing standards to ensure that the file is written without error. The following is an explanation of the writing rules obeyed by the STL file for binary and ASCII code storages.

How to Write STL File. An example to construct a triangular facet to illustrate how to write the STL file using ASCII code is given as follows:

solid file name // The name of the STL file
facet normal
outer loop // Start of definition
vertex A
vertex B
vertex C
endloop // End of definition
endfact // The overall patch definition ends
…… // repeat the definition in accordance with the above principles
end solid file name // It marks the end of the definition

"facet normal" is used to define the coordinate information of each directional components of the normal vector of a triangular facet. "vertex" is followed by the coordinate information of the vertexes of a facet. "endfacet" finishes all the information definitions of a facet, and a triangular facet has been successfully created.

The Binary Format of Facet. Shown in Fig. 6 is the binary format of a facet composed of a file header and triangular facet information. The file header is used for user identification and the major information is stored in the part of the facet information. The file header consists of 84 bytes, the major triangular facet information is 50 bytes, used to store the vertex and vector information, all information is stored in the format of real number, the three vertices and normal vector share four groups of 3-D data, each group data are 4 bytes, total 48 bytes are stored, the remaining two bytes are used to reserve other information. A file header includes 84 bytes, wherein the major is of 80 bytes, most of them are annotation contents including the creation time, the format, and the size of the file, and the software used to create the file, while the remaining four bytes recorded the total number of facets, which can be observed during the operation

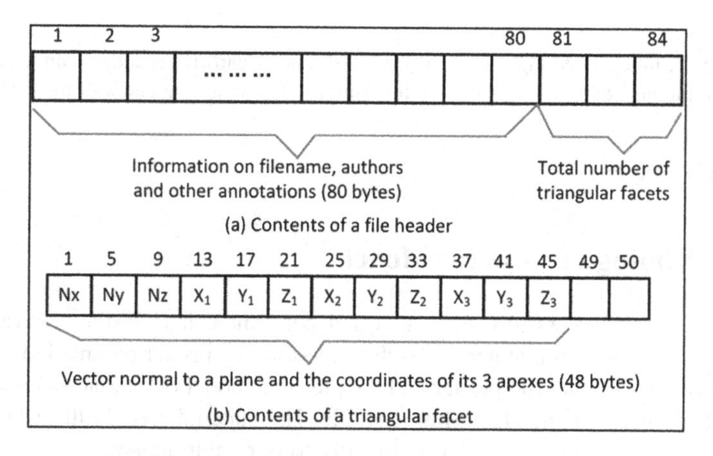

Fig. 6. The binary format of STL file.

to have a glance on the size of the file. If you create a model with N facets, the corresponding binary file size would be $84 + 50 \times N$ bytes.

Triangular Facet Filled Between Layers. The interlayer filling algorithm is concisely described as follows [2, 6].

Synchronous Marching Approach for Adjacent Contours. The basic idea underlying the approach is that, at the same time the above mentioned method is carried out, the connection of the endpoints on the contours in the adjacent slices is simultaneously realized. The approach is extensively applicable in a practice.

Maximizing for Minimal Interior Angles. The basic idea underlying the approach is to optimize the minimal internal angles at the positions of dense facets to maximize them.

Volume Rendering of STL File. In this work, we firstly use the isosurface function within Matlab to output the point and surface information of 3-D data, according to the rules of writing a STL file in ASCII code for the output of the triangular facet information reconstruction. In this way, we generate two STL files named Agent.stl and HeartMuscle.stl for the contrast agent in the cardiac cavity and the cardiac tissue,

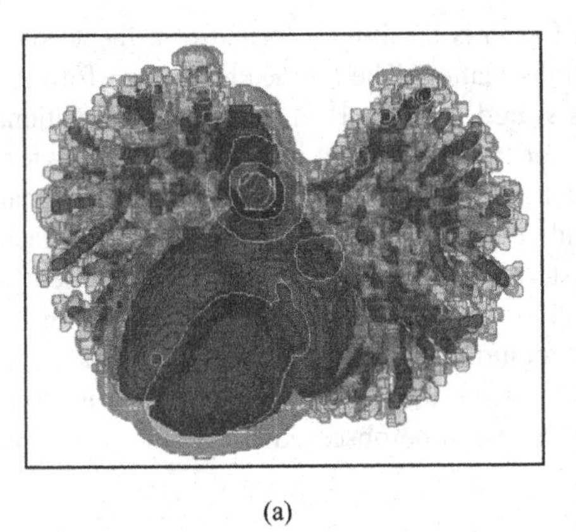

(a)

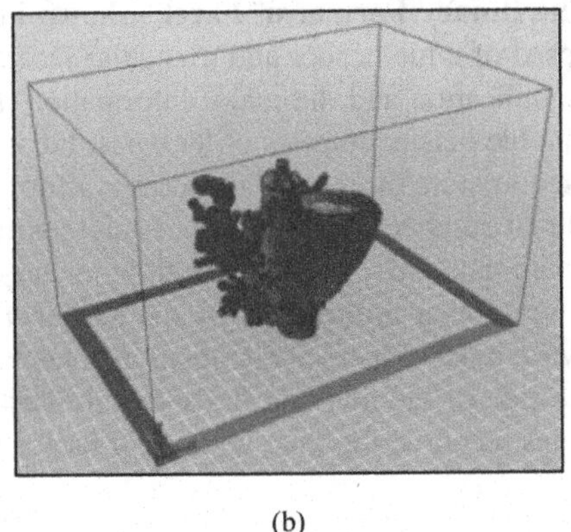

(b)

Fig. 7. (a) volume rendering of both Agent.stl and HeartMuscle.stl within 3-matic and (b) confirm the intactness of Agent.stl by the Flashprint software mounted in the 3-D printer.

respectively (see Fig. 7).

6 3-D Printing of Physical Model

After STL file is successfully generated and confirmed, it is further converted into G codes, which can be implemented by the table-top 3-D printer mounted in our lab. In Fig. 8(a) and (b) are the 3-D printer and the physical model of cardiac tissue produced by the 3-D printer with the HeartMuscle.stl file. The material used by the 3-D printer is a kind of degradable plastic, which is friendly to our environment.

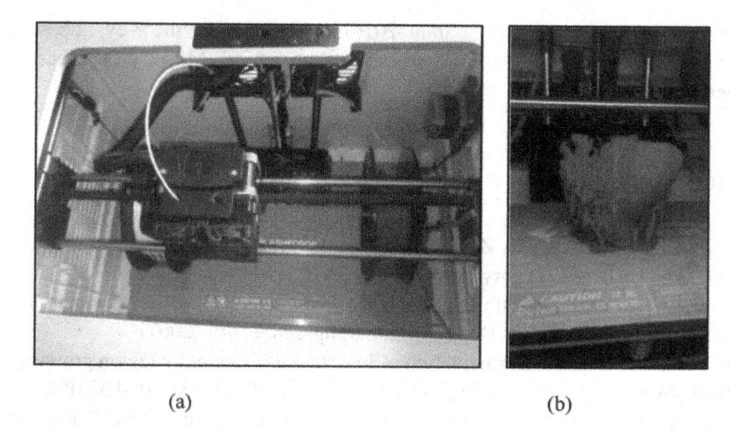

(a) (b)

Fig. 8. (a) The table-top 3-D printer mounted in our lab, and (b) the physical model of cardiac tissue produced by the 3-D printer in (a) with the HeartMuscle.stl file.

7 Design and Implementation of the System

7.1 Analysis on Demand of System Design

In the system design, we demanded some crucial functions such as the image data input, the conversion of data format, the semi-automatic image segmentation with the region growing technique, the 3-D modeling, and the generation of STL file. Meanwhile, the 3-D volume rendering, the image display and the data storage are also needed. The working flow chart of the developed system has been shown in Fig. 1.

7.2 System GUI

In the Matlab GUI designed for the prototype system as in [5], the left side is the DICOM data directory, the middle is the data information of each image, while the corresponding image is displayed on the right side. The "save data" button is used to store all.mat files consisting of DICOM data.

8 Discussions and Conclusions

In this work, we described how to develop and implement a prototype system of 3-D cardiac modeling and printing. The process included the image data input, the conversion of data format, the semi-automatic image segmentation with the region growing technique, the 3-D modeling, the 3-D volume rendering, the generation of STL file, the data storage, and so on. Preliminary experimental results showed that the prototype system developed by us performed well for the expected task. Validation on the accuracy of 3-D modeling and printing will be our works in the near future.

Acknowledgments. This work is supported by the New Star Team of XUPT, by the Department of Education Shaanxi Province, China, under Grant 15JK1673, by Shaanxi Provincial Natural

Science Foundation of China, under Grant 2016JM8034, and by the Key Lab of Computer Networks and Information Integration (Southeastern University), Ministry of Education, China, under Grant K93-9-2017-03.

References

1. Zeng, Z., Dong, F., Chen, X., Zhou, H., Zhou, J.: Three dimensions reconstruction of CT image by MATLAB. CT Theory Appl. **13**(2), 24–29 (2004)
2. Nie, X.: Study and achievement on method of generation STL files from industrial CT slicing image. Dissertation for Master Degree, Chongqing University (2007)
3. Peng, F., Bao, S., Zeng, B.: Segmentation of liver based on adaptive region growing. Comput. Eng. Appl. **46**(33), 198–200 (2010). doi:10.3778/j.issn.1002-8331.2010.33.056
4. He, B., Jin, Y., Li, Y.: Defining nuclear medical file format based on DICOM standard. Nucl. Electron. Detect. Technol. **21**(6), 440–443 (2001)
5. Yang, B.: Matlab-based GUI for medical imaging processing system. Dissertation for Master Degree, Jinan University (2014)
6. Hu, D., Li, Z., Li, D., Ding, Y., Lu, B.: Algorithm for rapid slicing based on geometric feature classification of STL model. J. Xi'an Jiaotong Univ. **34**(1), 37–40 (2000)
7. Xu, Y., Wu, X., Hu, Y.: 3D rebuilding of biology slice image by volume rendering in matlab. Comput. Eng. **27**(12), 114–115 (2001)

Design of Wisdom Home System for the Aged Living Alone

Ying Ma[1,2(✉)], Jianxing Li[1,2], and Lisang Liu[1,2]

[1] Fujian University of Technology, Fuzhou, China
may@fjut.edu.cn
[2] Key Laboratory of Digital Equipment, Fuzhou, Fujian Province, China

Abstract. According to the demand and development direction of China's industry and wisdom, the demand of social old-age home, this paper presents a PLC control Home Furnishing automation system technology, smart wearable body sensing technology and network sensing technology based on narrow band. Through the test, the system can realize the local, WEB and mobile terminal APP intelligent home control. According to the tracking of the number of steps, heart rate and moving path of the elderly, combined with the original health status and living habits of them, the home condition modification and health warning are given. And under abnormal circumstances, such as pop-up hints and mail alerts are given. The implementation of the system can effectively collect and manage the elderly state of home care, and have better market application prospects and system development value.

Keywords: Home care · Health assistance management · Smart home · Narrowband Internet of things

1 Introduction

With the aging trend of society becoming more and more serious, the problem of social support for the "empty nest" elderly is becoming more and more serious. Many elderly people do not pay enough attention to their health problems because they have no children to take care of, causing many social problems. The management of home life and health status of the "empty nest" elderly gradually attracted widespread attention. How to provide safe and effective service for the elderly in the old age is an important issue of concern for the government, the community and the elderly. Home Furnishing intelligent system integration technology of the original maturity, combined with the rise of wearable development somatosensory sensor technology and low-power narrowband Internet of things technology, gave birth to the people of the wisdom of the demand for home care system. Therefore, it is necessary to design a stable and reliable intelligent home for the aged. And the establishment of the system will provide a large

P. Krömer et al. (eds.), *Proceedings of the Fourth Euro-China Conference on Intelligent Data Analysis and Applications*, Advances in Intelligent Systems and Computing 682, DOI 10.1007/978-3-319-68527-4_20

data base for social pension, in order to provide policy and planning requirements. In this paper, the key technologies of smart home system and health assistant management system are introduced, and the system structure diagram is given. The home intelligent home system is designed by using S7-1200 and SCADA server, and the functions of health assistant management and mobile monitoring are realized. The system parameters are optimized by using the rule self-learning method [1].

2 System Architecture

The intelligent management of home hardware system, personalized service of mobile terminal and the collection of large data at home are the main focus of smart home. The base of smart home is the interconnection and interworking of home appliances and sensors. In order to realize the intelligent, safe and personalized Home Furnishing control, Home Furnishing with the temperature control system, humidity control system, light control system, and control system, safety monitoring system and health monitoring system. The temperature control system includes the control of ventilation equipment and air conditioning system. The humidity control system includes the control of humidifying equipment and air-conditioning system. The lighting system includes lighting and curtain system controls. The security monitoring system includes smoke sensing system, gas detection system, security sensor system and camera system. Health monitoring system includes kinds of wearable sensors and home health testing or fitness equipment. We connect all above kinds of signals through wired connection, RFID, WIFI or infrared communication to the field controller - S7-1200PLC. PLC achieves the collection of various types of signals and the implementation of control of household equipment output.

A separate intelligent home control system needs to be equipped with a SCADA server. As a management device it will manage and control the underlying control device PLC. Users can set their own home control according to their own needs in the system. The server will be based on user needs combined with the user's gender, age, physical health and the accumulation of user life during the operation of the data through self-learning to improve the control system to achieve humane intelligent comfortable user experience. At the same time SCADA server also provides WEB publishing capabilities to support multi-user mobile monitoring services. In addition the system provides mobile end APP services to meet the needs of different users of remote monitoring. Important information in the server through the NB-IOT upload the cloud through the different needs of large data screening calculation for the community the competent government departments and various service providers to provide data support [2, 3].

In this design idea we establish the control system structure as Fig. 1:

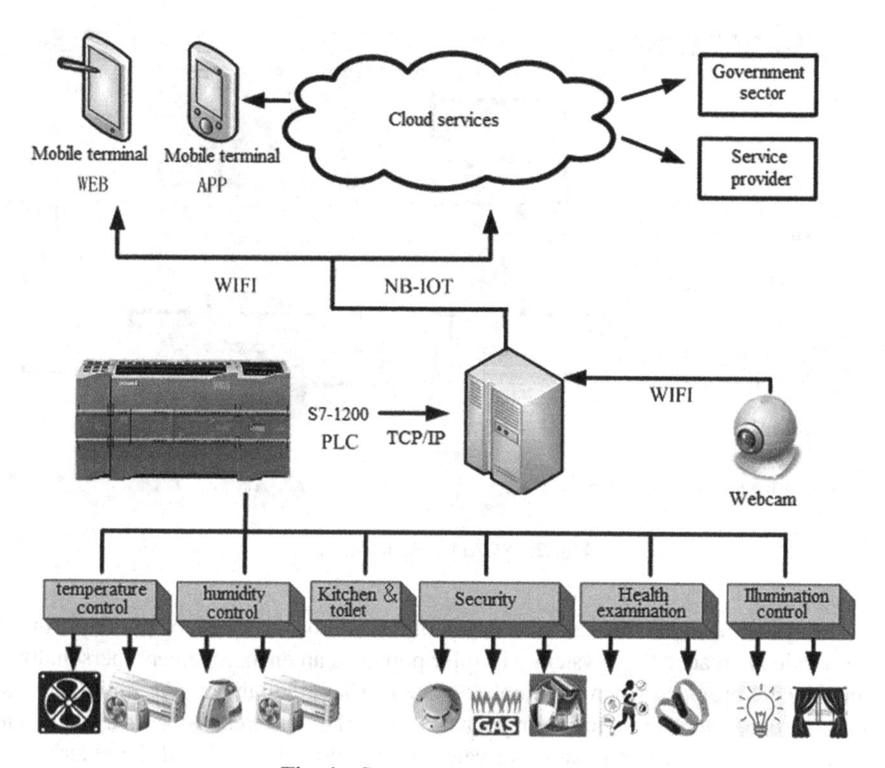

Fig. 1. System structure diagram

3 Designs and Implementation of Health Assistant Management Function

The initial operation of the system you can choose to enter the user's gender, age, basic medical history and daily habits. On the basis of this information the system will give a reasonable set of wisdom and initial health settings and health management recommendations. During the system operation the actual operation of the user will be gradually collected and the corresponding parameter values will be corrected step by step. Such as according to the regular pattern of sleep, the system will correct the automatic control of different periods of time lighting control; according to the daily bathing time it will adjust the water heater work; according to the user's daily movement and health basic states it will give appropriate sports reminders and warnings. The main interface of the system is shown in Fig. 2. If the system is not initially set up the system will operate in the basic mode. The health management proposal here will only give recommendations in the limit state. Such as the number of steps and mobile frequency ultra-high or ultra-low limit. The system will be in the course of self-learning to gradually sum up the user's living habits and thus gradually correct the corresponding parameter values and the corresponding health management recommendations.

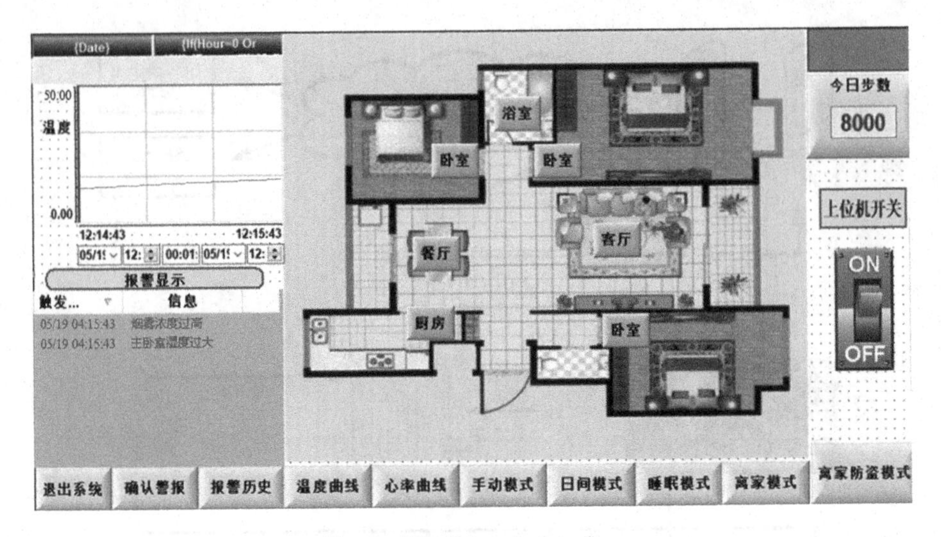

Fig. 2. SCADA main interface

The system designs the statistics shows of the number of the steps per day. And if the set value is reached the system will offer popup as an encouragement; personalized reminders are provided when the number of steps is less than the regular limit (such as: Proper aerobic exercise is good for your health!). The comprehensive analysis of the step number, trajectory and heart rate value information shows that if the elderly less activity trajectory and concentration range was small, or had abnormal heart rate. The system will determine that the status of the elderly should be abnormal, and the corresponding popup will be reminded on the SCSDA server and APP. And send email alerts to the appropriate service provider. In addition, if the security system of any security risks, the system will also carry out the necessary alarm services, as shown in Fig. 3.

Fig. 3. Mail receive alarm

4 Rule Self-learning Method

The system adopts the rule self-learning method based on the statistical analysis model to optimize the parameters. Using SCADA system to obtain the operation of each subsystem and the old people's work and rest information. After data classification and processing, log files are generated. Then extract the key fields from the log file. The working states of each subsystem in a certain time interval are calculated to form a set of rules. According to the threshold condition established, the existing control system parameters are revised to achieve the purpose of self-learning. [4, 5] Through the analysis of user's phased operation rules, the self-learning parameter setting rules are implemented, and the following steps can be adopted:

(1) Extract all work information from the log;
(2) It calculates the correction law of each work information at a given stage (which can be analyzed by weekly/monthly intervals);
(3) According to the difference between the statistical result and the current setting, it chooses the difference of different proportion according to the different object as the revised value of the original coefficient;

A day after the new rules are executed, the system asks the user for information. It judges whether the user likes the corrected home work state. Upon receipt of a positive reply and no reply, the user is deemed to have approved the amendment and will thus form a whitelist rule.

Take the light control system as an example, the frequency difference of the manual operation of the user before and after the rule learning method is revised as shown in Table 1.

Table 1. Manual frequency table.

Day	Morning	Noon	Night
1	2	0	3
2	1	1	2
3	1	0	1
4	0	0	1
5	0	0	0

From the data in the table, we can see that the parameters of the light control system revised by the rule learning method are more in line with the user's living habits. It can effectively reduce the frequency manually modified by the user. After the trial operation has reached a stable, the system will complete the user's customary fitting process, giving users a more comfortable home experience.

5 Mobile Terminal Services

As a smart home management system for the empty nest elderly, it is essential to provide mobile terminal user login service for the elderly children or other related service providers at home. The InduSoft software platform adopted by SCADA system has the function of WEB publishing. After the platform completes the WEB publishing settings, run the project in the server. After that, open any browser in different places, enter the IP address set in the server, and the application alias/picture name you set up before you can access the project and perform the operation of the corresponding user rights. The WEB functionality is in full agreement with the SCADA HMI. If the server uses the dynamic allocation of IP, then the IP address will not be able to login. Domain name resolution services can be used to map domain name addresses on the local router at the server side. Then the mobile client can login to the system only by using domain names.

Key information will be driven by event driven or timed drive by the server via a narrow network of things to communicate with the cloud. Then the APP program is developed to obtain the appropriate information from the cloud and realize remote control. When needed the system also provide webcam for authorized users to view security and the real-time situation of the elderly.

6 Conclusion

The social aged care problem is becoming more and more prominent, which leads to the rapid development of home aged care service. The progress of science and technology has led to higher pursuit of quality of life. In order to achieve a more humane way of life for the elderly, home care intelligent home system came into being. In this paper, through the comprehensive control of the home environment, to create a comfortable, safe and convenient home care environment. It can also provide timely warning of major security risks and the health of the elderly in the home. Smart home and home care are now in the ascendant sunrise industry. The system combines the two, and realizes the design and practice of cloud services by means of narrow band Internet of things. There is a broad market prospect.

Acknowledgement. In this paper, the research was supported by Scientific Research Fund of Fujian Provincial Education Department.

References

1. Chan, M., Estève, D., Escriba, C., Campo, E.: A review of smart homes present state and future challenges. Comput. Methods **44**(2), 79–84 (2008)
2. Peng, M.G., Yan, S., Zhang, K.S., et al.: Fog computing based radio access networks: issues and challenges. IEEE Netw. **30**(4), 46–53 (2015)
3. Bastug, E., Bennis, M., Debbah, M.: Living on the edge: the role of proactive caching in 5G wireless networks. IEEE Commun. Mag. **2014**(52), 82–89 (2014)
4. He, Q., Li, N., Luo, W.-J., Shi, Z.-Z.: A survey of machine learning algorithms for big data. PR&AI **27**(4), 327–336 (2014)
5. Ma, Y., Xie, J., Luo, J.: Study on optimal operation of reservoir based on directional self-learning genetic algorithm. J. Hydroelectric Eng. **28**(4), 43–55 (2009)

Decision Support System in Orthopedics Using Methodology Based on a Combination of Machine Learning Methods

Martin Radvansky[1](✉), Milos Kudelka[1], Eva Kriegova[2], and Regina Fillerova[2]

[1] Faculty of Electrical Engineering and Computer Science, VSB-Technical University of Ostrava, Ostrava, Czech Republic
martin.radvansky@vsb.cz

[2] Department of Immunology, Faculty of Medicine and Dentistry, Palacky University Olomouc, Olomouc, Czech Republic

Abstract. Analysis of data based on gene expressions characterizing serious disease is an area currently receiving high attention. The basic task is to classify patients, usually by searching for a small group of genes that provides sufficient classification power. However, very often, different gene combinations can describe different aspects of the problem being analyzed. In this paper, we present in a concrete example with one real dataset, a methodology that has repeatedly been successfully applied to different types of data. In addition to common statistical methods, this methodology combines methods such as a visualization of a dataset structure using networks, and feature-selection and neural network classification. The output of the application of the methodology is a system for decision support during the reoperation of patients with joint endoprosthesis.

Keywords: Classification · Feature selection · Gene expressions · Network analysis · Artificial neural network

1 Introduction

Biological data and in particular, human gene expression data, are increasingly investigated. The reason is new devices providing new and more accurate data resulting from the analysis of human samples. These samples can be associated with very complicated and often rare diseases. Therefore, very often there are only small groups of samples for analysis and the results can be affected by many factors unrelated to these diseases. Considering new approaches to analysis is necessary. These approaches are often combinations of statistic and machine learning methods. Increasingly, artificial neural networks are often used in this area which, unlike in traditional methods, are able to describe complex internal relations in data.

P. Krömer et al. (eds.), *Proceedings of the Fourth Euro-China Conference on Intelligent Data Analysis and Applications*, Advances in Intelligent Systems and Computing 682, DOI 10.1007/978-3-319-68527-4_21

The aim of this work is to design a tool for the classification of patients with and without prosthetic joint infection (PJI) based on selected gene expressions. The tool is planned to be used as a decision support system during operation. The organization of this paper is as follows. Section 2 introduces related work. In Sect. 3, there is the described methodology of gene expression data analysis. An example of using the introduced methodology, the dataset, and feature selection and classification are presented in Sect. 4. Section 5 concludes the paper.

2 Related Work

The analysis of data coming from biologic systems is not an easy task. There is no universal methodology about how to work with particular datasets. Although the structure of analyzed data is the same (one data table), unknown relations between measured values are often hidden. To find such relations, researchers must consider using a combination of data mining and machine learning methods. Together with the size of the dataset (in the meaning of very small or large-scale), the teams of researchers use many approaches.

Liu et al. [1] introduce a framework for working with combinational feature selection and ensemble neural network for the classification of gene expression data. He works with micro-array raw data which has been bootstrapped 100 times due to the small size of the dataset. Authors use PCA and extract the top 15 principal components; selection of important genes is made by the ranksum test. These components are used as the features for learning several neural networks, and the final classification is obtained by the majority voting. Moteghaed et al. [2] study the hybrid optimization algorithm and artificial neural networks (ANN) on micro-array data for cancer classification. Authors in this work improve the ability of the algorithm for the classification problem by finding a small group of biomarkers and also the best parameters of the classifier. They combine decision tree algorithms to find the relation between the biomarkers and use a hybrid of GA and PSO algorithms as a feature selection method. The fitness of each gene subset (chromosome) is determined by an artificial neural network classifier. The 10-fold cross-validation classification accuracy on the gene subset in the training and evaluation samples is evaluation criteria. For increased classification accuracy, authors use a decision tree classifier to see the relation between founded biomarkers and rule extraction. A different approach to the classification of gene expression data is used by Mehridehnavi and Ziaei [3]. Authors study minimal gene selection for classification and diagnosis prediction based on gene expression profile. They use a signal to noise ratio as the main tool in order to reduce data dimensionality. Authors try to classify training and testing data with the perceptron neural network. Due to a low number of data samples, and to achieve a more accurate result, they propose an approach based on ranked distances between the output of classifier and input data. Therefore, 14 germinal center and 12 activated like samples with the shortest distances are selected as a training set for an artificial neural network as the final classifier.

Our methodology is inspired by the previously mentioned works. The main difference is using network visualization as an exploratory analysis tool that helps in the first decisions and feature selection. Moreover, our final classifier is based on the parallel usage of artificial neural networks (ANN), and the output information includes not only their response but also their confidence. The next section describes the methodology in great detail.

3 Methodology and Tools

We have been working on several healthcare projects, and we have faced the problem of classifying patients in small datasets with relative gene expressions. Obtaining larger data is usually a problem because it is time and money consuming (the measurement of relative gene expression is done on special devices). Analysis of this kind of datasets should answer two questions:

1. Can we classify patients for a particular disease based on measured gene expressions?
2. Is there any combination of genes which is sufficient for disease detection?

To answer these questions, we introduce a methodology combining several methods and approaches. The application of this methodology is projected to the decision support system providing correct classification with high reliability (confidence). The following text introduces the methodology successfully used in several projects working with gene expression data and patient classification. The rest of the text in this section is focused on a brief description of used methods.

3.1 Methodology

The methodology of dataset analysis and classification can be divided into two main parts. The first part is focused on feature selection and making a model of the system (Fig. 1). The second one uses this model to create classifier subsequently used for the decision support system (Fig. 2).

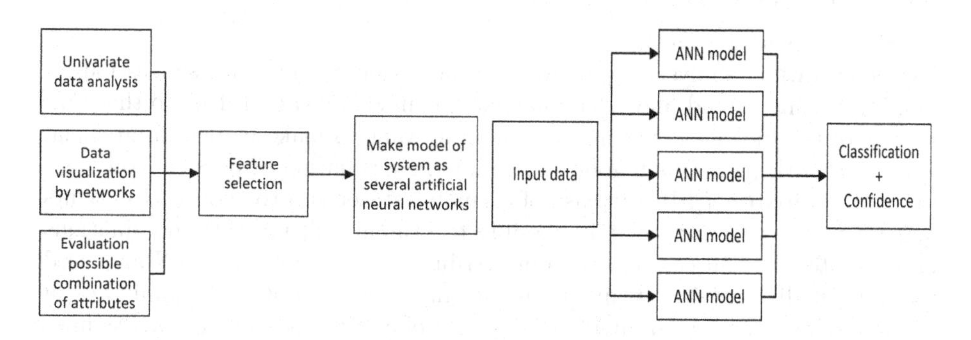

Fig. 1. Make model of system

Fig. 2. Classification

Model of the System

The first part of our methodology is making a model of the system. The measuring of gene expressions is a highly time and money consuming operation. This means we need to select the smallest possible number of genes within the range of input data. By our examination, using widely known methods for feature selection was usually not sufficient due to low confidence with the results. To have a reliable approach to get the required feature selection, it is necessary to combine common statistical methods, visualization and application of artificial neural networks.

Univariate Data Analysis

The first step in our evaluation process is univariate attribute analysis. This allows us to better understand the dataset we work with. We get information about data ranges and distributions, correlations, outliers, missing values, etc. Regarding the physicians, it must also be done visually; we use descriptive statistics, histograms and boxplots. Utilizing Receiver Operating Characteristic (ROC) curves, we can evaluate the ability of a binary classification system in case a threshold value is varied [4]. ROC curves and their confidence intervals [5] over all attributes can suggest which attribute should be a candidate for deeper examination.

Dataset Visualization Using Networks

The understanding of relations between attributes is one of the necessary steps during data analysis. A visualization is able to show us the hidden structure of data. For a visualization of the dataset, a transformation of data to the network (graph construction method) was applied. Network nodes represent individual patients (vectors of gene expression values); edges between the nodes represent a similarity of the corresponding patient vectors based on the Gaussian function. The edges were chosen to link the nearest neighbors (nodes having the highest similarity). The number of nearest neighbors for each node corresponds to node representativeness [6].

The networks were constructed for many combinations of attributes proposed by physicians and laboratory staff. The main goal of the visualization was to understand the internal structure of the dataset influenced by a selected combination of attributes.

Combinations of Attributes

By the univariate analysis and network visualization, we got some suggestions of attributes which could play an important role in classification. Due to the complexity of the biological system for which we want to make model, an artificial neural network (ANN) is a good choice for the evaluation of possible combinations of attributes [7]. In the case of a small dataset (up to thousand patients and up to ten attributes), it is possible to evaluate all possible combinations of attributes. For all combinations of attributes, we created an ANN and used the dataset dividing into training and testing sets. We obtained candidates of combinations of important attributes as a set of attributes for which ANNs have the smallest percentage of prediction and mean squared error (MSE).

Feature Selection
Since the initialization of neutral networks is made with randomly generated weights, we had to use the k-fold cross-validation procedure for all sets of candidate attributes. Then, for each set, we calculated the percentage of classification error and MSE. All attribute combinations were ordered by the classification error, and attribute combinations with the smallest value indicated the important set of attributes.

Classifier Based on ANN
The final step was to make a model of an examined biological system. Similarly to the feature selection phase, we used the ANN also for the model creation. There is no universal way to determine an optimum network topology just from the number of inputs and outputs. The internal structure of the network depends on the number of training examples and the complexity of the underlying system. Based on our evaluation, we suggested the feed-forward ANN with two hidden layers and back propagation as the training algorithm. This structure of the ANN has the ability to minimize output error on a small training dataset. In order to improve the reliability of the created model, we repeatedly used 10-fold cross-validation on important features. Finally, we used the top five trained ANNs with the smallest average error as a base for the model of the system. The number of ANNs used for the modeling of the system was selected by several experiments.

Classification
The model of the system will be used for the classification task. The unknown data we use as an input for our model containing five ANNs. Final classification of data is obtained as the average response value of these five ANNs (Eq. 1).

$$classification = \frac{1}{n}\sum_{i=1}^{n} answer(ANN_i) \tag{1}$$

where n is number of ANN used in model of system.

Although classification of input data is required information, it is necessary to know a confidence of this value. The response of the each ANN is in range zero to one. From five ANNs we obtained five network answers therefore we can calculate the confidence interval of these values. Final confidence of classification is calculated by Eq. 2.

$$confidence = 1 - \left(\frac{1}{n}\sum_{i=1}^{n}(answer(ANN_i) - classification)^2\right)^{\frac{1}{2}} \tag{2}$$

Both calculated values (classification and confidence) together give minimum necessary information for a decision about the proper classification of unknown gene expression data.

4 Methodology Application

In this section, our methodology introduced in Sect. 3 is presented in one example of a real-world dataset. The results are used as decision support for physicians to identify infection in the case of revision surgery.

4.1 Dataset

The dataset contains relative expressions of selected genes measured on the Xxpress cycler (BJS Biotechnologies, UK) system. It consists of 48 records with nine attributes; eight of which are measured gene expressions, and one is the PJI class. An example of the dataset is displayed in Table 1. Dataset is exported from Xxpress machine; there are no missing values.

Table 1. Example of gene expressions dataset

DEFA1	IL1B	LTF	TLR1	BPI	IFNG	TLR2	TLR4	PJI
0.005	0.001	0.098	0.085	0.366	0.028	0.277	2.219	0
0.203	0.218	0.189	4.595	0.047	4.000	2.144	9.190	1
0.277	0.014	0.342	0.319	0.121	0.056	0.901	0.732	1
.	.	.	.	.	.	.	.	.

4.2 Univariate Data Analysis

The first step in understanding the data is a univariate analysis. These methods help us find if a single gene with the capability to separate groups of patients exists. For this purpose, we use widely known boxplots and ROC curves. The dataset contains 23 patients classified as PJI and 25 non-PJI. We used boxplots to visualize distributions of patients by their classification for every single measured gene. The images of boxplots are depicted in Fig. 3 where green dots are non-PJI classified patients, and red colored ones are PJI positive. We tested hypotheses H_0 for the equality of distribution of both groups of patients at confidence level 95%. P-value lower than 0.05 reject hypothesis H_0, and we accepted the alternative hypothesis that there exists a difference between two groups of patients at confidence level 95% (values are in Table 2).

P-values and boxplots help us to suggest some genes which could be important in the dataset. There are six genes with a low P-value which are the main candidates to be important in the dataset. For the next evaluation of genes we can use ROC curves (see Fig. 4). Parameter AUC (area under the curve) should have suggested us decision power of single genes. Based on an AUC greater than 79%, we can select four attributes DEFA1, TLR1, LTF, IL1B that could be important in the dataset.

Table 2. P-values and AUC for single gene

Gene	P-value	AUC [%]
DEFA1	<0.0001	92.66
IL1B	0.0004	79.37
LTF	0.0003	79.66
TLR1	0.001	79.86
BPI	0.002	75.40
IFNG	0.451	44.94
TLR2	0.002	75.50
TLR4	0.358	56.65

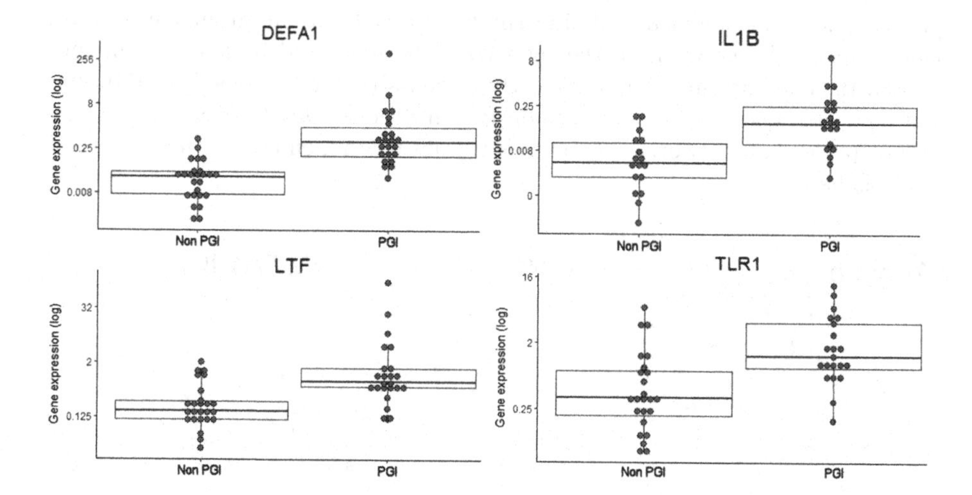

Fig. 3. Comparsion of data by single gene

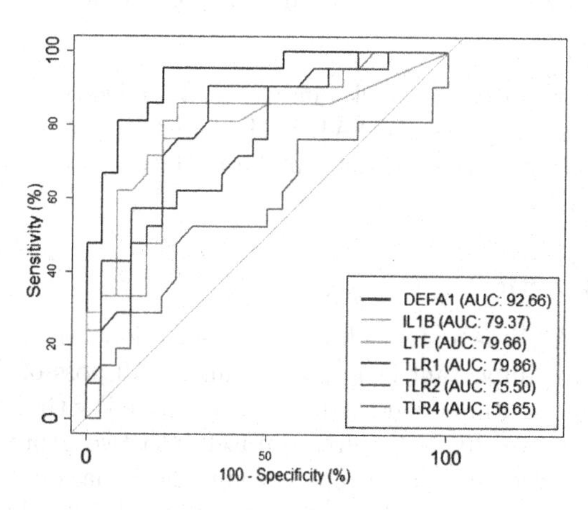

Fig. 4. ROC curves for selected genes

The selected genes are the result of our univariate analysis and need to be examined more. It has been reported that gene profiling in periprosthetic tissues may propose PJI biomarker TLR1 [8]; we take this into account in our next analysis. Explorative network analysis visualizing relations between selected genes is described in the next section.

4.3 Network Visualization

Network visualization shows us relations between similar patients. These relations are based on a selected sets of attributes. Together with univariate analysis, the networks give us a good overview of important genes. Figure 5 displays visualizations for different sets of genes. The left figure visualizes a network for all genes. A single gene network for TLR1 is in the middle. Five relatively dense parts of the network which could identify five (mixed) patient groups in the data can be seen. The network on the right could be separated by a single cut into two parts. The left part contains mostly patients with infections from PJI, and the right side contains patients of non-PJI infections. Due to lack of space, we do not present all networks created from genes recommended by the univariate analysis here.

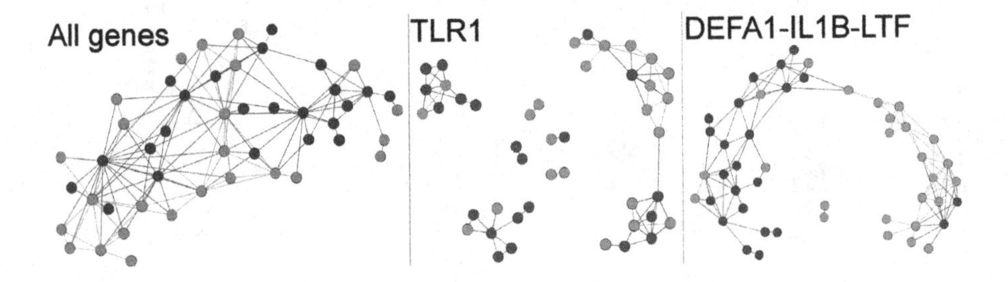

Fig. 5. Networks based on important genes

As result of explorative network analysis of visualization relations between genes, we selected three genes (DEFA1, IL1B, LTF) as candidates for important attribute combination in the dataset. In the next step, we verified these genes as the best choice.

4.4 Feature Selection

Next, we used an ANN as a method of feature selection. Although we have candidates for the most important genes, we evaluated all possible combinations of genes. The dataset contains eight genes so we worked with 256 possible combinations of genes. We divided source dataset into two groups. The first one contained 70% of patients as a learning set and the remaining 30% was used as a testing set. The obtained results were ordered by the classification error and MSE value. The top ten feature sets are shown in Table 3.

Table 3. Top 10 selected feature sets with smallest validation error

Candidates				10-fold cross-validation	
	Combination of genes	Validation error [%]	Validation MSE	Average error [%]	Average MSE
1	DEFA1, LTF	0	0.1	22	0.207
2	**DEFA1, IL1B, LTF**	14.29	0.18	**14**	**0.130**
3	DEFA1, IL1B, TLR2, LTF	14.29	0.15	34	0.298
4	DEFA1, IFNG, TLR2, TLR1	14.29	0.15	24	0.232
5	DEFA1, IFNG, BPI, LTF	14.29	0.15	22	0.178
6	DEFA1, IL1B, IFNG, LTF, TLR1	14.29	0.18	16	0.138
7	DEFA1, IFNG, TLR2, LTF, TLR1	14.29	0.10	26	0.227
8	DEFA1, IL1B, IFNG, TLR4, LTF, TLR1	14.29	0.19	20	0.187
9	DEFA1, IL1B, IFNG, BPI, LTF, TLR1	14.29	0.17	20	0.172
10	DEFA1, IFNG	28.57	0.18	16	0.143

Due to the random initialization of the neural network, we had to evaluate the top ten combinations by the 10-fold cross-validation. This method gave us reliable results, and therefore we can get a decision of important features combination in the dataset. Table 3 includes averaged values of error and MSE for 10-fold cross-validation.

4.5 Model of the System

We needed to create an ANN model of a biological system based on expressions of important genes. To make sure that our model can be used for the classification of patients, we utilized five parallel neural networks. For genes DEFA1, IL1B and TLR, we repeated the 10-fold cross-validation one hundred times. Top five neural networks with the smallest error were selected as a model of the system. The structure of the ANN was experimentally chosen as the network with two hidden layers, with 13 and 9 neurons, and, the training algorithm used back-propagation and logistic activation function.

Making the model was a necessary step for the classification task which was the main goal of our work. Description of the classification process follows in the next subsection.

4.6 Classification

In the classification task, we worked with the model of the system represented by the five neural networks. All of these ANNs are processed in parallel, and we obtain five results. These values are averaged (Eq. 1) to the final classification of input data. In some cases, although the mean value of the model response provides positive classification, we need to know what the relevance of the classification is. Positive classification of input data, in the case when network answers are near border value (0.5), is not very valuable because classification confidence can be low. For this purposes, our classifier has additional output calculated by the Eq. 2 and its value is the confidence of the classification. Classification and

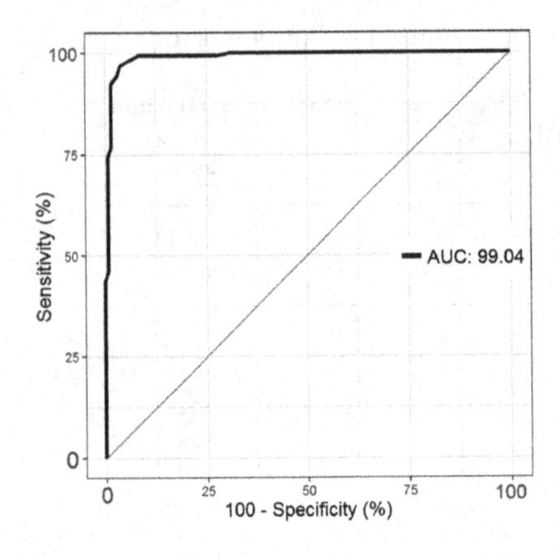

Fig. 6. ROC curve of classifier

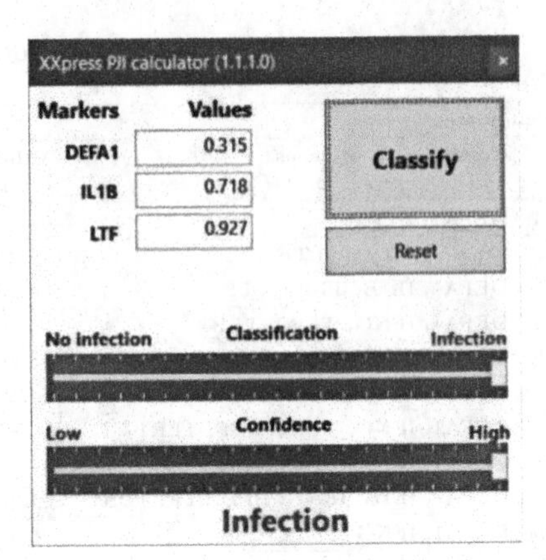

Fig. 7. Decision support tool

confidence is sufficient information for a physician as additional information for confirming or rejecting hypotheses about infection. Evaluation of our classification model was based on ROC curves. As a test group, we used 10 blinded records (patients not used in analysis). The ROC curve is depicted in Fig. 6, the AUC is 99.04%. In this evaluation, we correctly classified 100% patients while the average confidence of classification was 88.8%.

4.7 Decision Support Tool

In order to use our classifier as a decision support tool for physicians, a simple application was created (see Fig. 7); the application has a user-friendly MS Windows interface for physicians and laboratory medical staff. In the current state of knowledge, physicians have to decide on the correct treatment in case of PJI or non-PJIs during operations. Our PJI infection calculator can add extra information for physicians when he/she is not sure about the infection.

5 Conclusion

In our work, we presented a methodology for analyzing gene expressions data. In the proposed methodology, we use a combination of several approaches including traditional approaches (descriptive statistics, tools like boxplots and ROC curves), network visualization and machine learning methods, especially artificial neural networks. We described our application of artificial neural networks on feature selection and classification tasks. The presented methodology was used in several projects, and classification is compound into the electronic calculator used by medical staff as a decision support tool. There are several ways how to improve our methodology. Our main effort will be focused on automatic processing of the whole pipeline and the optimization of artificial neural network structures to get faster learning.

Acknowledgments. This work was supported by grant of Ministry of Health of Czech Republic (MZ CR VES16-31852A) and by SGS, VSB-Technical University of Ostrava, under the grant no. SP2017/85.

References

1. Liu, B., Cui, Q., Jiang, T., Ma, S.: A combinational feature selection and ensemble neural network method for classification of gene expression data. BMC Bioinform. **5**(1), 136 (2004)
2. Moteghaed, N.Y., Maghooli, K., Pirhadi, S., Garshasbi, M.: Biomarker discovery based on hybrid optimization algorithm and artificial neural networks on microarray data for cancer classification. J. Med. Signals Sens. **5**(2), 88 (2015)
3. Mehridehnavi, A., Ziaei, L.: Minimal gene selection for classification and diagnosis prediction based on gene expression profile. Adv. Biomed. Res. **2** (2013)
4. Hajian-Tilaki, K.: Receiver operating characteristic (ROC) curve analysis for medical diagnostic test evaluation. Caspian J. Intern. Med. **4**(2), 627–635 (2013)
5. Macskassy, S. Provost, F.: Confidence bands for ROC curves: methods and an empirical study. In: Proceedings of the First Workshop on ROC Analysis in AI, August 2004 (2004)
6. Zehnalova, S., Kudelka, M., Platos, J., Horak, Z.: Local representatives in weighted networks. In: 2014 IEEE/ACM International Conference on Advances in Social Networks Analysis and Mining (ASONAM), pp. 870–875. IEEE (2014)
7. Dayhoff, J.E., DeLeo, J.M.: Artificial neural networks: opening the black box. Cancer **91**(Suppl. 8), 1615–1635 (2001)
8. Cipriano, C., Maiti, A., Hale, G., Jiranek, W.: The host response: Toll-like receptor expression in periprosthetic tissues as a biomarker for deep joint infection. J. Bone Jt. Surg. Am. **96**(20), 1692–1698 (2014)

Sensor Technology and Signal Processing

Wireless Sensor Network Routing Protocol Research for High Voltage Transmission Line Monitoring System

Xuhong Huang(✉) and Sheng-hui Meng(✉)

School of Information Science and Engineering,
Fujian University of Technology, Fuzhou 350108, China
huangxuhl@163.com, menghui@fjut.edu.cn

Abstract. Wireless sensor network (WSN) was applied to high voltage transmission line monitoring system, according to the characteristics of the high voltage transmission line, put forward a kind of wireless sensor network Routing Protocol - Driven Clustering Routing Protocol DCRP (Driven Clustering Routing Protocol), that suitable for high voltage transmission line monitoring. And according to the characteristics of the different data that is collected in the high voltage transmission line monitoring, it is divided into two work modes—normal and emergency, and energy of the network is optimized. DCRP has good real-time performance, can effectively prolong the network life cycle, to meet the application requirement of high voltage transmission line monitoring system.

Keywords: WSN · High voltage transmission line monitoring · Routing algorithm · Energy optimization

1 Introduction

Serious natural disasters occur frequently in recent years. Especially lightning, debris flow, ice have seriously damaged on high voltage transmission line. Some criminals for profiteering risk cut and steal transmission cable. Distribution of high voltage transmission line is wide, the way by manual inspection alone, can't well realize monitoring. In order to ensure the safety of high voltage transmission line running effectively, it must establish an effective monitoring network for high voltage transmission line. Wireless sensor Network (WSN) is constituted by lots of cheap intelligent sensor node that is self-organization, also known as intelligent wireless sensor Network (Smart wireless sensor Network) [1]. The network is wireless network, can solve the problem that the high-voltage transmission lines are not allowed other devices to take electricity directly on it. Number of wireless sensor network nodes is enormous and nodes can be randomly distributed. Node not only has the ability to perceive and data processing [2], also has the function of routing. WSN can well solve the problem of high voltage transmission line wide distribution.

Wireless sensor network is Ad hoc network, with low power consumption of communication mode. But on the technical implementation, it is a big difference with the traditional Ad hoc network. WSN is easy deployment, high precision, high fault

P. Krömer et al. (eds.), *Proceedings of the Fourth Euro-China Conference on Intelligent Data Analysis and Applications*, Advances in Intelligent Systems and Computing 682, DOI 10.1007/978-3-319-68527-4_22

tolerance, large covering area, remote monitoring, but Center nodes need to do a lot of related data processing, integration, and the cache. It also has the characteristics of limited energy, topology changes frequently [3]. Limited energy refers to that the network nodes usually don't complement energy [4]. In addition, each node in wireless sensor network (WSN) is a random deployment, according to certain algorithm automatically organized into application oriented network. In the application of wireless sensor for high voltage transmission line monitoring, the comprehensive factors such as node closing, the change of transmitted power, line swing sharply in windy day, terrain and weather, etc. may change topology at any time. Therefore, it must design appropriate routing protocol. There are many routing protocols of different characteristics, but they have some limitations [5].

To meet the demand of the high voltage transmission line monitoring, according to the characteristics of them, suitable routing protocols of wireless sensor network for high voltage transmission line monitoring must researched, to raise the level of technology in high voltage transmission line monitoring.

2 The Network Mode of Wireless Sensor Network Monitoring System

The wireless sensor nodes for monitoring stress, vibration and temperature can be made into ring clasping on high voltage transmission lines. On transmission line one is to deploy from a distance, real-time monitors these data. The center node and video monitoring nodes can be fixed on the transmission tower. A large number of online annular monitoring sensor nodes and the sensor nodes on tower constitute self-organized network. The data from sensor nodes is gathered in the center node. After the data has been fused, it is send to the monitoring center. Online annular sensor nodes can be used by the exterior of bright colors, then can prevent the helicopter crashed the high tension line by mistake.

High voltage transmission line monitoring system has its features:

(1) high voltage transmission line network is huge, wide distribution, energy conservation requirements. Because the high voltage transmission line networks are distribution all over the country, include a lot of complex mountainous terrain, replacement the battery of nodes is difficult. So the problem of the network energy optimization is important.
(2) methods that collected different data in high voltage transmission line monitoring system are different. The environmental information, such as real-time stress, vibration and temperature is collect in normal times, and in other emergencies need to transmit a lot of data. The data of stress, vibration and temperature is collected by cyclical, and emergencies are random events.
(3) the emergencies will destruct some nodes, and the high voltage transmission line monitoring network will change dynamic.

Because above characteristics of the high voltage transmission line monitoring, and the span of high-voltage transmission lines is generally long, and at the same time many lines are parallel, and the sensor network scale is larger, the network use clustering network structure. The whole wireless sensor network is divided into several region. One region is between two poles. Each region is divided into several clusters. Each cluster has a cluster head, with one hop communication between members of the cluster and the cluster head. The cluster head communication with the center node by multiple hops communication. The data fused from cluster head will be sent to center node, so as to reduce the traffic.

Sensor nodes can be set up for four kinds of state, respectively, receiving state, sending state idle state, and the sleep state. Through transformation among the four states, it tries to save energy consumption of nodes, prolong the network lifetime.

At ordinary times, most of the base station of high voltage transmission line monitoring network is a fixed mode, but when emergency occurs, the network structure is random variation. Therefore, the network can work in two kind modes, normal and emergency. The data quantity of monitoring normal stress, vibration and temperature is small, the sensor nodes can be driven by the time of periodic. When emergencies and stolen transmission line event occurs, network must send a lot of data by event driven type.

3 Routing Algorithm for High Voltage Transmission Line Monitoring Network

In order to make the routing protocol has higher extensibility, make load more evenly, this paper proposes a wireless sensor network routing protocol that is suitable for high voltage transmission line monitoring - driven routing protocol (DCRP).

High DCRP uses asynchronous clustering method. The communication between members of the cluster and the cluster head is by one hop. The communication between cluster head and the center node is by multiple hops. The date fused from cluster head will be sent to the data fusion center node (Sink node), so as to reduce the traffic. The method of asynchronous replace cluster head nodes can cause energy consumption equilibrium.

Cluster head nodes in network constitute the backbone network. Based on the minimum hop routing algorithm, it establish routing lines for each cluster head. Then the trunk link of cluster head nodes is form.

Routing algorithm is as follows:

(1) ‘Father_id’, the parents of all the cluster head nodes and ‘Min_hop’, the minimal hop for reaching the Sink node, initialized to 0;
(2) The Sink node broadcast ‘Hop_Msg’ message that is constructed of message identifier, sending node ID and send the minimum hop count of ‘Min_hop’ adding 1;
(3) ‘Father_id’ of cluster head nodes with receiving ‘Hop_Msg’ is set to the Sink, ‘Min_hop ‘is set to 1. At the same time, the ‘Min_hop’ and ‘Father_id’ are update, ‘Hop_Msg’ message continue to broadcast to its neighbor nodes;
(4) The cluster head nodes who received ‘Hop_Msg’ message check ‘Min_hop’ from this message.If it is less than its own ‘Min_hop’, it updates its Father_id and

Min_hop, then continue to broadcast ‘Hop_Msg’ message to its neighbor nodes, otherwise does not handle;

(5) repeat step (4), processed until all nodes in the network have handled.

The algorithm flow chart is shown in Fig. 1.

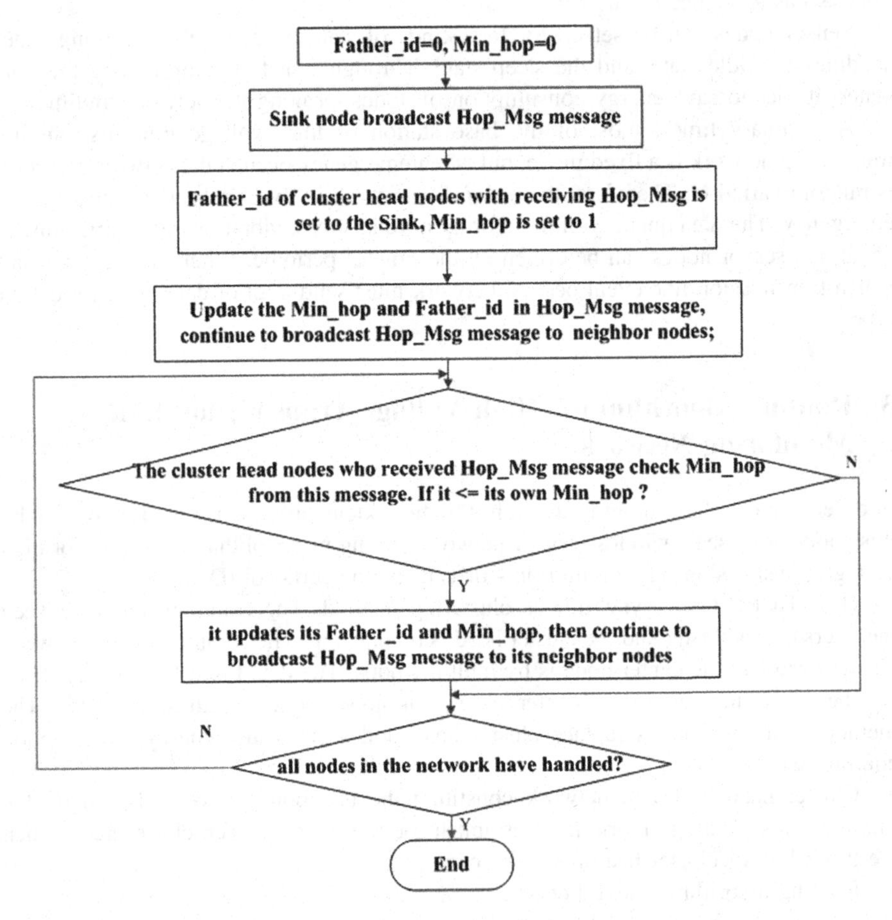

Fig. 1. Composition block diagram of wireless sensor nodes.

The cluster head node communication with Sink nodes by multiple hops, transmit monitoring data. The cluster head nodes and Sink nodes form the backbone. The backbone uses the minimum hop routing algorithm, routing lines is established for each of the members. Each cluster head node can obtain the minimum hop and information of father node, form a minimum hop field. Energy consumption of cluster head is the biggest. For balancing energy consumption within the cluster, asynchronous mode of changing the cluster head is used. When energy of cluster head is lower than the threshold value, the ‘CH_change’ message is broadcasted within the cluster. After other nodes within cluster receiving the message, they transmit their own location and information about energy to the cluster head. According to these information with the energy and location optimization principle, Cluster head select the new cluster head and broadcasted information of new cluster head. New cluster head inherit information about the original father of cluster head nodes and the minimum hop. The other nodes modify the routing table. The situation of Change cluster head is shown in Fig. 2.

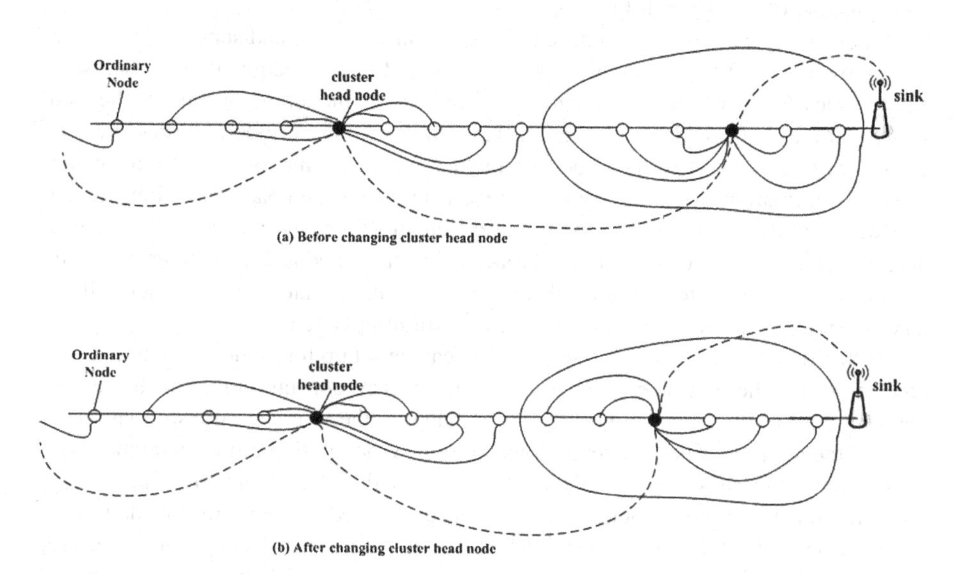

Fig. 2. Situation of change cluster head 4 network energy optimization.

4 Network Energy Optimization

Wireless sensor network node is usually deployment only ones, independent. The sensor nodes on the high voltage transmission line are serviced inconvenient. The energy of node is limited, and supplement for energy is inconvenient. In order to meet the demands of stability monitoring, decreasing power consumption of network supplement to extend network life is necessary, and must save cost as much as possible. Most of the base station in monitoring network of high voltage transmission line, at ordinary times, is fixed mode. when the unexpected events occur, the network structure is random variation. DCRP network has two work modes, ordinary and sudden event. Ordinary, monitoring data quantity of line stress, vibration and temperature is small, the sensor nodes can be driven periodic to collect data, transmission periodically. When emergencies event occurs, it is to send a lot of data, it use event driven type.

High In high voltage transmission line monitoring, different methods for collecting information are use in different environmental. Usually, real-time stress, vibration and temperature sensor is needed. When other emergencies event occur, remote video monitoring sensors are driven for data transmission. Data collection of stress, vibration and temperature is cyclical, data collection of emergencies is random events. At ordinary times, stress, vibration and temperature sensor nodes are driven by periodic time. These sensor nodes for stress, vibration and temperature distribution on high voltage transmission line, sent the network information, and information of stress, vibration and temperature with small data volume to the cluster head periodically. Cluster head sensor nodes rapid fuse data and forward the data to center node. As a result of these information data volume is not big, high transmission rate is not need, and consume of energy is little.

When emergencies event occur, a lot of data must be send and sensor nodes can be driven by event. The threshold is set into periodic data acquisition mode. when emergencies event occur, its numerical is more than induction threshold, the node sensing at the first broadcast awaken package to the adjacent node and preserve local event information. The awaken package include node information, event information, relevant information. After nearby nodes receiving the awaken package, it immediately test, determine correlation, decided to accept or reject awaken package. After cluster head receiving the awaken package, broadcasting it immediately. After receiving the broadcast, video monitoring node determine correlation, start the video surveillance and transmit monitoring information to the monitoring center.

At the same time, because the sensor node can be set up four kinds of states, such as receive, send, idle and sleep. Sleep state little energy consumption, and the energy consumption of idle state is relatively low. Changing the four state transformation can achieve the goal of reducing energy consumption. When periodic time and emergency aren't come, the sensor nodes are try to make in the sleep state to reduce energy consumption. When nodes detect incidents occur or need periodic transfer daily data, node is from state sleep to send state; When node is in a state of sleep, if it is received effective signal, it is from sleep into the receiving state; When nodes sending or receiving is end, it get into idle state; if, in a certain period of time or need to send and receive data, it send or receive in; During this period, if no information must be sent and received, the sensor nodes change into sleep state.

5 Simulation

The DCRP protocol is simulated with NS2 and compared with the MHC [6] routing protocol, which evaluates the network energy consumption and the transmission delay.

The number of nodes in the simulation were set to 100, these nodes randomly distributed in the area of 200 m * 1 m, each node has the initial energy of 1 j, Sink node is located in the coordinates (x = 0, y = 0). The total time of the simulation is 500 s, and the monitoring node sends the usual packet at 384 bits per every other 5 s. Let's say that when 300 s, a sudden event occurs, the packet is 2166338.

Figure 3 compares the two protocols in the total energy consumption of the network nodes. As you can see from the diagram, the MHC routing protocol runs to 413 s and the node energy runs out while the DCRP still has energy in the 500 s. During the entire run time, the DCRP protocol node total energy consumption is less than the MHC routing protocol.

Figure 4 compares two protocols in the data transfer latency while the network is running. As you can see from the diagram, the DCRP protocol data transfer delay is almost the same as the MHC routing protocol in normal times. But the DCRP protocol

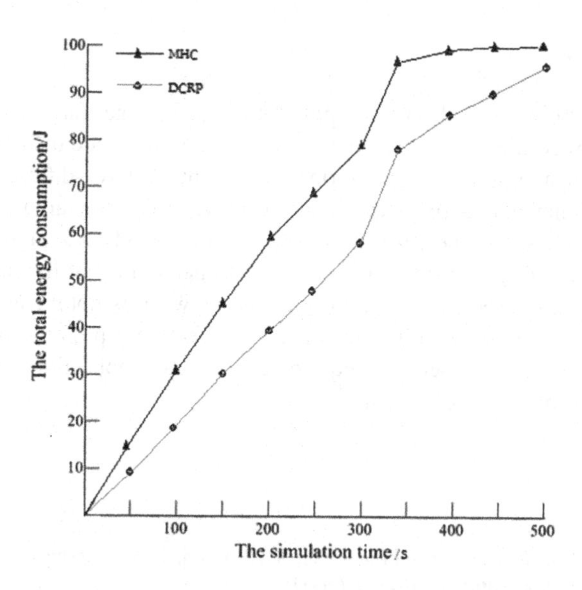

Fig. 3. Network node total energy consumption and time.

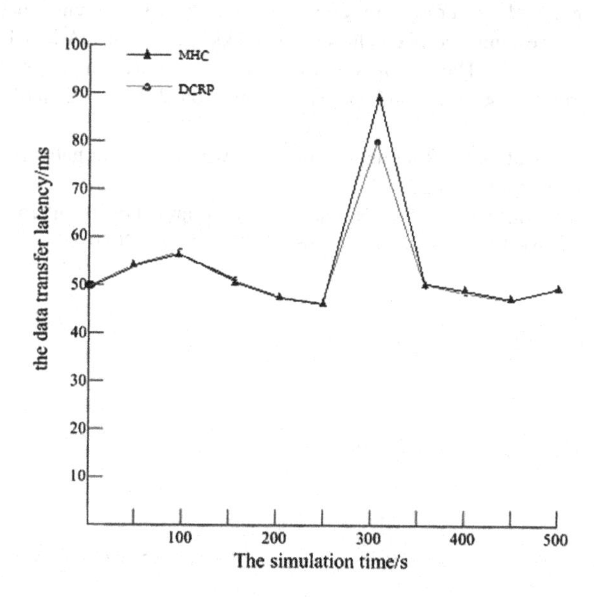

Fig. 4. The data transfer latency.

data transfer latency is less than the MHC routing protocol when an unexpected event occurrences. Therefore, the DCRP protocol applies to the characteristics of high-voltage transmission line monitoring system.

6 Conclusion

Wireless sensor network (WSN) was applied to high voltage transmission line monitoring system. According to characteristics of a large scale, wide distribution, and that high voltage transmission line monitoring system muse have different methods for different environmental and the emergencies event will cause the failure of some nodes, a suitable wireless sensor network routing protocol, DCRP, for high voltage transmission line monitoring is proposed. According to peculiarity of the data collected in the high voltage transmission line, energy of the network is optimization. DCRP has good real-time, high energy utilization rate, can effectively prolong the life cycle of monitoring network, and meet to requirement of the application in high voltage transmission line monitoring system.

References

1. Akyidiz, I.F., Su, W., Sankarasubramanian, Y., Cayirci, E.A.: Survey on sensor networks. IEEE Commun. Mag. **40**(8), 102–114 (2002)
2. Sun, L., Li, J.: Wireless Sensor Network (WSN). Tsinghua University Press, Beijing (2005)
3. Akkaya, K., Younis, M.: A survey on routing protocols for wireless sensor networks [EB/OL] (2003). http://www.es.umbc.edit/kemall/mypapers/Akkaya_YounisjoAdHocRevised.pdf
4. Li, H., Jiang, S., et al.: The wireless sensor network application oriented modeling and analysis of information accuracy and energy consumption. J. Sens. Technol. **20**(2), 408–412 (2007)
5. Al-Karaki, J.N., Kamal, A.E.: Routing techniques in wireless sensor network: a survey. IEEE Wirel. Commun. **11**(6), 6–28 (2004)
6. Duan, W.-f., Qi, J.-d., Zhao, Y.-d.: Research on minimum hop count routing protocol in wireless sensor network. Comput. Eng. Appl. **46**(22), 88–90 (2010)

Near/Far-Field Polarization-Dependent Responses from Plasmonic Nanoparticle Antennas

Hancong Wang[1,2,3](✉), Wenbin Zheng[2,3], Shihao Huang[3], Haiyun Zhang[3], Deyao Lin[3], Jian Chen[2,3], Qingzhou Ye[3], and Chi Hu[2,3]

[1] The Key Laboratory for Automotive Electronics and Electric Drive of Fujian Province, Fujian University of Technology, Fuzhou 350108, China
whcuser@163.com

[2] Fujian Provincial Key Laboratory of Digital Equipment, Fujian University of Technology, Fuzhou 350108, China

[3] School of Information Science and Engineering, Fujian University of Technology, Fuzhou 350108, China

Abstract. The localized surface plasmons of metallic nanoparticles are able to concentrate light into small volumes, which lead to a variety of fundamental studies and practical applications in plasmonics. For example, by strong coupling between metallic nanoparticles, plasmonic antennas are able to concentrate and re-emit light in a controllable way. A variety of structures of optical antennas have been investigated in in the past decade. The near- and far-field responses of the plasmonic nanoantennas for example, the intensity and phase distributions, and the emission polarization state are found to be sensitive to polarization. This sensitivity is determined to arise from structural properties including particle size, shape, spacing, relative positions and symmetry of nanoparticles. In this review, we will discuss our recent advances in plasmonic nanoparticle antennas from the polarization point of view, i.e., control of the incident polarization-dependent near-field enhancement, control of the (far-field) polarization of elastic or inelastic scattering light, and outlook the corresponding impacts in understanding physics and nanophotonic devices applications.

Keywords: Plasmonics · Nanoantennas · Polarization · Nanoparticle · Generalized Mie theory

1 Introduction

Optical nanoantennas (or plasmonic antennas) as an analogue of microwave antennas at the nanoscale are of great interest due to the unique ability of controlling absorption and emission at visible and infrared region [1, 2], such as focusing optical fields to sub-diffraction limited volumes [3], enhance the excitation and emission of molecules [4, 5] and quantum emitters [6]. Propagating light can be converted into nanoscale enhanced near field [7–9], and vice versa, a localized excitation can be coupled to directed radiation. The plasmonic response from nanoantennas depends on the incident

P. Krömer et al. (eds.), *Proceedings of the Fourth Euro-China Conference on Intelligent Data Analysis and Applications*, Advances in Intelligent Systems and Computing 682, DOI 10.1007/978-3-319-68527-4_23

polarization, and also the emission polarization states can be tuned by its shape, material, dimension, geometry, and operation frequency. Most of these coupled antenna such as dimers [10, 11] and trimers [4] are based on metallic nanostructures that support surface plasmons (SP). Single and coupled nanoantennas have been investigated thoroughly by far-field spectroscopy, exploiting two-photon luminescence and near-field scanning microscope. Abundant applications have been found in surface enhanced Raman scattering [3, 12], optical manipulation [11, 13], biosensing [14] and integrated photonic devices [15, 16], etc.

Here we provide an overview of the polarization effects in coupled nano-aggregates antenna system. Polarization properties such as the incident polarization-dependent near field, and the polarization of elastic and inelastic emission, can be controlled effectively by different geometries of nanoantennas, and will be discussed respectively.

2 Control of Incident Polarization-Dependent Near-Field Enhancement

The coupling of single nanoparticle aggregates are able to concentrate light into small volumes, which are responsible for the electromagnetic field enhancement particularly pronounced in nanogaps between coupled nanoparticles, named "hot spots" [3]. The intensity and phase distributions of near field can be tuned by the shapes and relative positions of nanoparticles. The control of far-field polarization by nanoparticle trimer had been fully investigated, and mainly focused on the linear incident polarization [4].

Actually, a trimer can be the simplest asymmetric structure, and its near-field enhancement is also sensitive to circular polarization light (CPL). In Fig. 1, we use the excitation of left-hand circularly polarized light (LCP) and right-hand circularly polarized light (RCP) to theoretically investigate the near-field enhancements in the gap of asymmetric trimer. The generalized Mie theory (GMT) is a relatively accurate method to calculate near field which is the constructive/destructive interference [17] of two near-field parts induced respectively by two orthogonal electric field components constituting CPL. We found that LCP and RCP led to very different enhancements and near-field interference by investigating the phases of hot spots in Ag nanoparticle trimer ($R_1 = R_2 = 40$ nm, with the gap distance of 1 nm) for parallel and perpendicular polarized at 532 nm. Just as the introduction of circular dichroism (CD) to the realm of metamaterials, in the plasmonic trimer, hot spots also has CD effect, that is, to be "hot" or "cold" (enhance or diminish) depending on the polarization state of the excitation. Similar to the definition of CD signals for far-field transmission or reflection, the CD factor of hot spots describing near-field response could be defined by $\rho = log_{10}[(M_R - M_L)/(M_R + M_L)]$, where M_L or M_R are the enhancement factors for LCP or RCP excitation, respectively. The enhancement factor is often defined as $log_{10}M = log_{10}(|E_N/E_i|^2)$. A giant CD is obtained at 532 nm ($\rho = 0.9986$). This significant effect would be "Incident circular polarization (ICP)" Raman optical activity (ROA) [18] if there are individual Raman scattering molecules to probe the hot spots. This study could pave the way towards ultra-strong polarization-dependent light-matter interaction on the nanoscale plasmonic devices, with possible applications such as, SERS or ROA substrate, plasmonic nanoantennas and sensor, and polarization sensitive devices, etc.

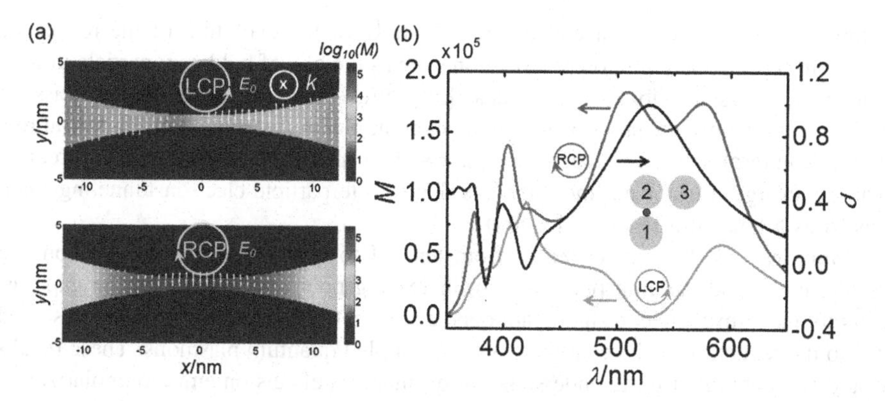

Fig. 1. (Reproduced from [19]) Local CD response of the right-angle trimer. (a) Electric field intensity distributions around the interparticle junction in the right-angle trimer excited by LCP and RCP at 532 nm, respectively. (b) Local field intensity and local CD response ρ (black curve, to the right axis) at the hotspot in the interparticle junction as a function of the incident wavelength. The green and red curves are for LCP and RCP excitations, respectively. The CD response reaches its maximum value of 0.998 at 532 nm. (Color figure online)

3 Control of Emission Polarization

The emission polarization states from nanoparticle antenna can also be control by the plasmonic gap. The quantum effects can also drastically change the coupling strength as the feature size approaches atomic scales [20]. As shown in Fig. 2, we present a

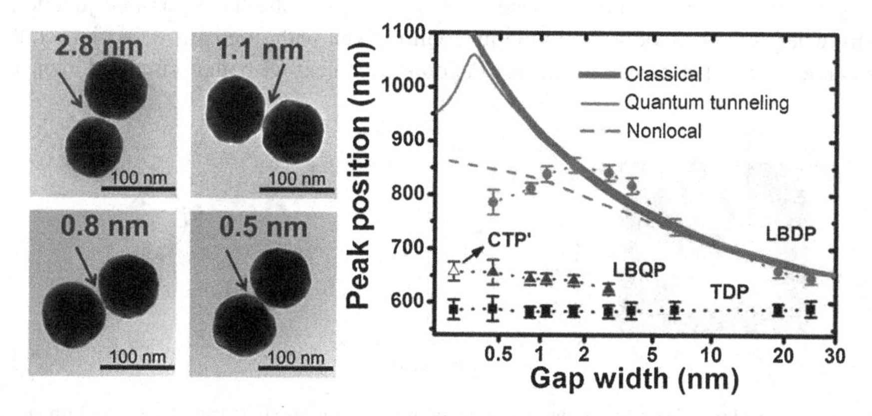

Fig. 2. (Reproduced from [21]) Evolution of scattering peak positions of dimer antennas. (a) TEM images of gold dimers. Averaged diameter of particle is 80±2 nm. Scale bars are 100 nm. (b) Evolution of the LBDP (red circles), LBQP/CTP′ (blue triangles/ hollow triangle), and TDP (black squares) as a function of the gap width. The gray curve is the classical prediction for gold nanoparticle dimers with the diameter D = 80 nm. The broadening of the curve shows the influence of diameter variation ΔD = 2 nm on the resonant peak. Red solid and dashed curves are quantum theoretical results incorporating electron tunneling and nonlocal effect by QCM and LAM, respectively. The dotted lines are plotted to guide eyes. The uncertainties of peak positions are caused by the noise level of spectra when finding the maximum of scattering intensity. (Color figure online)

comprehensive experimental and theoretical study of the evolution of the resonance (scattering) peak and emission polarization state as the gap of gold nanoparticle dimer–antenna narrows to subnanometer scale [21]. We clearly can identify the classical regime (> ~3 nm), the crossover regime where nonlocal screening plays a central role with a feature of saturated resonance peak (~ 1–3 nm) [22], and the quantum regime where a charge transfer plasmon appears due to interparticle electron tunneling manifested as a blue-shift (~ 0.8–1.1 nm).

Moreover, as the gap decreases from tens of to a few nanometers, the bonding dipole mode tends to emit photons with increasing polarizability. When the gap narrows to quantum regime, a significant depolarization of the mode emission is observed due to the reduction of the charge density of coupled quantum plasmons. These results would be beneficial for the understanding of quantum effects on emission polarization of nanoantennas and the development of quantum-based photonic nanodevices.

4 Control of Polarization of Inelastic (Raman) Scattering

Not only the elastic scattering from metal nanoparticle, but also the light polarization from an emitter such as Raman scattering (RS) of molecules in the gap between nanoantennas can be manipulated significantly [4, 23]. Compared with the symmetric response in dimer antenna, a drastically new behavior is obtained from a trimer structure. These properties of such a trimer with the SEM image shown in Fig. 3a can be simulated by treating the nanoparticles as spheres. Here the low concentration of molecules used ensures that each aggregate contains no more than a single molecule [24]. In order to simulate this situation of trimer, calculation was performed by assuming that the molecule is placed in turn in each of the three possible junctions. Only when the molecule is set in the junction marked with a red arrow in Fig. 3a, the calculated and experimental results are in good agreement for both normalized intensity

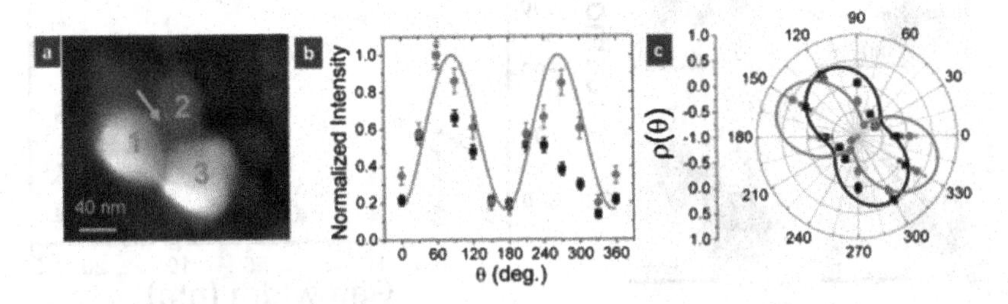

Fig. 3. (Reproduced from [4]) Polarization response of a nanoparticle trimer. (a) SEM image of the trimer. (b) Normalized SERS intensity at 555 nm (black squares) and 583 nm (red circles) as a function of the angle of rotation by the λ/2 wave-plate. The intensities at both wavelengths show approximately the same profile, but the maximal intensity is observed at ~75^0, which does not match any pair of nanoparticles in the trimer. The green line the result of a calculation assuming that the molecule is situated at the junction marked with red arrow in SEM image. (c) Depolarization ratio (ρ) measured at 555 nm (back squares) and 583 nm (red circles). The black and red lines show the result of calculations at the two wavelength, assuming that the molecule is situated at the junction marked with red arrow in SEM image. (Color figure online)

and depolarization, which also confirms the assumption that only one molecule in the junction, contributes to the signal.

The intensity profile in Fig. 3b is maximal at an angle of $\sim 75^{\circ}$, which is close to the axis of 1st and 2nd nanoparticles. The depolarization ρ in Fig. 3c is defined as $\rho = (I_{//}+I_{\perp})/(I_{//}+I_{\perp})$, where $I_{//}$ and $I_{\perp}$ are RS signals with orthogonal polarization. Whereas, the depolarization ratio profiles do not coincide with each other and in addition they are both rotated with respect to the intensity profile. The depolarization pattern of the 555 nm light is rotated by $\sim 45^{\circ}$, while the 583 nm light is rotated by $\sim 75^{\circ}$. What should be note is this counter-intuitive wavelength-dependent polarization rotation is not an accident. The rotation only exists in the cases with the number of the particles are larger than two.

5 Summary

Compare with single structures, more important nanoantennas are coupled systems which induce near-field enhancement and far-field scattering. Plasmonic antennas can be used to manipulate light properties at the nanoscale. With the help of SP coupling, the near-field light intensity, and the far-field emission can be well tailed. The polarization-related effects in plasmonic nanoparticle (dimer and trimer) antennas have been introduced here. These results could pave the way towards polarization-dependent light-matter interaction on the nanoscale plasmonic and quantum devices, with possible applications such as, SERS or ROA substrate, plasmonic nanoantennas [25] and sensor, polarization sensitive devices, compact optical diodes and switches, etc.

Acknowledgements. This work was supported by the National Natural Science Foundation of China (61604041), the Natural Science Foundation of Fujian Province of China (2016J05147), the Startup Foundation of Fujian University of Technology (GY-Z160049), the Mid-youth Project of Education Bureau of Fujian Province (JAT160331), and the Fujian Provincial Major Research and Development Platform for the Technology of Numerical Control Equipment (2014H2002).

References

1. Alu, A., Engheta, N.: Theory, modeling and features of optical nanoantennas. IEEE Trans. Antennas Propag. **61**(4), 1508–1517 (2013)
2. Giannini, V., Fernandez-Dominguez, A.I., Heck, S.C., Maier, S.A.: Plasmonic nanoantennas: fundamentals and their use in controlling the radiative properties of nanoemitters. Chem. Rev. **111**(6), 3888–3912 (2011)
3. Xu, H., Bjerneld, E.J., Käll, M., Börjesson, L.: Spectroscopy of single hemoglobin molecules by surface enhanced raman scattering. Phys. Rev. Lett. **83**(21), 4357–4360 (1999)
4. Shegai, T., Li, Z., Dadosh, T., Zhang, Z., Xu, H., Haran, G.: Managing light polarization via plasmon–molecule interactions within an asymmetric metal nanoparticle trimer. Proc. Natl. Acad. Sci. U.S.A. **105**(43), 16448–16453 (2008)

5. Crozier, K.B., Zhu, W., Wang, D., Lin, S., Best, M.D., Camden, J.P.: Plasmonics for surface enhanced raman scattering: nanoantennas for single molecules. IEEE J. Sel. Top. Quantum Electron. **20**(3), 152–162 (2014)
6. Curto, A.G., Volpe, G., Taminiau, T.H., Kreuzer, M.P., Quidant, R., van Hulst, N.F.: Unidirectional emission of a quantum dot coupled to a nanoantenna. Science **329**(5994), 930–933 (2010)
7. Li, Z., Hao, F., Huang, Y., Fang, Y., Nordlander, P., Xu, H.: Directional light emission from propagating surface plasmons of silver nanowires. Nano Lett. **9**(12), 4383–4386 (2009)
8. Li, Z., Zhang, S., Halas, N.J., Nordlander, P., Xu, H.: Coherent modulation of propagating plasmons in silver-nanowire-based structures. Small **7**(5), 593–596 (2011)
9. Li, Z., Bao, K., Fang, Y., Guan, Z., Halas, N.J., Nordlander, P., Xu, H.: Effect of a proximal substrate on plasmon propagation in silver nanowires. Phys. Rev. B **82**(24) (2010)
10. Shegai, T., Miljković, V.D., Bao, K., Xu, H., Nordlander, P., Johansson, P., Käll, M.: Unidirectional broadband light emission from supported plasmonic nanowires. Nano Lett. **11** (2), 706–711 (2011)
11. Svedberg, F., Li, Z., Xu, H., Kall, M.: Creating hot nanoparticle pairs for surface-enhanced Raman spectroscopy through optical manipulation. Nano Lett. **6**(12), 2639–2641 (2006)
12. Wang, W., Li, Z., Gu, B., Zhang, Z., Xu, H.: Ag@SiO2 core-shell nanoparticles for probing spatial distribution of electromagnetic field enhancement via surface-enhanced Raman scattering. ACS Nano **3**(11), 3493–3496 (2009)
13. Li, Z., Kall, M., Xu, H.: Optical forces on interacting plasmonic nanoparticles in a focused Gaussian beam. Phys. Rev. B **77**(8) (2008)
14. Lal, S., Clare, S.E., Halas, N.J.: Nanoshell-enabled photothermal cancer therapy: impending clinical impact. Acc. Chem. Res. **41**(12), 1842–1851 (2008)
15. Ozbay, E.: Plasmonics: merging photonics and electronics at nanoscale dimensions. Science **311**(5758), 189–193 (2006)
16. Kirchain, R., Kimerling, L.: A roadmap for nanophotonics. Nat. Photon. **1**(6), 303–305 (2007)
17. Rodríguez-Fortuño, F.J., Marino, G., Ginzburg, P., O'Connor, D., Martínez, A., Wurtz, G. A., Zayats, A.V.: Near-field interference for the unidirectional excitation of electromagnetic guided modes. Science **340**(6130), 328–330 (2013)
18. Chuntonov, L., Haran, G.: Maximal Raman optical activity in hybrid single molecule-plasmonic nanostructures with multiple dipolar resonances. Nano Lett. **13**(3), 1285–1290 (2013)
19. Wang, H., Li, Z., Zhang, H., Wang, P., Wen, S.: Giant local circular dichroism within an asymmetric plasmonic nanoparticle trimer. Sci. Rep. **5**, 8207 (2015)
20. Tame, M.S., McEnery, K.R., Özdemir, Ş.K., Lee, J., Maier, S.A., Kim, M.S.: Quantum plasmonics. Nat. Phys. **9**(6), 329–340 (2013)
21. Yang, L., Wang, H., Fang, Y., Li, Z.: Polarization state of light scattered from quantum plasmonic dimer antennas. ACS Nano **10**(1), 1580–1588 (2016)
22. Luo, Y., Fernandez-Dominguez, A.I., Wiener, A., Maier, S.A., Pendry, J.B.: Surface plasmons and nonlocality: a simple model. Phys. Rev. Lett. **111**(9), 093901 (2013)
23. Li, Z., Shegai, T., Haran, G., Xu, H.: Multiple-particle nanoantennas for enormous enhancement and polarization control of light emission. ACS Nano **3**(3), 637–642 (2009)
24. Le Ru, E.C., Meyer, M., Etchegoin, P.G.: Proof of single-molecule sensitivity in surface enhanced Raman scattering (SERS) by means of a two-analyte technique. J. Phys. Chem. B **110**(4), 1944–1948 (2006)
25. Li, Z., Xu, H.: Nanoantenna effect of surface-enhanced Raman scattering: managing light with plasmons at the nanometer scale. Adv. Phys. X **1**(3), 492–521 (2016)

Car Collision Warning System for Cornering on Mountain Roads

S.H. Meng[1,2(✉)], S.B. Hu[3(✉)], A.C. Huang[4], T.J. Huang[5], J.J. Jia[1], and Xuhong Huang[1]

[1] School of Information Science and Engineering, Fujian University of Technology, Fuzhou 350118, Fujian, China
menghui@fjut.edu.cn

[2] Key Laboratory of Big Data Mining and Application, School of Information Science and Engineering, Fujian University of Technology, Minhou, Fuzhou 350118, Fujian, China

[3] Institute of Electro-Optical Science and Engineering, National Cheng Kung University, Tainan 70101, Taiwan

[4] Department of Electrical Engineering, National Sun Yat-Sen University, Kaohsiung, Taiwan

[5] Kaohsiung Municipal ChungShan Senior High School, Kaohsiung, Taiwan

Abstract. This study proposes an ultrasound-based collision avoidance/warning safety system for vehicles cornering on mountain roads. The design includes a hardware and a software system. The hardware system consists of a single-chip microcontroller with a minimum system board, a power module, and an automotive radar system. The radar system includes a stc98c52 control module, LCD 1602 display module, and hc-sr04 ultrasonic detection module. The software was programmed using C on the Keil uVision4 platform.

Keywords: Ultrasound · Early warning · Obstacle

1 Background and Research Motivation

Driving safety has become a significant problem in China with the rapid increase in the number of cars [1]. An ultrasonic cornering radar can remind the driver of an upcoming corner, effectively reducing the possibility of accidental collision [2]. When a vehicle is cornering, the automotive cornering radar measures the distance in front of the vehicle using ultrasound, and compares this with the pre-set safety distance. If the threshold is exceeded, the radar warns the driver so that he/she can react in a timely manner to avoid a potential accident.

2 System Operation Overview

The hardware system consists mainly of an ultrasonic transmitter module, ultrasonic receiver module, control module, buzzer, and display module, as shown in Fig. 1. During operation, the ultrasonic transmitter module sends out ultrasonic waves, which

P. Krömer et al. (eds.), *Proceedings of the Fourth Euro-China Conference on Intelligent Data Analysis and Applications*, Advances in Intelligent Systems and Computing 682, DOI 10.1007/978-3-319-68527-4_24

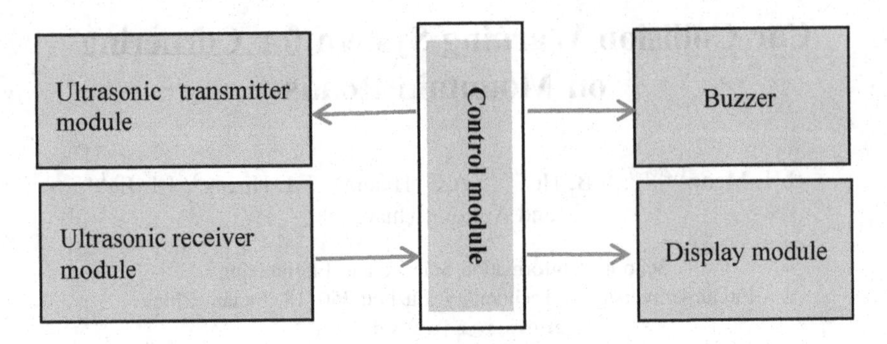

Fig. 1. Hardware system structure.

are then reflected and received by the receiver module. The microcontroller times this process and computes the distance between the vehicle and obstacle based on the calculated time. The calculated distance is then compared to the threshold pre-set by the driver, and displayed on the display module. If it is below the threshold, the buzzer will be triggered and a warning message will be displayed on the screen to notify the driver of the obstacle around the corner.

According to a previous publication, "The effect of driver's response time on driving safety and detecting system", the average driver reaction time is 766 ms [3, 4]. The Road Traffic Safety Law of the People's Republic of China specifies a maximum vehicle cornering speed of 30 km/h [5]. Based on these figures, a vehicle would travel 6.38 m from the moment at which the driver notices the obstacle, to the moment of applying the break. Another formula often used by Chinese traffic police to estimate a vehicle's initial breaking speed [6] is as follows:

$$\text{Velocity2} = 254 * u * s \tag{1}$$

where u is the ground friction coefficient and s is the vehicle skid mark length. Unusually, u is 0.7 for concrete roads and 0.6 for asphalt roads. A vehicle driving at 30 km/h would therefore have a coasting distance of 5.06 m after breaking. In total, after the driver notices the obstacle, the vehicle would normally travel 11.44 m before fully stopping. Therefore, if the radar detection range is equal to or greater than 11.44 m, it should leave enough reaction time for the driver to reduce the likelihood of an accident.

2.1 Operation of the Ultrasound Module

The TRIG port is the start port of the ultrasound module. When a high level is input, the ultrasound module starts working, and when the reflected ultrasonic waves are collected, the ECHO port outputs a high level. After receiving this, the microcontroller stops the timer and calculates the distance to the obstacle based on the calculated time. (Fig. 2)

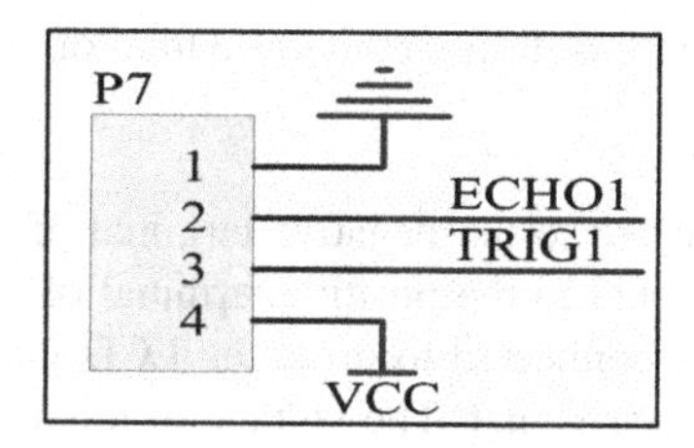

Fig. 2. Ultrasound module circuit diagram.

2.2 Control Module

The motor controller chip serves as the control module. EN3 and EN4 are motor enabling ports, and the motor can only operate after a high level has been input. IN5 and IN6 are motor control ports. When IN5 is on low and IN6 on high, the motor works in forward mode; otherwise it works in reverse mode. This is shown in Fig. 3.

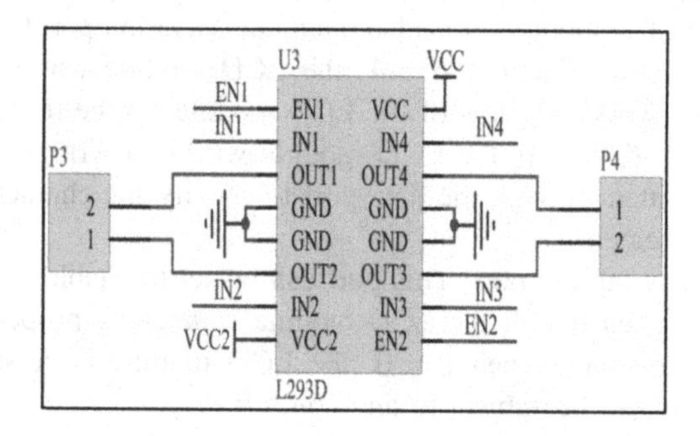

Fig. 3. Circuit diagram of the control module.

2.3 LCD Module

With the emergence of various new semiconductor materials, the liquid crystal display devices market is extremely competitive. Among them, LCD and OLED devices are very popular among consumers and developers due to their excellent performance. For this design, there is a single requirement from the LCD module; it need only display the prompt information. For this reason, the inexpensive LCD1602 module is selected as the display device for the entire system. The LCD1602 module has the following features:

1. Able to display 16 * 2 characters. This fully meets our normal display needs;
2. Low power consumption. The LCD1602 module requires only a 2 mA driving current to read and write the LCD screen, greatly saving energy and increasing the service life of other system components.
3. Simple driving circuit. The SCM can easily set the internal register of the LCD module through data analysis, without stringent requirements on the IO ports.
4. Adjustable working voltage within a certain range based on need. This ensures that it will not fail to display characters or display garbled characters when the voltage decreases.

The LCD1602 module in this design contains a total of 16 pins, and the function of each pin is shown below:

VSS: No. 1 pin, connected to the negative terminal of the power supply;

VDD: No. 2 pin, connected to the positive terminal of the power supply. Usually the 5 V power is connected to drive the LCD module;

VL: No. 3 pin, LCD bias signal. The voltage of this port can be set to adjust the display brightness of the LCD screen. Usually this port is connected with the sliding rheostat by VCC. Users can slide the rheostat to change the resistance of the sliding rheostat, and thereby change the voltage of the LCD screen port.

RS: No. 4 pin, data and command selecting port. It selects an action object based on the TTL level of the port. When RS = 1, the data register of the LCD module is operated; when RS = 0, the command register of the LCD module is operated.

R/W: No. 5 pin, read and write selecting port. This port can be set to read and write the LCD module. It is also the main operation port to enable the liquid crystal display. When R/W = 1, the LCD module works in reading state, where the working status of the LCD module can be read through the data port; when R/W = 0, the LCD module works in writing state, where data can be written through the data port to specify the characters displayed on the LCD screen.

E: No. 6 pin, enabling port. This port can be set to enable or disable the LCD module. When E = 1, the LCD module is working properly and ready for character display; when E = 0, the LCD module is closed and the LCD screen will not be subject to any control;

D0-D7: No. 7–14 pins, 8-bit data ports. Data can be written to data ports through any one of SCM ports P0, P1, P2, and P3.

BLA: No. 15 pin, the positive pole of backlight. Usually it is connected to the positive pole of a 5 V voltage to supply power for backlight;

BLK: No. 16 pin, the negative pole of backlight. Usually it is connected to the negative pole of a 5 V voltage to supply power for backlight

2.4 Main Program Flow and Principle

The main program primarily involves transmission and receipt of the hr-sr041 ultrasound module, the timing, whistle, LCD display, and the safety distance setting, using a keyboard.

When the system is started, After the ultrasound module receives the echo signal, the system will calculate the distance based on the result from the timer and check if the radar alarm is on. If it is not, directly display the calculated distance on the LCD, where the radar status is also “off”. If the radar alarm function is on and the calculated distance is smaller than the pre-set safety distance, the buzzer will whistle to remind the driver.

3 System Test and Debugging Data

The 1st set of data were obtained from a test case where the obstacle was set 300 cm in front of the vehicle, and the system warning safety distance was set to 0.5 m. The system successfully triggered the alarm in the 255–270 cm range, as shown in Table 1.

Table 1. Test data: obstacle set in front of vehicle with a safety distance of 0.5 m

Preset safety distance	Number of times	Vehicle stopping distance (cm)
0.5 m	1	264
0.5 m	2	260
0.5 m	3	264
0.5 m	4	270
0.5 m	5	260

In the 2nd test case, the obstacle was set 300 cm in front of the vehicle and the system warning safety distance was set to 1 m. The system successfully triggered the alarm in the 196-210 cm range, as shown in Table 2.

Table 2. Table captions should be placed above the tables.

Preset safety distance	Number of times	Vehicle stopping distance (cm)
1 m	1	201
1 m	2	201
1 m	3	196
1 m	4	196
1 m	5	196

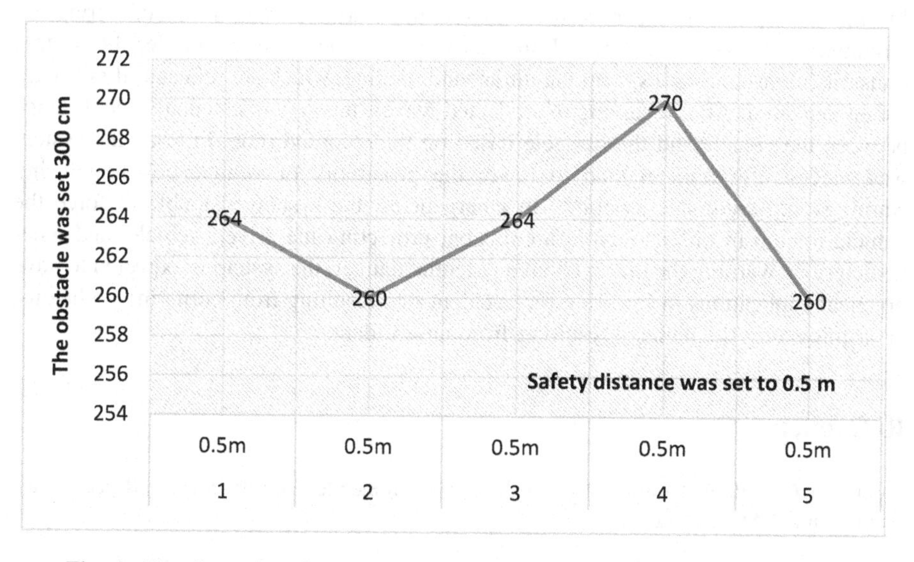

Fig. 4. Test data: obstacle set in front of vehicle with a safety distance of 0.5 m.

The data from the system test and debugging shown in Tables 1 and 2 are displayed in the line charts below. In this study, Figs. 4 and 5 show that the sensing success rates for all test cases reached almost 100%.

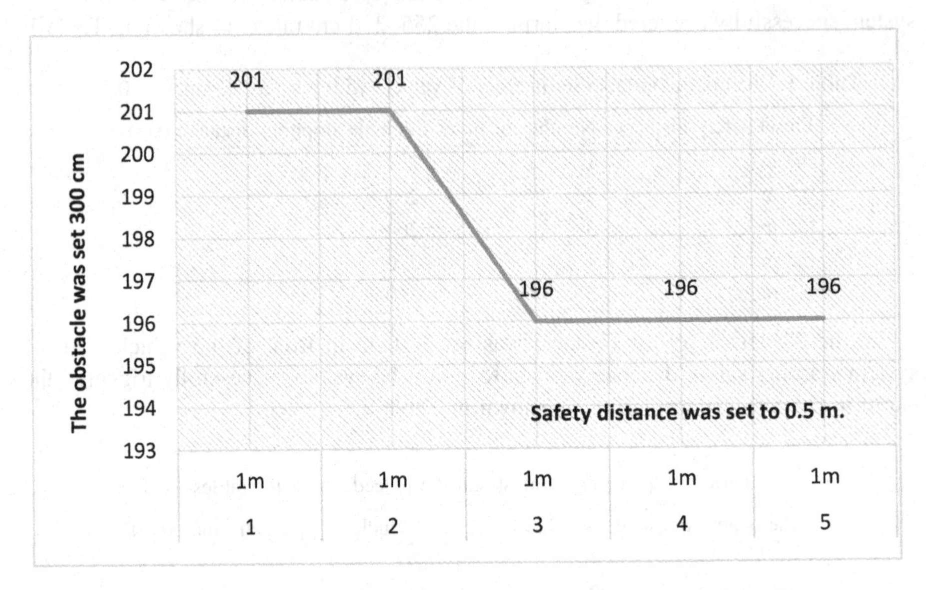

Fig. 5. Test data: obstacle set front of vehicle with a safety distance of 1 m.

4 Conclusion

In this study, a smart car collision avoidance/warning system for cornering on mountain roads was developed. Using stc89c52 as the control module, the system transmits ultrasonic waves from the ultrasound module, which are reflected if obstacles are encountered. After receiving the reflected waves, the system calculates the distance between the vehicle and the obstacle based on the recorded time. In tests, the system was successfully implemented to lower the possibility of vehicle collision, using warnings and automatic stopping. In emergencies, the system was able to force the vehicle to stop in order to avoid a collision, protecting the driver, vehicle, and other vehicles. By warning the driver about a potential danger, the system is expected to have practical applications in lowering the accident rate resulting from cornering collisions, or in mitigating the damage resulting from an accident.

References

1. Zhiyan Consulting Group. 2016–2022 China Automobile Industry Market Research and Investiment Strategy Report

2. Meng, S.H., Huang, A.C., Huang, T.J., Cai, Z.M., Ye, Q.Z., Zou, F.M.: Ultrasonic ranging-based vehicle collision avoidance system. In: Proceedings of the 12th International Conference on Intelligent Information Hiding and Multimedia Signal Processing (IIH-MSP-2016), 21–23 November 2016, pp. 13–19. Springer Heidelberg (2016). doi:10.1007/978-3-319-50212-0_26. ISBN: 978-3-319-50212-0
3. Cheng, Y.: Automobile Driving Psychology. Chinese People's Liberation Army Publishing House, Beijing (1989)
4. Lanxin, G., Songnan, L.: Research on reaction time of driver. Energy Conserv. Environ. Prot. Transp. **2**, 25–29 (2015)
5. Regulation on the Implementation of the Road Traffic Safety Law of the People's Republic of China. New Laws Regul. (6), 20–28 (2004)
6. Pengfei, G., Jianbang, D.: Skid mark-based vehicle speed calculation method and its application in traffic accident management. J. Jiangsu Police Officer Coll. **18**(3), 182–185 (2003)

Relay Node Selection Strategy for Wireless Sensor Network

Lingping Kong[1], Jeng-Shyang Pan[1,2(✉)], Shu-Chuan Chu[3], and John F. Roddick[3]

[1] Innovative Information Industry Research Center, Harbin Institute of Technology Shenzhen Graduate School, Shenzhen, China
jengshyangpan@fjut.edu.cn

[2] Fujian Provincial Key Laboratory of Big Data Mining and Applications, College of Information Science and Engineering, Fujian University of Technology, Fujian, China

[3] School of Computer Science, Engineering and Mathematics, Flinders University, Adelaide, Australia

Abstract. In the past few years, people have experienced advanced interest in the potential use of wireless sensor network in applications such as environment surveillance, military field protection and medical treatment. Usually hundreds even thousands of sensors are scattered randomly in remote environment. In generally, for the scalability of sensor network, cluster techniques are often used to group nodes into several sets. And in this paper, we proposed a new method called GF_CENTER to select the positions of centers, which is also a kind of *k*-center problem. And a new fitness function is presented for optimize the resolution. We compare our method with two other algorithms. GF_CENTER method minimizes the number of centers in the network. From the experiments, GF_CENTER find smaller number of centers than the other methods, which lower the construction fee of network.

Keywords: Wireless sensor network · Genetic algorithm · Farthest first traversal

1 Introduction

Wireless sensor networks (WSNs) are networks of sensor nodes, those nodes are randomly or customized scattered over a field for the purpose of monitoring certain environment, detecting area pollution. Usually there is one Sink node in the center of network, which is responsible for collecting and processing data. In the sensor network, all the sensor nodes work cooperatively for sensing interest packet and re-transmitting the packet to Sink node through one or more steps. Every sensor node performs measurements, sense and communicating over a certain equipped device. For most sensor network, this many-to-one communication pattern is more common [1], but the vital flaw is hard to resolve. The inner nodes cost more energy than the outer nodes during the same period time, because the inner nodes need to send their own packet and they are also required to re-send the outer packets to Sink node due to network

P. Krömer et al. (eds.), *Proceedings of the Fourth Euro-China Conference on Intelligent Data Analysis and Applications*, Advances in Intelligent Systems and Computing 682, DOI 10.1007/978-3-319-68527-4_25

transferring structure. This condition easily causes running out of energy resulting in the death of inner node. The problem is hard to avoid, part of the reason is that most of the sensor nodes in network are powered by a battery, which cannot be re-charged or re-placed. The solar power or other natural energy is not stable for energy harvesting strategy [2]. The worst is that all the outer nodes will be forced to stop working even if they are full of energy, if there are no inner nodes available, because there is no route for transmitting outer event packets [3]. For part of this reason, the hierarchical wireless sensor network is evolving. It is very important for network scalability and stability. Clustering technique is to group sensor nodes into several sets, every sensor node is distributed to its closest cluster and all the interest packets are transmitted to the cluster head instead of Sink node. In the past years, a lot of papers about cluster techniques are proposed. Ameer presents a taxonomy and general classification of published clustering schemes, and the author also compares to the previous clustering algorithms based on metrics such as convergence rate, cluster stability, cluster overlapping, supporting for node mobility and location awareness characteristics [4].

The rest of the paper is organized as follows. In Sect. 2 we discuss some of related work, and introduce a few *k*-center problem resolution strategies and their related researches. Section 3 presents the proposed algorithm GF_CENTER in detail. In Sect. 4 several experiments are tested to compare the performance of these simulated algorithms. Finally the conclusion part will be summarized in Sect. 5.

2 Related Works

There is a similar problem to cluster technique, which is *k*-center problem. Instead of group nodes into several clusters, *k*-center problem is to pick *k* positions from a set of pivots that these *k* pivots could cover all the rest of pivots in the area. And the number of *k* should be as few as possible. This is the original *k*-center problem. Recently, some generalizations of the *k*-center problem are considered in the literature [5–7]. One generalization for the *k*-center problem, named as capacitated *k*-center problem, was introduced in [8]. In this kind of problem, it is required to locate *k* centers in a certain area, and assign other points into its closest center, but the condition is that each center should have a limited number of dominated points. Andreas considered the more general variant of the *k*-center problem. The author shows both approximation and fixed-parameter hardness results, and also provides similar fixed-parameter approximations for the weighted *k*-center problems. As in the paper [9], the author consider a restricted covering problem, in which a convex polygon G with n vertices and an integer k are given, the objective is to cover the entire region of G using k congruent disks of minimum radius r, centered on the boundary of G. In the paper [10], the author considers the widely used *k*-center clustering problem and its variant used to handle noisy data, *k*-center with outliers. The algorithm is fast, memory efficient and matches their sequential counterparts in distributed settings.

In this paper, GF_CENTER is designed to locate up to k nodes that are work as cluster center, and every other common node connects at least one center. The aim is to minimize the number of center node and maximum the distance between centers and common nodes. There are many algorithm are proposed to solve point selection problem. One of the widely and popular algorithms used is farthest first traversal (*FF*), which is introduced by Gonzalez [11]. The first center is chosen by randomly, and the successive point centers are selected by the rules, which have the longest distances with the previous centers. This is the brief of farthest first traversal idea. The other one is a variant of minimum dominating set (*DO*) [12].

3 GF_CENTER Implement

This part we introduce a new proposed method for center selection problem, which is called GF_CNETER based on genetic algorithm and Farthest-first traversal. Genetic algorithm is a search heuristic in the field of artificial intelligences [13, 14], which generate solutions to optimization problems using process like in natural evolution, such as selection, mutation, crossover and inheritance.

3.1 The Annotation of Symbol

Table 1 summarizes the symbols used in the paper.

Table 1. Annotation of symbol

Symbol	Annotation
N	The number of total sensor nodes
M	The population of chromosomes
F_N	The chosen nodes of farthest first traversal
r	Mutation rate
$\alpha_1 \alpha_2$	Constant parameter
Pick_N	The number of value '1' in one chromosome
Cov_N	The number of nodes covered by value '1' representation nodes in one chromosome
D_ns	The total distance between all nodes to Sink node

3.2 The Process GF_CENTER

- First, sort all the nodes and give each node an order number from '1' to '*N*', then pick a digit randomly, use this node which the digit presented as the start point of farthest first traversal to generate one solution. In the farthest first traversal operation, we choose the next center that has fixed ratio distance from the previous centers. Mark the solution as *F_N*. Repeat the process *M* times with *M* different random digit, then we can get *M* set of *F_N*.
- Initialization: Initialize *M* chromosomes, where each chromosome is a *N*-bit binary string with value '0' and '1, which *N* is the number of sensor nodes. Each bit in the chromosome represents a sensor node in the network, the value '0' bit represents a common node, and value '1' bit denotes that this bit represented node is a center. The best chromosome with the optimization fitness value is the final result. One

F_N generates one chromosome, and the value '1' bits are just the nodes from *F_N*. See Fig. 1.

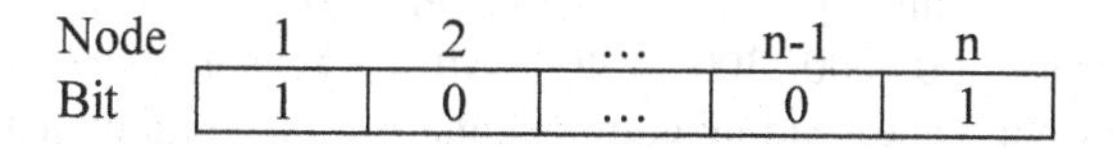

Fig. 1. A chromosome structure

- Evaluation. Get the fitness values of *M* chromosomes using Eq. 1. Store the best one with the best fitness value.
- Parents selection techniques. Simply divides the population into two sets, Set *A* and Set *B*, Set *A* has those chromosomes with better fitness values. The other half are stored in Set *B*. The father and mother chromosomes are from different sets.
- Crossover process. Let *C1* (mother) $C_{i,1}, .., C_{i,n}$ and *C2* (father) $C_{j,1}, .., C_{j,n}$ be the parents chromosomes. Use one middle-point operator, the child *C3* strings of *C1* and *C2* is

$$C_3 := C_{i,1}, C_{i,2}, ..C_{i,\lceil n/2 \rceil}, C_{j,\lceil n/2 \rceil +1}, C_{j,\lceil n/2 \rceil +2}, .., C_{j,n}$$

- Mutation process. Mutation process [15] works by inverting a bit value in the chromosome with a small probability. Mutation rate is 0.02. The following Fig. 2 shows the transformation.

Before	Node	1	2	...	n-1	n
mutation	Bit	1	0	...	0	1
After	Node	1	2	...	n-1	n
mutation	Bit	1	1	...	0	1

Fig. 2. Mutation process

3.3 Fitness Function

A fitness function is an objective function that is used to evaluate the evolution solution. This function includes *Pick_N*, *Cov_N* and *D_ns* parameters. *Pick_N* is the number of value '1' bits. *Cov_N* is the summation of value '0' bits, and those value '0' bits need to satisfy the following conditions: One, each value '0' bits is only counted once. Two, those bits represented nodes can connect to at least one value '1' bit represented node. *D_ns* is the sum of distance from all the nodes to the Sink. The two parameters α_1, α_2 are constants are 0.5 and 0.5.

$$F = \alpha_1 \times Cov\_N + \alpha_2 \times \frac{D\_ns}{Pick\_N} \quad (1)$$

4 Experiment Analysis

The size of simulation wireless sensor network is a 200 × 200 square area, all sensor nodes are randomly scattered in the environment. The number of sensor nodes in network differs from 100, 200, 300 to 400. All the sensor nodes equipped with the same sensing, receiving and sending components. But the communicating radius keeps shorter after the growing number of sensor nodes deployed in network. 40 units communicating radius for 100-nodes network, and 30 units radius for 200-node network, 25 units and 20 units is for 300-node and 400-nodes network corresponding. The simulation chromosome population number is 20.

The Table 2 shows the result of the amount of centers by three methods, farthest first traversal and Dominating set and GF_CENTER. For 100-node network, GF_CENTER only uses 16 centers to cover the rest of nodes in network, and the other methods use 18 centers at least.

Table 2. The number of center results of three algorithms

Algorithm	Number of nodes			
	100	200	300	400
GF_CENTER	16	27	42	62
Farthest_First	19	29	47	67
Dominating set	18	29	44	64

The experiments run 42 times in the same condition except for the different random seed to test the performance of GF_CENTER. Table 3 lists the maximum value (Max), minimum value (Min) and standard deviation (StD) and average value (AVG) of our scheme. In the best case, it only uses 13 centers to build a full coverage sensor network, and 16 centers for the worst case. However even for the worst case, it is still better than the other two methods, FF needs 19 and DO needs 18. Table 4 shows the results of fitness value in three methods, and if the fitness value is bigger, it is better.

Table 3. The stability of GF_CENTER algorithm

Value	Number of nodes			
	100	200	300	400
Max	16	27	42	62
Min	13	25	37	57
AVG	15.119047	26.571428	41.071428	60.190476
StD	0.748826	0.630248	1.197413	1.292343

The following figures are the simulation configuration of GF_CENTER routing structure. The node in blue in the center is Sink node. Those nodes in red are centers and they can communicate with Sink node directly and collect the data from common sensing node. The rest of nodes in black are common sensing nodes (Fig. 3).

Table 4. Fitness values of three methods

Algorithm	Fitness values			
	100	200	300	400
GF_CENTER	341.61	401.23	437.10	442.60
Farthest_first	244.47	355.96	367.08	397.14
Dominating set	256.30	355.96	384.98	409.45

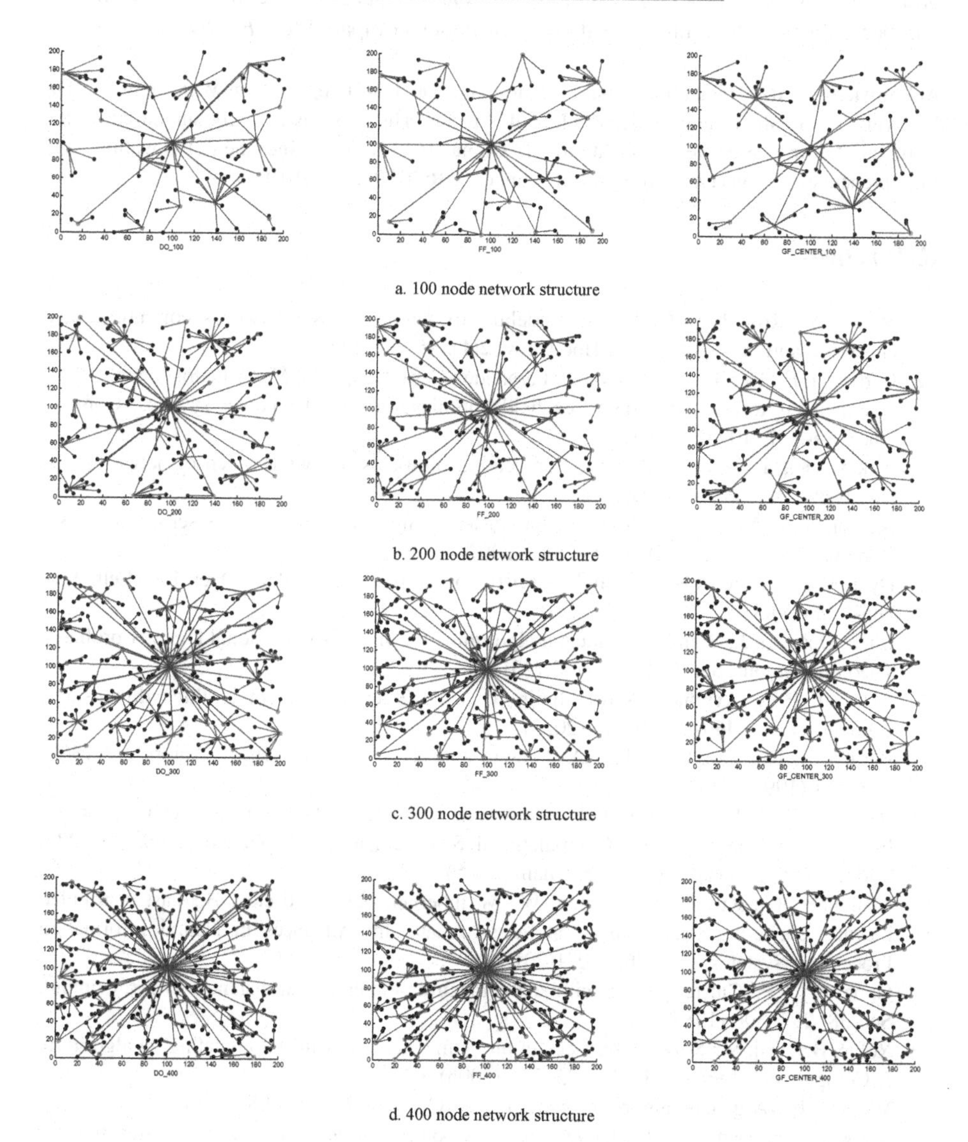

Fig. 3. Nodes network structure

5 Conclusion

This paper proposed a new method for solving node selection problem, and applying it into wireless sensor network for clustering the sensor nodes. GF_CENTER works based on genetic algorithm and farthest first traversal, the special design of GF_CENTER lower the times of repeated fitness function evaluation for complex problems. The experimental results show that GF_CENTER could find a smaller number of centers to connect all the rest of nodes compared to *FF*, *DO* methods.

Acknowledgment. The authors would like to thank Dr. Ivan Lee, Dr. Tien-Wen Sung and Dr. Pei-Wei Tsai for their valuable feedback to this paper. This work is partially supported by the National Natural Science Foundation of China (61371178), Shenzhen Innovation and Entrepreneurship Project with the project number: GRCK20160826105935160.

References

1. Mhatre, V., Rosenberg, C.: Design guidelines for wireless sensor networks: communication, clustering and aggregation. Ad Hoc Netw. **2**(1), 45–63 (2004)
2. Carrizosa, E., Domínguez-Bravo, C., Fernández-Cara, E., Quero, M.: A heuristic method for simultaneous tower and pattern-free field optimization on solar power systems. Comput. Oper. Res. **57**, 109–122 (2015)
3. Akkaya, K., Younis, M.: A survey on routing protocols for wireless sensor networks. Ad Hoc Netw. **3**(3), 325–349 (2005)
4. Abbasi, A.A., Younis, M.: A survey on clustering algorithms for wireless sensor networks. Comput. Commun. **30**(14), 2826–2841 (2007)
5. Dyer, M.E., Frieze, A.M.: A simple heuristic for the p-centre problem. Oper. Res. Lett. **3**(6), 285–288 (1985)
6. Chen, D.Z., Li, J., Wang, H.: Efficient algorithms for the one-dimensional k-center problem. Theoret. Comput. Sci. **592**, 135–142 (2015)
7. Agarwal, P.K., Avraham, R.B., Sharir, M.: The 2-center problem in three dimensions. Comput. Geom. **46**(6), 734–746 (2013)
8. Barilan, J., Kortsarz, G., Peleg, D.: How to allocate network centers. J. Algorithms **15**(3), 385–415 (1993)
9. Basappa, M., Jallu, R.K., Das, G.K.: Constrained k-center problem on a convex polygon. In: International Conference on Computational Science and Its Applications, pp. 209–222. Springer International Publishing, Cham (2015)
10. Malkomes, G., Kusner, M.J., Chen, W., Weinberger, K.Q., Moseley, B.: Fast distributed k-center clustering with outliers on massive data. In: Advances in Neural Information Processing Systems, pp. 1063–1071(2015)
11. Gonzalez, T.F.: Clustering to minimize the maximum inter-cluster distance. Theor. Comput. Sci. **38**, 293–306 (1985)
12. Robič, B., Jurij, M.: Solving the k-center problem efficiently with a dominating set algorithm. J. Comput. Inf. Technol. **13**(3), 225–234 (2005). CIT
13. Whitley, D.: A genetic algorithm tutorial. Stat. Comput. **4**(2), 65–85 (1994)
14. Thi-Kien, D., Tien-Szu, P., Nguyen, T.-T.: A compact articial bee colony optimization for topology control scheme in wireless sensor networks. J. Inf. Hiding Multimed. Sig. Process. **6**, 297–310 (2015)
15. Güçyetmez, M., Çam, E.: A new hybrid algorithm with genetic-teaching learning optimization (G-TLBO) technique for optimizing of power flow in wind-thermal power systems. Electr. Eng. **98**(2), 145–157 (2016)

Adaptive Signal Processing of Fetal PCG Recorded by Interferometric Sensor

Radek Martinek[1(✉)], Radana Kahankova[2], Jan Nedoma[1], Marcel Fajkus[1], Homer Nazeran[3], and Jana Nowakova[4]

[1] Department of Cybernetics and Biomedical Engineering, Faculty of Electrical Engineering and Computer Science, VSB-Technical University of Ostrava, Ostrava, Czech Republic
radek.martinek@vsb.cz

[2] Department of Telecommunications, Faculty of Electrical Engineering and Computer Science, VSB-Technical University of Ostrava, Ostrava, Czech Republic

[3] Department of Electrical and Computer Engineering, University of Texas El Paso, 500 W University Ave, El Paso, TX 79968, USA

[4] Department of Computer Science, Faculty of Electrical Engineering and Computer Science, VSB-Technical University of Ostrava, Ostrava, Czech Republic

Abstract. This paper is focused on the design, implementation, and verification of an adaptive system for processing of the fetal phonocardiogram (*f*PCG) recorded by the novel interferometric sensor. The main interference to be suppressed in the abdominal signal is the maternal phonocardiogram (*m*PCG). In this article, adaptive methods based on Least Mean Square and Recursive Least Square algorithms are used for the elimination of the maternal component. Evaluation of the filtration quality is provided using the objective parameters (Signal Noise to Ratio, Sensitivity, and Positive Predictive Value).

Keywords: Interferometer · Non-invasive measurements · Fetal heart rate (*f*HR) · Maternal heart rate (*m*HR) · EMI-free · Adaptive system · Least mean squares (LMS) algorithm · Recursive least squares (RLS) algorithm · Fiber optic sensors · Fetal Phonocardiography (*f*PCG)

1 Introduction

Fetal Phonocardiography (*f*PCG) was discovered in the 17th century [1, 2]; however, the interest in this research has only occurred over last few years. It was also enabled by the introduction of the Electronic Fetal Monitoring (EFM). In today's obstetric, EFM is used in most of the labors and besides *f*PCG, it includes fetal Electrocardiography (*f*ECG), Pulse Oximetry, Magnetocardiogram (*f*MCG), and Cardiotocography (CTG) [3–5].

Currently, *f*PCG is used only as a secondary tool for diagnosing the fetal heart rate (*f*HR). The main reason is that the acquired signal is a mixture of acoustic and pressure components from the fetus, the mother, and other noise sources. Therefore, it is

P. Krömer et al. (eds.), *Proceedings of the Fourth Euro-China Conference on Intelligent Data Analysis and Applications*, Advances in Intelligent Systems and Computing 682, DOI 10.1007/978-3-319-68527-4_26

generally very noisy. Moreover, the recorded signal is highly dependent on the location of the data acquisition, gestational age, fetal and maternal position, etc. [6, 7].

This article focuses on the processing of the noisy signal in order to suppress the most significant unwanted component, i.e. maternal PCG (*m*PCG). Since it is possible to measure the undesired signal (by using thoracic lead), the adaptive algorithm can be used for suppression of such signal. Additionally, the paper deals with processing of a signal recorded by a novel approach – using the interferometric sensor. Consequently, this signal differs from a conventional PCG signal recorded by means of a microphone. Since it is based on optical, not electrical principle, it can be used even in environments where the usage of conventional methods is limited, e.g. magnetic resonance or in case of water birth. This is an early stage of the research, and thus synthetic data were used for the first experiments. After a complicated legislative process and approved clinical tests, the novel sensor can be utilized for the measurement on the pregnant women. These signals will be used for the verification of the results obtained using synthetic data.

2 Methods

The aim of this paper is to design an adaptive system for elicitation of the maternal component from the abdominal signal. We used synthetic data to assess the filtration quality using objective methods, where the reference signal is needed. Synthetic data utilized for the experiments were created on the base of [8] and also by means of modified *f*ECG signal generator introduced in [9, 10], which is now dynamical. To ensure the most accurate synthetic data, the measurements on non-pregnant women were carried out. This way, physiological and pathological data were created, both 600 s long.

The main positive of utilization of the synthetic data is that they allow obtaining the reference (ideal) *f*PCG signal for an objective verification of the designed system. The research is in its initial phase must go through a complicated legislative process and approved clinical tests. Using real data for the verification is therefore not possible at present.

2.1 Non-invasive Interferometric Measurement Sensor

Proposed non-invasive interferometric sensors is comprised of Mach-Zehnder fiber optic interferometer formed by 1×2 and 2×1 power couplers with an even split ratio. Interferometric sensors belong to the highest-performance group of optical sensors as they are capable of measuring even tiny differences in the fiber core refractive index and in the optical fiber length. The optical laser source (with bandwidth $\approx$0.01 nm and output power $\approx$1 mW is divided into two fiber arms forming the reference and the measurement arms. The measurement arm is encapsulated into polymer polydimethylsiloxane (PDMS), see Fig. 1. The PDMS polymer does not affect the function of the fiber optic sensors, see literatures [11, 12]. The reference arm must stay in a stable environment.

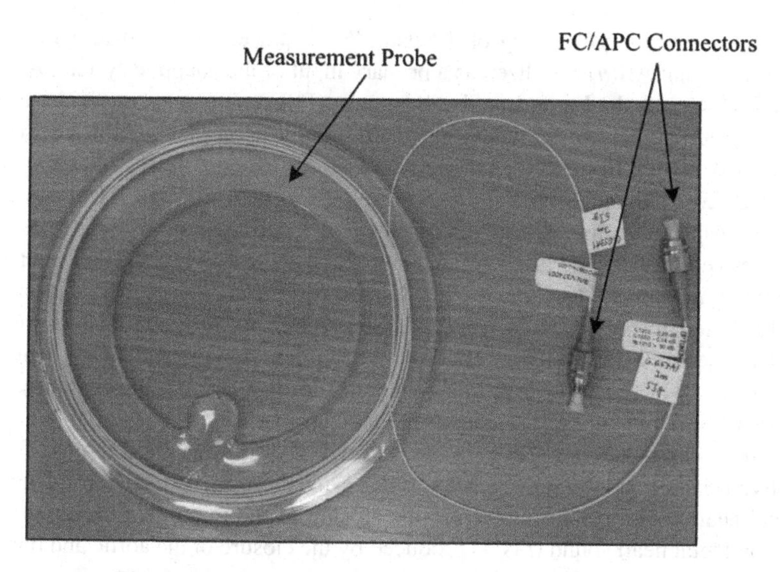

Fig. 1. Non-invasive interferometric measurement probe.

3 Experimental Setup

Figure 2 shows an adaptive system for *f*PCG extraction. It consists of two interferometric sensors – IS_T and IS_A. IS_T placed on the maternal thorax, the recorded signal TM(n) is preprocessed and then used as a reference input of the adaptive system since it is considered to be completely maternal. Optical interrogator system and DSP unit consists of Optical Spectrum Analyzer (OSA) with the sampling frequency of 1 kHz and Digital control circuits. The further processes such as digitization, amplification and filtering are realized in the preprocessing part. Preprocessing part is realized by means of the well-known techniques, which is thoroughly described in literature [13–16]. The digitization and amplification of the fetal and maternal signals are realized by means of the card NI-USB 6210, which has an 8-bit Analog-to-Digital Convertor with a built-in amplifier enabling the choice of amplification gains in steps of 1, 10 and 50. The associated software support development environment of this card, LabVIEW,

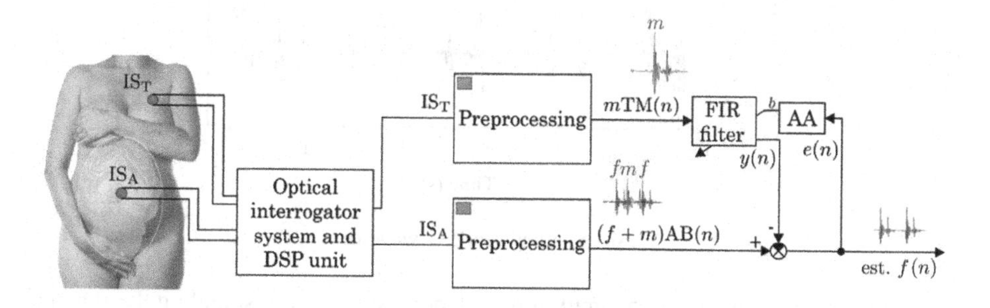

Fig. 2. Non-invasive interferometric measurement scheme.

enables further digital processing of the data. IS_A is placed on the maternal abdomen and records signal AB(n) is utilized as a primary input of the adaptive system. Adaptive algorithm (AA) sets the Finite Impulse Response (FIR) coefficients on the base of the backpropagated signal $e(n)$. The output of the filter $y(n)$ is then subtracted from the preprocessed signal from IS_A, AB(n), which contains both fetal (f) and maternal (m) component. The output signal $f(n)$ corresponds to the estimated *f*PCG signal.

In this paper, two representatives of adaptive algorithms were tested. LMS as a representative of stochastic gradient adaptation and RLS as a representative of recursive optimal adaptation [17–20]. The main difference between these two approaches is that RLS-based methods recursively find the coefficients that minimize a weighted linear least squares cost function relating to the input signals, which are considered deterministic, while LMS-based methods aim to reduce the mean square error (MSE) and the input signals are considered stochastic.

Figure 3 shows the ideal *f*PCG waveform, which serves as the reference for both subjective (visual) and objective (using SNR) evaluation of the filtration quality. The first fetal heart sound (fS_1) results from the closing of the mitral and tricuspid valves. The second fetal heart sound (fS_2) is produced by the closure of the aortic and pulmonic valves.

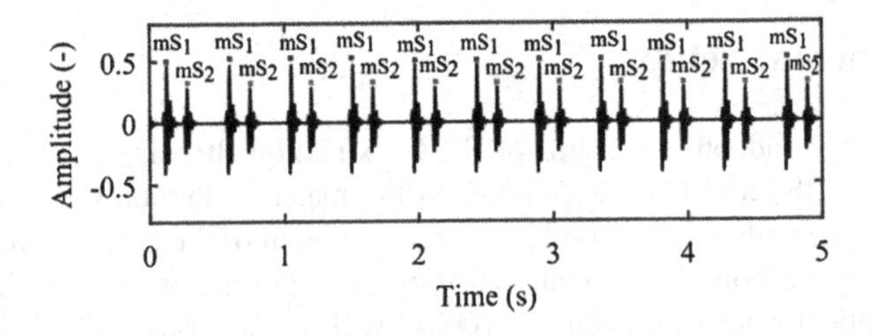

Fig. 3. Ideal *f*PCG waveform.

Figure 4 shows the ideal *m*PCG signal measured by thoracic electrodes. According to Fig. 2, it is the reference input of the adaptive system which is considered to be only maternal, i.e. does not include fetal component. Maternal first and second heart sounds are denoted as mS_1 and mS_2, respectively.

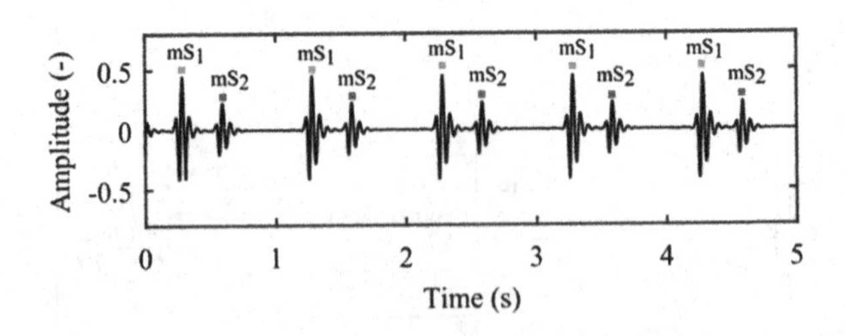

Fig. 4. Synthetic *m*PCG waveform.

Figure 5 shows the primary input of the adaptive system corresponding the scheme in Fig. 2. The signal is recorded by abdominal lead; the signal is composed of the maternal and fetal component. It is important to note that the maternal component differs from the *m*PCG recorded by thoracic electrodes due to its passing through the maternal body. It influences the signal, and therefore the maternal component cannot be suppressed by a simple subtraction of the reference *m*PCG from composed signal. For determining the fetal heart rate (*f*HR), it is necessary to detect fS_1 which is difficult without the filtration. All of the signals were plotted for the tested signal, where $SNR_{IN} = -4$ dB (see Table 1).

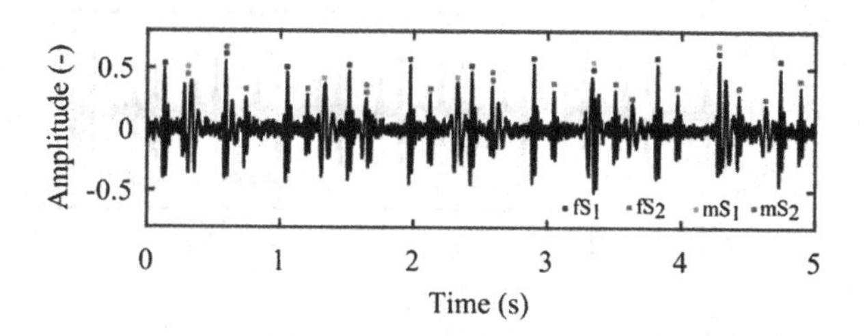

Fig. 5. Abdominal PCG recorded by IS_A (*a*PCG = *f*PCG + *m*PCG, $SNR_{IN} = -4.00$ dB).

Table 1. Results of the tested algorithm.

	LMS			RLS		
SNR_{IN} (dB)	SNR_{OUT} (dB)	S^{+}_{fHR} (%)	SNR_{IN} (dB)	SNR_{OUT} (dB)	S^{+}_{fHR} (%)	SNR_{IN} (dB)
−1.00	3.248	97.16	−1.00	3.248	97.16	−1.00
−2.00	3.237	96.71	−2.00	3.237	96.71	−2.00
−3.00	3.019	96.40	−3.00	3.019	96.40	−3.00
−4.00	1.847	96.12	−4.00	1.847	96.12	−4.00
−5.00	1.057	96.04	−5.00	1.057	96.04	−5.00
−6.00	1.246	94.86	−6.00	1.246	94.86	−6.00
−7.00	0.957	91.91	−7.00	0.957	91.91	−7.00
−8.00	0.315	91.47	−8.00	0.315	91.47	−8.00

Figure 6 shows the output of the adaptive system based on LMS algorithm. It is clear that the maternal component was successfully suppressed. However, the filtration also affected the fS_1, i.e. the fetal component was suppressed as well. That influences

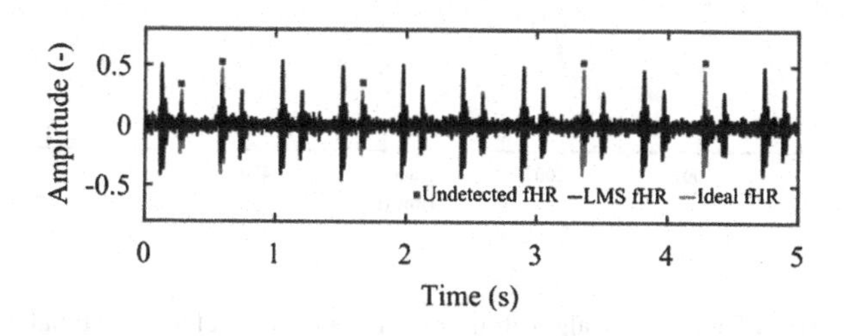

Fig. 6. Output of the adaptive system based on LMS algorithm.

the determination of *f*HR, which was decreased. The filter parameters (step size and filter length) were set empirically (considering the value of SNR) on the base of authors' experiences ($\mu = 0.025$, $N = 31$).

Figure 7 shows the output of the adaptive system based on RLS algorithm. Compared to LMS algorithm, it is more computationally intensive but also more accurate since it does not suppress the desired fetal component (fS_1). Based on the

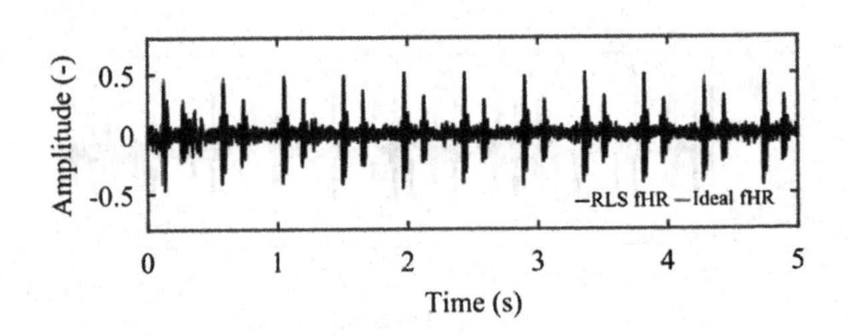

Fig. 7. Output of the adaptive system based on RLS algorithm.

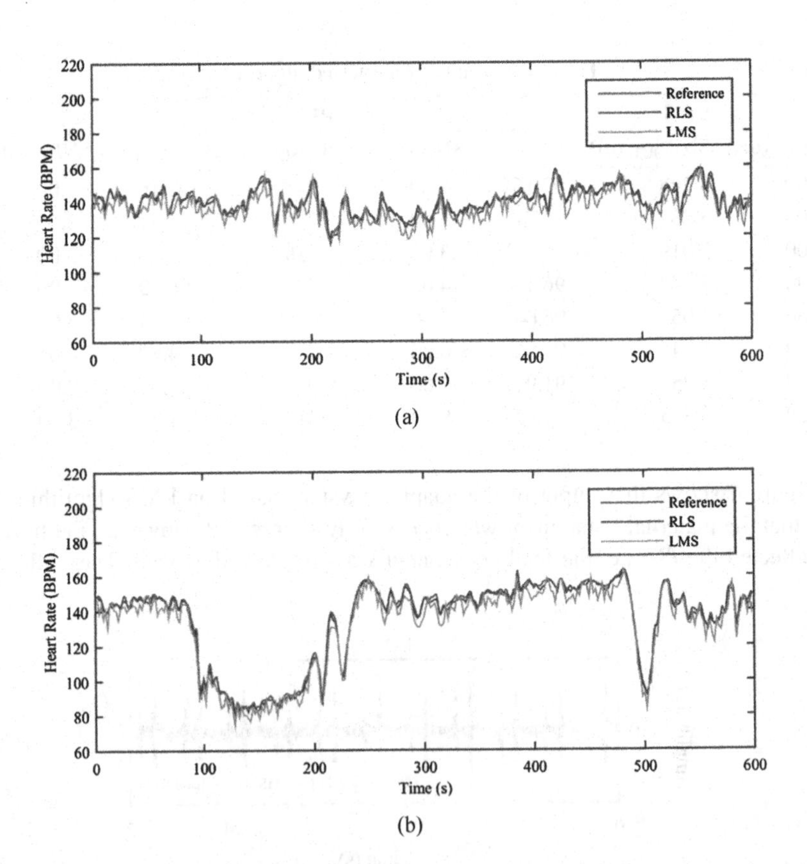

Fig. 8. Comparison of the tested algorithms outputs with the reference signal for (a) physiological and (b) pathological type of data.

authors' experience, the forgetting factor and the filter length was set as $\lambda = 0.9479$, $N = 47$, respectively.

Finally, Fig. 8 shows the results of the determined *f*HR of both tested algorithms in comparison with the ideal (reference) *f*PCG signal. The results confirm that the LMS algorithm causes a decrease of the estimated *f*HR since it eliminates some of the desired fS_1 besides undesired maternal sounds. Moreover, Fig. 8a shows that for RLS algorithm, the estimated *fHR* does not significantly differ from the reference one. Authors also tested the algorithms on the pathological data (see Fig. 8b). The results show that the type of data does not affect the filtration quality. Noting that the detection was performed according to [19].

Table 1 shows the results of the experiments for LMS and RLS algorithm, respectively. The performance was accessed by the difference between input and output SNR, Sensitivity (S^{+}_{fHR}) and the Positive Predictive Value (PPH_{fHR}) that are associated with the success in detection of fHR. When considering the values of SNR_{OUT}, the algorithms achieve approximately the same results. Nevertheless, the values of S^{+}_{fHR} and PPH_{fHR} confirm that the RLS algorithm overcomes the LMS algorithm since it was more accurate in the *f*HR detection. It is also obvious that the higher is the SNR, the lower the performance of the LMS algorithm. In contrast, RLS algorithm is effective even for noisy signals.

4 Conclusion

It this paper, an adaptive system was designed and implemented for processing the PCG signal recorded utilizing a novel interferometric sensor. The systems based on LMS and RLS algorithms were tested for the *f*PCG extraction. The system based on RLS algorithm significantly overcame the LMS algorithm; therefore, it is more suitable for suppressing the maternal component. The algorithms were tested on synthetic data. In the future research, the data from clinical practice should be utilized. Moreover, the future research should involve the influence of the sensor placement, the fetal position, and the gestation age on the quality of the filtration. Another important part of the research is the optimization of the filter settings, which were only set empirically for the experiments in this paper.

Acknowledgements. This article was supported by the Ministry of Education of the Czech Republic (Projects Nos. SP2017/128 and SP2017/79). This research was partially supported by the Ministry of Education, Youth and Sports of the Czech Republic through Grant Project no. CZ. 1.07/2.3.00/20.0217 within the framework of the Operation Programme Education for Competitiveness financed by the European Structural Funds and from the state budget of the Czech Republic.

References

1. Sartwelle, T.P.: Electronic fetal monitoring: a bridge too far. J. Legal Med. **33**, 79–313 (2012). doi:10.1080/01947648.2012.714321

2. Tan, B.H., Moghavvemi, M.: Real time analysis of fetal phonocardiography. In: Proceedings of the IEEE, Region 10 Annual International Conference, New York, pp. II-135–II-140. IEEE Press (2000)
3. Adithya, P.C., Sankar, R., Moreno, W.A., Hart, S.: Trends in fetal monitoring through phonocardiography: challenges and future directions. Biomed. Sig. Process. Control **33**, 289–305 (2017). doi:10.1016/j.bspc.2016.11.007
4. Liu, C., Springer, D., Li, Q., Moody, B., Juan, R.A., Chorro, F.J.: An open access database for the evaluation of heart sound algorithms. Physiol. Measur. **37**(12), 2181–2213 (2016). doi:10.1088/0967-3334/37/12/2181
5. Tang, H., Li, T., Qiu, T., Park, Y.: fetal heart rate monitoring from phonocardiograph signal using repetition frequency of heart sounds. J. Electr. Comput. Eng. **2016**, 2404267 (2016). doi:10.1155/2016/2404267
6. Chourasia, V.S., Tiwari, A.K., Gangopadhyay, R.: A novel approach for phonocardiographic signals processing to make possible fetal heart rate evaluations. Digit. Sig. Proc. **30**, 165–183 (2014). doi:10.1016/j.dsp.2014.03.009
7. Jezewski, J., Roj, D., Wrobel, J., Horoba, K.: A novel technique for fetal heart rate estimation from Doppler ultrasound signal. Biomed. Eng. Online **10**(1), 92 (2011). doi:10.1186/1475-925X-10-92
8. Cesarelli, M., Ruffo, M., Romano, M., Bifulco, P.: Simulation of foetal phonocardiographic recordings for testing of FHR extraction algorithms. Comput. Methods Programs Biomed. **107**(3), 513–523 (2012). doi:10.1016/j.cmpb.2011.11.008
9. Martinek, R., Kelnar, M., Koudelka, P., Vanus, J.: A novel LabVIEW-based multi-channel non-invasive abdominal maternal-fetal electrocardiogram signal generator. Physiol. Measur. **37**(2), 238–256 (2016). doi:10.1088/0967-3334/37/2/238
10. Martinek, R., Kelnar, M., Vojcinak, P., Koudelka, P.: Virtual simulator for the generation of patho-physiological foetal ECGs during the prenatal period. Electron. Lett. **51**(22), 1738–1740 (2015). doi:10.1049/el.2015.2291
11. Nedoma, J., Fajkus, M., Vasinek, V.: Influence of PDMS encapsulation on the sensitivity and frequency range of fiber-optic interferometer. In: Optical Materials and Biomaterials in Security and Defence Systems Technology XIII, p. 99940P. SPIE, Edinburgh (2016). doi:10.1117/12.2243170
12. Nedoma, J., Fajkus, M., Bednarek, L., Frnda, J., Zavadil, J., Vasinek, V.: Encapsulation of FBG sensor into the PDMS and its effect on spectral and temperature characteristics. Adv. Electr. Electron. Eng. **14**(4-Special Issue), 460–466 (2016). doi:10.15598/aeee.v14i4.1786
13. Zazula, D.: Application of fibre-optic interferometry to detection of human vital sign. J. Laser Health Acad. **2012**(1), 27–32 (2012). doi:10.13140/2.1.2033.4726
14. Brandstetter, P., Klein, L.: Second order low-pass and high-pass filter designs using method of synthetic immittance elements. Adv. Electr. Electron. Eng. **11**(1), 16–21 (2013). doi:10.15598/aeee.v11i1.800
15. Donghui, Z.: Wavelet approach for ECG baseline wander correction and noise reduction. In: 27th Annual International Conference of the Engineering in Medicine and Biology Society, Shanghai, pp. 1212–1215. IEEE Press (2005)
16. Martinek, R., Kahankova, R., Nazeran, H., Konecny, J., Jezewski, J., Janku, P., Fajkus, M.: Non-invasive fetal monitoring: a maternal surface ECG electrode placement-based novel approach for optimization of adaptive filter control parameters using the LMS and RLS algorithms. Sensors **17**(5), 1154 (2017). doi:10.3390/s17051154
17. Fajkus, M., Nedoma, J., Martinek, R., Vasinek, V., Nazeran, H., Siska, P.: A non-invasive multichannel hybrid fiber-optic sensor system for vital sign monitoring. Sensors **17**(1), 111 (2017). doi:10.3390/s17010111

18. Kim, S., Hwang, D.: Murmur-adaptive compression technique for phonocardiogram signals. Electron. Lett. **52**(3), 183–184 (2015). doi:10.1049/el.2015.3449
19. Naseri, H., Homaeinezhad, M.R.: Detection and boundary identification of phonocardiogram sounds using an expert frequency-energy based metric. Ann. Biomed. Eng. **41**(2), 279–292 (2013). doi:10.1007/s10439-012-0645-x
20. Martinek, R., Nedoma, J., Fajkus, M., Kahankova, R., Konecny, J., Janku, P., Nazeran, H.: A phonocardiographic-based fiber-optic sensor and adaptive filtering system for noninvasive continuous fetal heart rate monitoring. Sensors **17**(4), 890 (2017). doi:10.3390/s17040890

A Stereo Audio Steganography by Inserting Low-Frequency and Octave Equivalent Pure Tones

Hung-Jr Shiu[1], Bor-Shing Lin[2], Bor-Shyh Lin[3], Wei-Chou Lai[1], Chien-Hung Huang[4(✉)], and Chin-Laung Lei[1]

[1] DCNS Lab., Graduate Institute of Electrical Engineering, National Taiwan University, Taipei 10617, Taiwan
[2] Department of Computer Science and Information Engineering, National Taipei University, New Taipei City 23741, Taiwan
[3] Institute of Imaging and Biomedical Photonics, National Chiao Tung University, Tainan 71101, Taiwan
[4] Department of Computer Science and Information Engineering, National Formosa University, Huwei 63201, Yunlin County, Taiwan
chhuang@nfu.edu.tw

Abstract. Audio steganography is a technique for hiding messages in audio signals such that no one except the sender and intended recipient suspects the existence of the messages. This paper proposes a new method for stereo audio steganography that exploits some characteristics of the human auditory system (HAS). Messages are embedded by inserting low-frequency and octave-equivalent pure tones into different channels. Hidden messages are extracted by comparing the frequency domain data of the left channel with those of the right channel. Experimental results reveal that the quality of the message-hiding audio that is generated by the method is only very slightly less than that of the host audio, so attackers cannot perceive the hidden messages.

Keywords: Audio steganography · Human auditory system (HAS) · Low-frequency pure tone insertion · Octave equivalence

1 Introduction and Related Work

Covert communication has important applications, such as in battlefield communications and bank transactions. These applications must deny access to unauthorized persons to important information that is being transmitted. By hiding information in a cover medium such as an image or audio, the existence of that information is concealed during transmission and the attention of malicious hackers or intruders will not be drawn to any secret messages. Steganography is used to embed messages into many host media, such as image, audio and video. Among them audio steganography is especially challenging since the human auditory system (HAS) is more sensitive than other human perceptual systems [1], so a small modification to an audio signal can cause perceptible distortions. Although the HAS is highly sensitive, certain phenomena can be exploited to achieve audio steganography, including auditory masking, octave

P. Krömer et al. (eds.), *Proceedings of the Fourth Euro-China Conference on Intelligent Data Analysis and Applications*, Advances in Intelligent Systems and Computing 682, DOI 10.1007/978-3-319-68527-4_27

equivalence and the relationship between loudness and frequency. Octave equivalence refers to the fact that, if the frequency of a pitch is 2^n times higher or lower than the frequency of another pitch, then these two pitches are considered to be 'the same'. Accordingly, if a tone with a low frequency and small amplitudes are several octaves below a stronger tone in the host audio, it will be imperceptible. In recent years, a number of audio steganographic methods have been proposed and demonstrated, including least significant bit (LSB) techniques [2, 3], parity coding [3], the spread spectrum method [3], phase coding [3], echo hiding [4] and other methods [5]. LSB is one of the simplest techniques for embedding messages. Inaudibility is achieved by replacing the least significant bits in some bytes of host audio with high precision (e.g., 16 bits quantization). The spread spectrum (SS) method attempts to spread secret information across the frequency spectrum of the audio signal. Phase coding relies on the fact that the phase components of a sound are not as perceptible to the human ear as is noise. It works by replacing the phase of an initial audio segment with a reference phase that represents the secret information. The phases of the remaining segments' are adjusted to preserve the relative phase between segments. Echo hiding embeds data by introducing a few echoes into the audio signal. Data are hidden by varying the initial amplitude, decay rate, and offset of the echo. If the delay is small enough, then the human ear cannot distinguish the original signal from the echo [4]. Fallahpour and Megías. presented a method that modifies the FFT magnitudes that are in a band of frequencies between 5 and 14 kHz [5]. This low-complexity method can achieve high capacity, but it causes perceptual distortions. This paper introduces a new, stereo-two-channel-audio, steganographic method that is based on the above phenomena. This method embeds messages by inserting low-frequency pure tones (sinusoids) into segments of a host stereo audio such that, for each segment, the inserted tone and the loudest tone in the host audio segment are octave-equivalent. Experimental results indicate that the scheme generates a message-hidden audio with nearly no degradation of quality.

2 Methods and Procedures

The encoding procedure divides the host audio into segments and embeds one bit in each segment. The decoding procedure segments the data-embedded audio, extracts the bit that is hidden in each segment by calculating the difference in the transferred domain data between the left channel and the right channel.

2.1 Octave Equivalence

In music theory, an interval is the difference between two pitches. An octave is the special interval between one musical pitch and another with half or double its frequency. It is a natural phenomenon that has been referred to as the "basic miracle of music". The human ear tends to hear both notes as being essentially "the same", because of their closely related harmonics. This phenomenon is called octave equivalence [6].

2.2 Data Embedding Procedure

The encoding process requires the transferred domain data of the host stereo audio. The discrete Fourier transform (DFT) is used herein. The sequence of N complex numbers $x_1, x_2, \ldots, x_N$ is transformed into an N-periodic sequence of complex numbers $X_1, X_2, \ldots, X_N$ using the DFT formula [7],

$$X_k = \sum\nolimits_{n=1}^{N} x_n \cdot e^{\frac{-i2\pi k(n-1)}{N}} \tag{1}$$

The frequency of an inserted tone must not be less than a pre-defined boundary $frequency_{min}$ in Hertz (Hz), and must be less than $2 \cdot frequency_{min}$ Hz. Let x be the host stereo audio that is sampled at SR Hz, yielding L samples. First, x is split into k non-overlapping segments, each with N samples:

$$x = \{x_1, x_2, \ldots, x_k\}, N = \left\lceil \frac{SR}{frequency_{min}} \right\rceil \tag{2}$$

k is computed from L and N, which is the number of bits that can be hidden in the host audio. For example, for a 3 s host audio, sampled at $SR = 48000$ Hz and $frequency_{min} = 30$ Hz, N is 1600 and k is 90:

$$k = \left\lfloor \frac{L}{N} \right\rfloor \tag{3}$$

Every segment x_i is composed of left channel signals x_{Left_i} and right channel signals x_{Right_i}.

$$x = \left\{ \left\{ x_{Left_1}, x_{Right_1} \right\}, \left\{ x_{Left_2}, x_{Right_2} \right\}, \ldots, \left\{ x_{Left_k}, x_{Right_k} \right\} \right\}$$

For every segment x_i, $i = 1, 2, \ldots, k$, an N-points DFT is applied to x_{Left_i} and x_{Right_i} yielding results that are denoted as y_{Left_i} and y_{Right_i}:

$$y_{Left_i} = DFT\left(x_{Left_i}\right), y_{Right_i} = DFT\left(x_{Right_i}\right) \tag{4}$$

Then Sum_{Left_i} and Sum_{Right_i}, the sums of absolute values of coefficients $y_{Left_i}(j)$ and $y_{Right_i}(j)$ of sinusoids with frequency $freq(j) = \frac{(j-1)}{N}$ (Hz), are calculated from y_{Left_i} and y_{Right_i}:

$$Sum_{Left_i} = \sum_{j=1}^{N} abs(y_{Left_i}(j)) \cdot f(j), \quad Sum_{Right_i} = \sum_{j=1}^{N} abs(y_{Right_i}(j)) \cdot f(j)$$

$$f(j) = 1, if\ frequency_{min} \leq \frac{(j-1)}{N} \cdot SR < 2 \cdot frequency_{min}$$
$$f(j) = 0, otherwise \tag{5}$$

For the i-th segment, if $Sum_{Left_i} \geq Sum_{Right_i}$, then that segment represents bit '1'. If $Sum_{Left_i} < Sum_{Right_i}$, then the segment represents bit '0'.

Let the bit string message to be hidden m of length k. For every x_i in x, the encoding process determines whether or not a low-frequency pure tone is inserted into x_i by $m(i)$ and Sum_{diff_i}, which is the difference between Sum_{Left_i} and Sum_{Right_i}:

$$Sum_{diff_i} = Sum_{Left_i} - Sum_{Right_i} \tag{6}$$

The four situations are as follows.

(a) $m(i) = 0$, and $Sum_{diff_i} \geq 0$. (b) $m(i) = 0$, and $Sum_{diff_i} < 0$. (c) $m(i) = 1$, and $Sum_{diff_i} \geq 0$. (d) $m(i) = 1$, and $Sum_{diff_i} < 0$.

In (b) and (c), no operations are performed on x_i. In (a), a low-frequency pure tone h_i is inserted into right-channel x_{Right_i} to reduce Sum_{diff_i} such that after the insertion $Sum_{diff_i} < 0$. In (d), h_i is inserted into x_{Left} to make $Sum_{diff_i} \geq 0$.

h_i is required, it is created by the following steps. The generation of h_i in situation (d) is described here. The equivalent process in situation (a) involves changing y_{Left_i} to y_{Right_i} and y_{Left_i} to y_{Right_i} in the first step, and inserting h_i into the right channel instead of the left channel.

First, the $frequency_{hide_i}$ of h_i must be determined from $y_{Left_i}(p)$. $y_{Left_i}(p)$ has the largest absolute value in y_{Left_i} and $\frac{(p-1)}{N}$ is the frequency of the loudest tone in the segment. $frequency_{hide_i}$ is several octaves below $\frac{(p-1)}{N}$:

$$y_{Left_i}(p) = max\left(abs(y_{Left_i})\right),\ frequency_{hide_i} = \frac{(p-1)}{N} \cdot SR \cdot \frac{1}{2^{q_i}} \tag{7}$$

Note that q_i is a non-negative integer, implying

$$frequency_{min} \leq frequency_{hide_i} < 2 \cdot frequency_{min}$$

Then, a sinusoid h'_i with $frequency_{hide_i}$ can be generated:

$$h'_i(t) = \sin\left(2\pi \frac{t-1}{SR} \cdot frequency_{hide_i}\right), t = 1, 2, \ldots, N \tag{8}$$

If $h'_i(t)$ is directly inserted into the input signals, then it will be perceptible and annoying, since discontinuities will be present around segments' boundaries, and the embedding of data may be suspected. Therefore, the last step is to reshape $h'_i(t)$ by applying the following reshaping function $S_i(t)$.

$$S_i(t) = \alpha_i \cdot \cos\left(\frac{t\pi}{N-1} - \frac{\pi}{2}\right) = \alpha_i \cdot \sin\left(\frac{t\pi}{N-1}\right), t = 1, 2, \ldots, N, \tag{9}$$

where α_i is the average amplitude of x_i.

Finally, the inserted tone h_i and the output segments z_{Left_i} and z_{Right} are obtained:

$$h_i = \left\{h'_i(t) \cdot S_i(t)\right\}, t = 1, 2, \ldots, N$$

$$z_{Left_i} = x_{Left_i} + h_i, z_{Right_i} = x_{Right_i}, \tag{10}$$

and the output stereo audio z is,

$$z_i = \{z_{Left_i}, z_{Right_i}\}, i = 1, 2, \ldots, k, z = \{z_1, z_2, \ldots, z_k\} \tag{11}$$

Algorithm 1 describes the control flow of the encoding procedure.

Algorithm 1: Data embedding procedure
Input: host stereo audio x, message m
Output: stereo z
Step 1: divide x into k segment
Step 2: for $i = 1$ to k
    apply DFT to x_{Left_i} and x_{Right_i}, get y_{Left_i} and y_{Right_i}
    compute Sum_{Left_i} and Sum_{Left_i} from y_{Left_i} and y_{Right_i}
    compute Sum_{diff_i}
    if $Sum_{diff_i} \geq 0$
        if $m_i == 0$
            insert a low-frequency tone to right channel
    else
        if $m_i == 1$
            insert a low-frequency tone to left channel
Step 3: return z

.

2.3 Data Extracting Procedure

The received stereo audio z with hidden data is firstly divided into k non-overlapping segments. For every segment $z_i = \{z_{Left}, z_{Right}\}$, an N-point DFT is applied to z_{Left_i} and z_{Right_i}, yielding y'_{Left_i} and y'_{Right_i}, respectively. Then, Sum'_{Left_i}, Sum'_{Right_i} and Sum'_{diff_i} are calculated from y'_{Left_i} and y'_{Right_i}. If $Sum'_{diff_i} \geq 0$, then the hidden bit m_i is 1; otherwise it is 0. Algorithm 2 presents the extracting procedure.

Algorithm 2: Data extraction procedure
Input: embedded audio z
Output: message m
Step 1: divide z into k segments
Step 2: for $i = 1$ to k
    apply DFT to x_{Left_i} and x_{Right_i}, get y_{Left_i} and y_{Right_i}
    compute Sum_{Left_i} and Sum_{Left_i} from y_{Left_i} and y_{Right_i}
    compute Sum_{diff_i}
    if $Sum_{diff_i} \geq 0$
        $m_i == 1$
    else
        $m_i == 0$
Step 3: collect all m_i to form m and return m

.

3 Experiments and Results

The first experiment evaluates the quality of the output audio. The second elucidates the relationship between the hiding bit rate and quality degradation. The last demonstrates that if the host audio is properly chosen, then the bit rate can be increased without perceptual distortions. Eighteen randomly selected and uncompressed (.wav) audio clips of music with different styles are used as host audio. These 18 audio clips are grouped in to six groups, as shown in Table 1. All of these audio clips are in stereo and 10 s- duration. They are sampled at 48000 Hz and quantized with 16 bits.

Table 1. Host audio used in the experiments.

Host audio	Music style	Host audio	Music style
$A_{01}-A_{03}$	Classical	$A_{10}-A_{12}$	Piano
$A_{04}-A_{06}$	Jazz	$A_{13}-A_{15}$	Pop
$A_{07}-A_{09}$	Opera	$A_{16}-A_{18}$	Rock

To measure the perceptual quality, the Perceptual Evaluation of Audio Quality (PEAQ) algorithm, is used [8]. The PEAQ algorithm compares the quality of the host audio with its data-hiding counterpart and returns a parameter called the Objective Difference Grade (ODG), which ranges from −4 to 0. The definition of ODG is listed as Table 2. In the first experiment, $frequency_{min} = 30$ Hz is set, so the number of bits that can be hidden in each clip is 300. The messages to be hidden are m_i, $i = 1, 2, \ldots, 18$, which are randomly generated binary strings; the output audio clips are A'_i, and the ODG of each output A'_i is given in Table 3. Table 4 clearly indicates almost no quality degradation in any of the output audio clips. The worst ODG is in A'_{13} (Pop), and is −0.366, which is between imperceptible and perceptible but not annoying. Over 70% of our outputs are graded 0 by PEAQ, revealing that the human ear would have great difficulty in perceiving the embedded tones. Furthermore, as presented in Figs. 1 and 2, only a few unusual peaks appeared in the time-domain signals of the output audio A'_{13}, so attackers will not suspect the embedded messages. In the first experiment, A'_{18} has the best ODG while A'_{13} has the worst, indicating that different host audio clips may exhibit different degrees of quality degradation. In this experiment A_{13} and A_{18} are used and our scheme is applied using $frequency_{min}$ from 30 Hz to 120 Hz; then the ODGs are calculated, yielding the results in Fig. 3.

Table 2. Definition of ODG.

Impairment description	ITU-R Grade	ODG
Imperceptible	5.0	0.0
Perceptible, but not annoying	4.0	−1.0
Slightly annoying	3.0	−2.0
Annoying	2.0	−3.0
Very annoying	1.0	−4.0

Table 3. ODG of each output audio.

Output	ODG	Output	ODG	Output	ODG	Output	ODG	Output	ODG	Output	ODG
A'_{01}	0	A'_{04}	0	A'_{07}	0	A'_{10}	−0.059	A'_{13}	−0.366	A'_{16}	0
A'_{02}	−0.171	A'_{05}	0	A'_{08}	0	A'_{11}	−0.060	A'_{14}	0	A'_{17}	0
A'_{03}	0	A'_{06}	−0.319	A'_{09}	0	A'_{12}	−0.051	A'_{15}	0	A'_{18}	0

Table 4. ODGs of different data rates.

Host audio	ODG of 30 Hz	ODG of 60 Hz	ODG of 120 Hz
A'_{10} A'_{11}	–0.059 –0.060	–1.044 –1.377	–1.806 –2.245
A'_{12} A'_{13}	–0.051 –0.366	–1.008 –2.739	–2.025 –3.518
A'_{16} A'_{17} A'_{18}	0 0 0	–0.7 –0.99 –0.13	–1.58 –1.67 –0.95

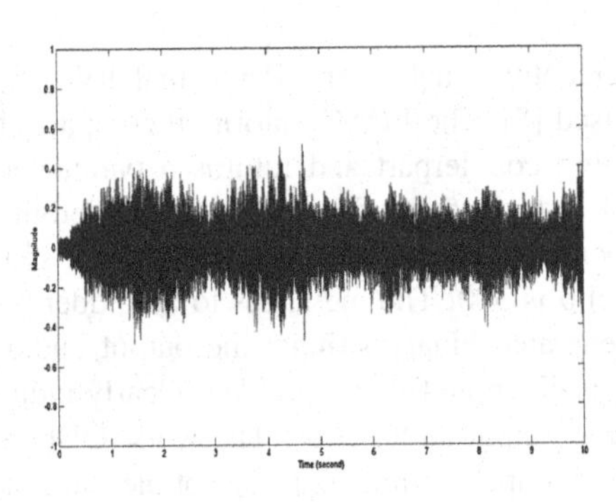

Fig. 1. Left channel signals of A'_{13}.

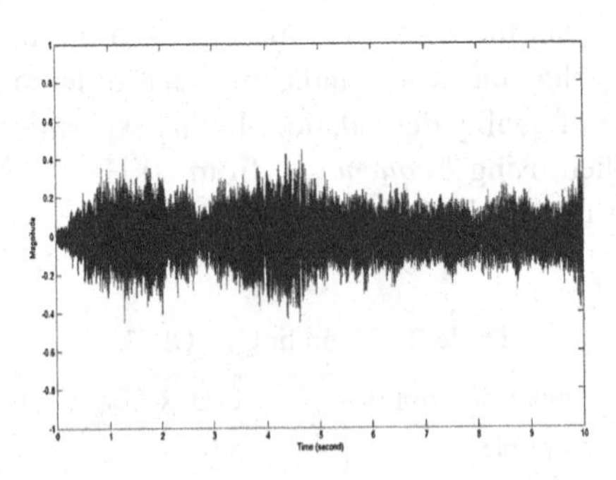

Fig. 2. Right channel signals of A'_{13}.

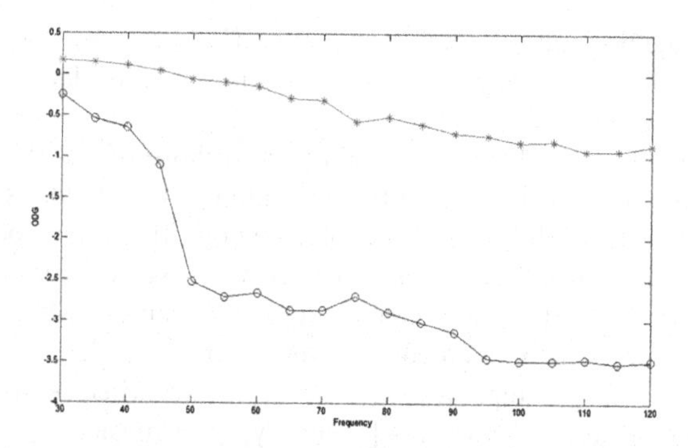

Fig. 3. ODG result of using different $frequency_{min}$.

4 Discussion

The value of ODG decreases dramatically starting from 40 Hz of A_{13} while A_{18} still has good ODG value when $frequency_{min}$ is 120 Hz. This difference is explained by frequency masking. As presented in Figs. 4 and 5, the magnitudes of the 30 to 120 Hz components of A_{13} are low, so if a low-frequency tone is inserted into A_{13}, it will be

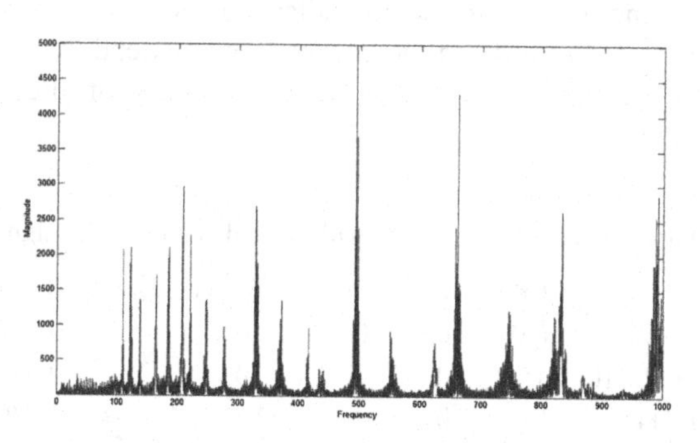

Fig. 4. Frequency spectrum of A_{13}.

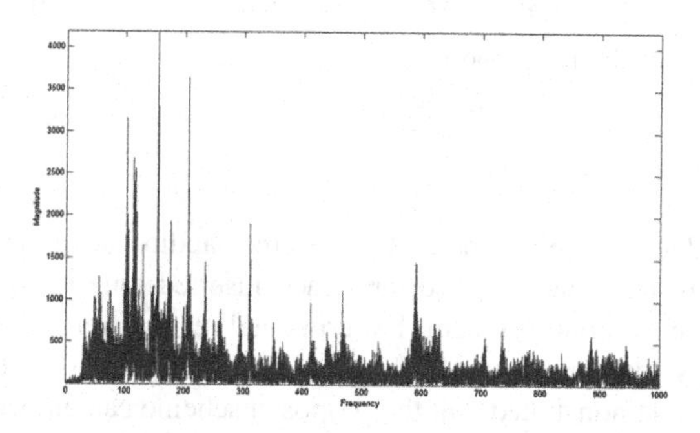

Fig. 5. Frequency spectrum of A_{18}.

detected easily since no frequency masking occurs. However, A_{18} provides a favorable environment for low-frequency insertion because it has larger low-frequency magnitudes.

The experimental results herein are compared to those of Fallahpour and Megías [5], who used the modification of FFT magnitudes to encode secret messages. Although our bit rate is much less than that obtained by [8], our method yielded almost no quality degradation; over 70% of our obtained ODG scores were 0, and none was less than −0.4. The ODGs that were obtained by Fallahpour and Megías [5] are almost −1 indicating the insertions are perceptible. Moreover, the experiments of Fallahpour and Megías [5] used only trumpet and violoncello music, whereas our scheme can be applied to a wider range of instruments [5] and types of music.

The bit rate of our scheme equals the input parameter $frequency_{min}$. For example, if $frequency_{min}$ is 30 Hz, then the number of bits that can be embedded in an one second audio clip is 30. The time complexity is affected by $frequency_{min}$. In our encoding and decoding procedures, the input audio is split into $frequency_{min}$ segments and DFT is applied to every segment. The DFT functions, whose time complexity is $O(n \log n)$ where n is the number of samples in a segment, is the bottlenecks. As a result, the time complexity is $O(n \log n)$, n is $frequency_{min} \cdot O(n \log n)$. Recent years, authors proposed robustness under several signal processings [9–14]. Table 5 presents the comparisons of bit error ratio between related works [10–14] under different signal processings. LPE is low pass filter to allow signals survive lower than 3 kHz; DC is padding a direct current noise on the stego-audio; mp3 is the process of a compression and decompression with 64 kbps; Re-Quantization is to extend the bits of the sampled value from 16 to 32 bits; Re-sampling is the process of digital-to-analog and analog-to-digital sampling.

Table 5. The bit error ratio comparisons with related works under signal-processings.

Method	LPF	DC	mp3	Re-quantization	Re-sampling
Proposed	0.0%	0.0%	0.0%	0.0%	0.0%
Akhaee et al. [10]	15.0%	0.0%	10.2%	0.0%	0.0%
Wu et al. [11]	∅	∅	4.3%	∅	9.87%
Chen and Wu [12]	∅	4.5%	6.5%	11.9%	6.4%
Shiu et al. [13]	100%	0.0%	100%	0.0%	0.0%
Shiu et al. [14]	0.0%	0.0%	0.0%	0.0%	0.0%

∅: The authors did not propose.

5 Conclusion

This work introduced a new method of stereo audio steganography in which low-frequency pure tones are inserted, and each inserted tone is an octave of a particular tone in the host audio segment. Experimental results were presented to explain why low-frequency tones were used. The data-hiding audio was tested using PEAQ, and the results thus demonstrated that the proposed scheme can embed data into stereo audio with nearly no perceptual distortion. Also, robustness performs better than some related works considering on a few of attacks.

Acknowledgment. This research is supported by the Ministry of Science and Technology, Taiwan, under Contract Nos. MOST 105-2221-E-150-063 and MOST 106-2221-E-150-056.

References

1. Neubauer, C., Herre, J., Brandenburg, K.: Continuous steganographic data transmission using uncompressed audio. In: Information Hiding 1998, LNCS, vol. 1525, pp. 208–217. Springer, Heidelberg (1998)
2. Gudla, S., Reyya, S., Kotyada, A.: Key based least significant bit (LSB) insertion for audio and video steganography. Int. J. Comput. Sci. Eng. **3**(1), 60–69 (2013)
3. Nosrati, M., Karimi, R., Hariri, M.: audio steganography: a survey on recent approaches. World Appl. Program. **2**(3), 202–205 (2012)
4. Kondo, K.: Evaluation of a stereo audio data hiding method using inter-channel decorrelator polarity. In: 2010 International Conference of Acoustics Speech and Signal Processing, Dallas, pp. 277–280. IEEE (2010)
5. Fallahpour, M., Megías, D.: High capacity method for real-time audio data hiding using the FFT transform. In: International Conference on Information Security and Assurance 2009, pp. 91–97. Springer, Heidelberg (2009)
6. Deutsch, D.: The psychology of music, 3rd edn. Elsevier, San Diego (2013)
7. Bracewell, R.N.: The Fourier transform and its applications. McGraw-Hill, New York (1986)
8. ITU-R Recommendation BS.1387: Method for Objective Measurements of Perceived Audio Quality. International Telecommunication Union – Radiocommunication Sector, November 2001
9. Shiu, H.J., Ng, K.L., Fang, J.F., Lee, R.C.T., Huang, C.H.: Data hiding methods based upon DNA sequences. Inf. Sci. **180**(11), 2196–2208 (2010)
10. Akhaee, M.A., Saberian, M.J., Feizi, S., Marvasti, F.: Robust audio data hiding using correlated quantization with histogram based detector. IEEE Trans. Multimed. **11**, 834–842 (2009)
11. Wu, S., Huang, J., Huang, D., Shi, Y.Q.: Efficiently self-synchronized audio watermarking for assured audio data transmission. IEEE Trans. Broadcast. **51**, 69–76 (2005)
12. Chen, O.T.C., Wu, W.C.: Highly robust, secure, and perceptual-quality echo hiding scheme. IEEE Trans. Audio Speech Lang. Process. **16**, 629–638 (2008)
13. Shiu, H.J., Tang, S.Y., Huang, C.H., Lee, R.C.T., Lei, C.L.: A reversible acoustic data hiding method based on analog modulation. Inf. Sci. **273**, 233–246 (2014)
14. Shiu, H.J., Lin, B.S., Cheng, C.W., Huang, C.H., Lei, C.L.: High-capacity data-hiding scheme on synthesized pitches using amplitude enhancement-a new vision of non-blind audio steganography. Symmetry **9**(6), 92 (2017)

Energy Control and Management

Sensor Data Mining for Gas Station Online Monitoring

Zhihao Wei[1,2,3], Kebin Jia[1,2,3](✉), and Zhonghua Sun[1,2,3]

[1] College of Electronic Information and Control Engineering, Beijing University of Technology, Beijing, China
kebinj@bjut.edu.cn
[2] Beijing Laboratory of Advanced Information Networks, Beijing, China
[3] Beijing Advanced Innovation Center for Future Internet Technology, Beijing, China

Abstract. Aiming at the geography and manpower inconvenience during the monitoring and inspection of the gas station, a remote online monitoring system for the gas station based on the sensor network is proposed in this paper, and the early warning analysis of the abnormal state of the gas station is proposed based on the data mining technology.

Firstly, a gas station senor dataset is built based on the sensor network of the gas station. Then, based on the B/S architecture, a gas station online monitoring system is built. Finally, based on the sensor dataset data mining, an abnormal state of the gas station analysis method is proposed.

Experiments show that the classifier method proposed in this paper has the generalization ability, it can analysis and alarm the abnormal state of the gas station which improve the intelligence and convenience of the gas station monitoring.

Keywords: Sensor network · Data mining · Gas station · B/S architecture

1 Introduction

Gas station is an important municipal infrastructure facilities, it is mainly used to storage oil and gas. With the addition of oil and gas recovery stations, more than 95% of the oil and gas in the pipeline and tank closed operation, more than 95% of the basic oil and gas emissions, greatly reducing environmental pollution [1].

In order to ensure the safety of the operation of the gas station, it is necessary to regularly monitor and maintain the key parameters of the storage tank, pipeline, refueling gun and recovery processing device. However, due to the differences in the geographical distribution of the gas stations themselves, it is necessary to spend a lot of human resources to carry out routine equipment testing and environmental monitoring on the gas stations distributed everywhere. By using the gas station sensor network and data mining analysis process, the daily maintenance and monitoring of the gas station can cost less manpower resources [2].

In this paper, an online monitoring technology of the gas station is proposed. The remote access of the gas station real-time data is realized through the sensor network

P. Krömer et al. (eds.), *Proceedings of the Fourth Euro-China Conference on Intelligent Data Analysis and Applications*, Advances in Intelligent Systems and Computing 682, DOI 10.1007/978-3-319-68527-4_28

architecture of the gas station. The B/S architecture is used to build the online monitoring system of the gas station. Based on the data mining Technology, the gas station real-time data classification and identification is researched, and thus to achieve the abnormal state of the gas station analysis and early warning.

2 Gas Station Remote Online Monitoring System Construction

2.1 Data Acquisition and Transmission Architecture of Gas Station Based on Sensor Network

The data and transmission architecture of the remote online monitoring system of the gas station mainly consists of three layers, as shown in Fig. 1.

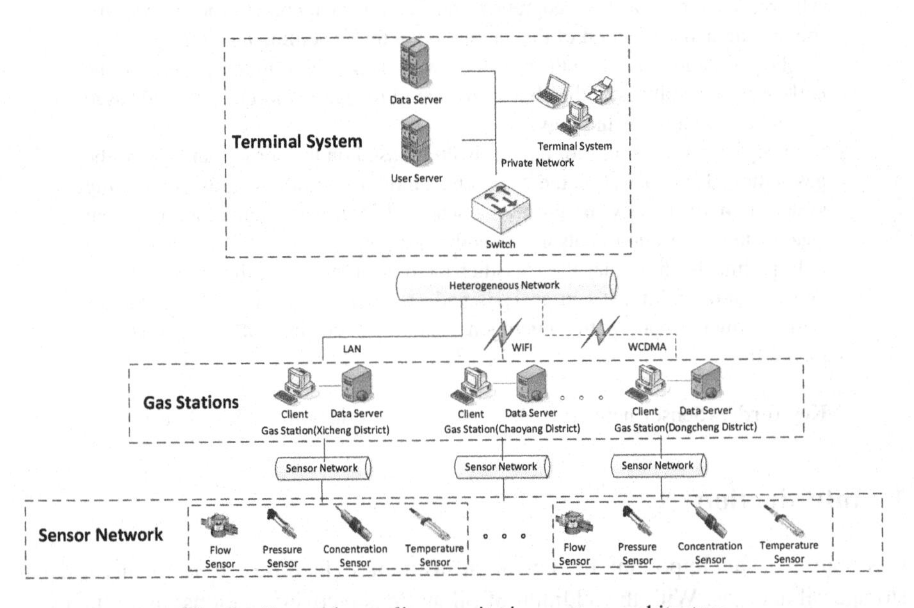

Fig. 1. Gas station online monitoring system architecture

In the above structure, the functions of each layer are as follows: The sensing layer is a data acquisition layer, which is composed of temperature sensor, pressure sensor and flow sensor set in the key part of the gas station, and can realize the real-time gas station comprehensive information And the middle layer is the local pretreatment layer of the gas station, which is composed of the computer set in the gas station and realizes the pretreatment of the sensor data, thus reducing the total amount of data transmitted to the upper layer and improving the efficiency of data transmission. The application layer is the terminal layer, which is composed of the server group and the user interaction device set up in the equipment maintenance center and the environmental monitoring center to realize the excavation analysis of the sensor data of the gas station, and provide the reference information for the decision maker.

2.2 Online Monitoring System Based on B/S Architecture

The gas station online monitoring system is able to visualize the gas station sensor data received by the terminal, providing users with a way to remotely view the gas stations distributed in various geographical locations.

The system is based on B/S architecture design, and the application of JavaScript technology and the .NET Framework technology, to achieve the gas station remote online monitoring system structures. In the user management and other data storage, the system uses a database based on Mysql. The data display system interface is shown in Fig. 2.

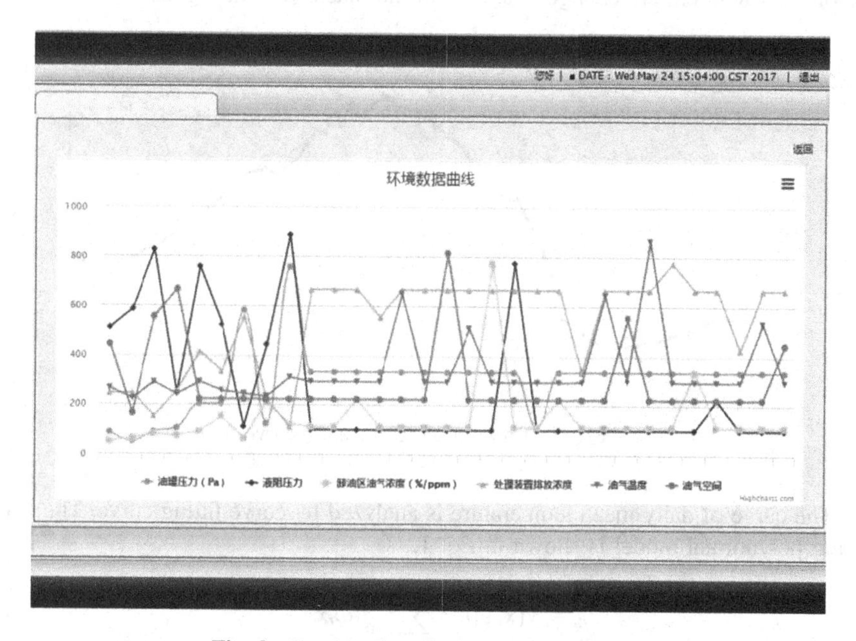

Fig. 2. Sensing data display system interface

3 Abnormal State Monitoring Based on Data Mining

3.1 Abnormal State Monitoring and Alarming

By analyzing the data of the gas station acquired via the network transmission, it is possible to judge the operation of the critical equipment of the gas station to see if the equipment is in an abnormal state. For the occurrence of large abnormalities of the device, you can use the alarm mechanism to achieve abnormal and timely feedback failure.

Gas station key equipment alarm mechanism to monitor the object mainly include storage tanks, fuel guns. Through the sensor data for data cleaning, classifier modeling and other steps to achieve the key equipment monitoring alarm.

3.2 Oil Tank Monitoring Based on Cumulative Temperature Variation

The storage tank is the core of the petrol filling station, it is usually set below the gas station surface, with sensors like the temperature sensor, pressure sensor and other sensor group. With the analysis of the sensor data at the storage tank, the current situation of the storage tank can be proved and the abnormal state can be early warning.

Due to the temperature of the oil storage tank is affected by the outdoor temperature, the traditional single temperature threshold warning mode has a large instability [3].

Based on a 72 h temperature change data of the storage tank, the average curve of the oil tank temperature change is shown in the black lines in Fig. 3.

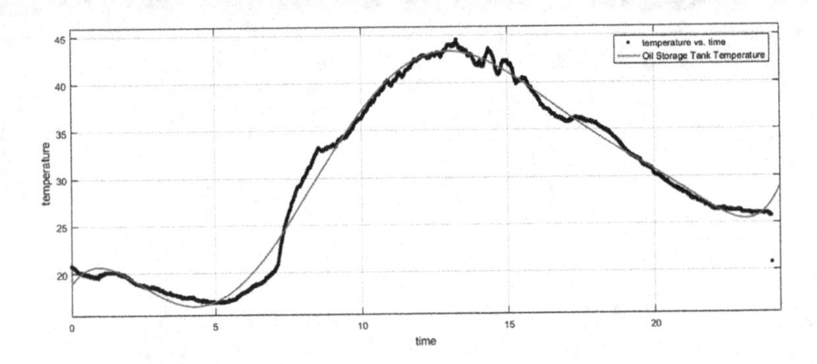

Fig. 3. Daily mean temperature curve

The curve of daily mean temperature is analyzed by curve fitting curve. The curve fitting polynomial model is shown in Eq. 1.

$$y(x, \omega) = \sum_{i=0}^{i=M} \omega_i x^i \quad (1)$$

Through the curve fitting, the recent temperature cycling curve of the oil storage tank is obtained, as shown by the blue line in Fig. 3. According to the curve fitting results, set the dynamic temperature alarm threshold T with time, as shown in Eq. 2.

$$T = y(t) \pm A \quad (A \geq 0) \quad (2)$$

where T is the dynamic temperature alarm threshold, y(t) represents the conventional tank temperature at time t, and A represents the dynamic range. Through this process, the accuracy of the tank temperature alarm can be improved.

3.3 Refueling Gun Abnormal Monitoring Based on Sensor Group

A variety of sensors located in the tanker area enable analysis of the state of the fuel gun and its delivery piping. Through the analysis of the five kinds of sensing data, such as the duration of fueling, the amount of return gas, etc., the sealing analysis of the pipeline is realized [4]. Among them, the fueling duration, recovery of oil and pipe seal the distribution of water as shown in Fig. 4.

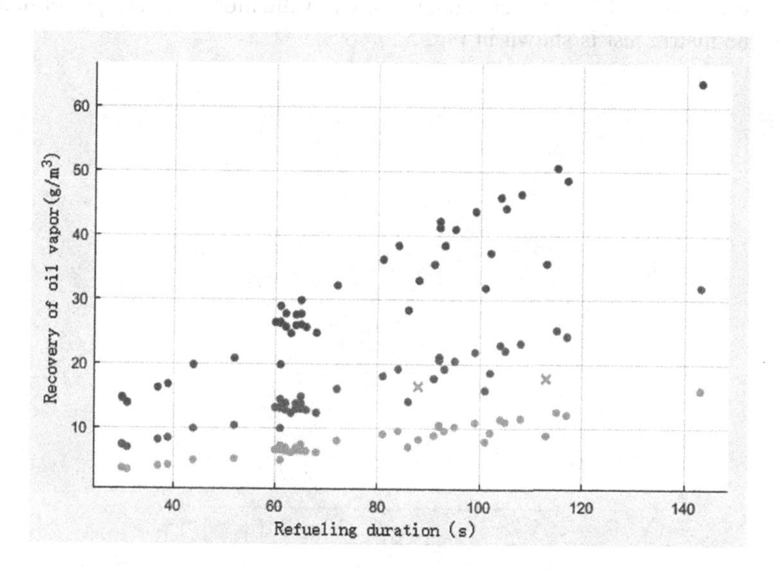

Fig. 4. Partial fuel gun sensor data distribution diagram (Color figure online)

The Y-axis is the corresponding recovery oil and gas volume, the color indicates the sealing state of the pipeline, the blue indicates the normal state, the yellow indicates the warning state, and the red indicates the alarm state.

This paper chooses to model the sensor data based on the classifier, and then analyzes the relationship between the fueling duration and the recovery of oil. The mainstream classifier model includes SVM classifier, decision tree classifier [5]. In this paper, the SVM classifier model is selected as the reference model. Through the fuel gun sensor data for data cleaning and normalization step, and feature matrix build step, and then to achieve abnormal state analysis [6].

4 The Experimental Results and Analysis

The study in this paper is based on the senor network data of gas station. In accordance with the methods in Sect. 3.3, Result of the fuel gun based data experimental is shown as follow.

The 5–dimensional feature data matrix Dataset1 containing 1740 samples was constructed by the data analysis process in Sect. 3.3. As shown in Table 1.

Table 1. Results of Dataset1

Flag	Type	Quantity
1	Normal	580
2	Alert	580
3	Alarm	580

Based on the SVM classifier model, 5 cross validation tests are performed. The truth value matrix test is shown in Fig. 5.

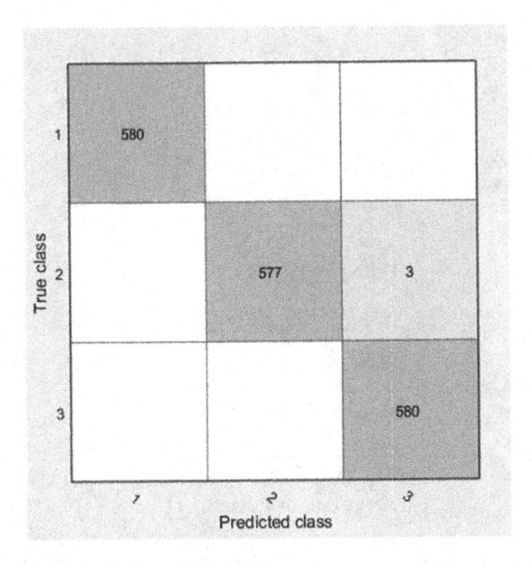

Fig. 5. Truth value matrix based on fuel gun sensor data classifier

By analyzing the 5-cross validation truth matrix shown in Fig. 5, it can be seen that the classifier has the ability to analyze the state of the refueling gun.

The generalization capability of the SVM classifier based on Sect. 3.3 is then further tested. Select the fuel gun data set. Data 2 data set includes a total of 240 fuel gun sensor data, including 80 normal data, 80 early warning data, 80 alarm data (Table 2).

Table 2. Results of generalization

Correct data	Miss data	Correct Rate
238	2	99.16%

It can be seen from the analysis that the SVM classifier in Sect. 3.3 has the generalization ability of pipeline sealability identification.

5 Conclusion

This paper mainly aims at the data mining research based on the analysis and alarm of the key parts of the gas station based on the sensor data of the gas station. A set of on - line monitoring system for gas stations based on three - layer sensor data acquisition and transmission architecture is designed and implemented for the daily limit of daily maintenance and environmental monitoring of gas stations. Then, based on the data mining of the sensor data, the curve fitting prediction based on the temperature of the oil storage tank is carried out. The oil and gas transportation pipeline is closed based on the fuel gun sensor data Feature matrix construction. Finally, a classifier for environmental state alarm and oil and gas pipeline closed alarm is constructed by using the established feature matrix. Through the experimental analysis, the SVM classifier designed in this paper have the ability of state recognition. Through the use of unfamiliar data on the classifier to test, proved that the generalization ability of the classifier to meet the requirements.

Acknowledgments. This paper is supported by the Project for the Key Project of Beijing Municipal Education Commission under Grant No. KZ201610005007, Beijing Postdoctoral Research Foundation under Grant No. 2015ZZ-23, China Postdoctoral Research Foundation under Grant Nos. 2016T90022, 2015M580029, Computational Intelligence and Intelligent System of Beijing Key Laboratory Research Foundation under Grant No. 002000546615004, and The National Natural Science Foundation of China under Grant No. 61672064.

References

1. Lee, S., Choi, I., Chang, D.: Multi-objective optimization of VOC recovery and reuse in crude oil loading. Appl. Energy **108**, 439–447 (2013). doi:10.1016/j.apenergy.2013.03.064
2. Singh, A., Thakur, N., Sharma, A.: A review of supervised machine learning algorithms. In: 2016 3rd International Conference on Computing for Sustainable Global Development, pp. 1310–1315 (2016)
3. Zou, X.L., Tan, Z.M., Qian, C., Tang, et al.: Fitting of daily variation curve of road surface temperature. J. Chang'an Univ. (Nat. Sci. Ed.), (03), 40–45 (2015)
4. Murphey, Y.L., Luo, Y.: Feature extraction for a multiple pattern classification neural network system. In: International Conference on Pattern Recognition, pp. 220–222 (2002). doi:10.1109/ICPR.2002.1048278
5. Abdul Rahim, N., Paulraj, M.P., Adom, A.H.: Adaptive boosting with SVM classifier for moving vehicle classification. Procedia Eng. **53**(7), 412–418 (2013). doi:10.1016/j.proeng.2013.02.054
6. Bennett, K.P., Bredensteiner, E.J.: Duality and geometry in SVM classifiers. In: Seventeenth International Conference on Machine Learning. Morgan Kaufmann Publishers Inc., pp. 58–63 (2000)

Research on Energy Consumption of Building Electricity Based on Decision Tree Algorithm

Yaxue Pang[1,2,3], Xinhua Jiang[2,3(✉)], Fumin Zou[1,2,3], Zhenhua Gan[2,3], and Junmin Wang[1,2,3]

[1] Beidou Navigation and Smart Traffic Innovation Center of Fujian Province Fuzhou, Fuzhou 350118, Fujian, China
yaxuepang@163.com
[2] Fujian University of Technology, Fuzhou 350108, Fujian, China
xhjiang@fjut.edu.cn
[3] Fujian Key Laboratory for Automotive Electronics and Electric Drive, Fujian University of Technology, Fuzhou 350118, Fujian, China
fmzou@fjut.edu.cn

Abstract. The consumption of building energy was increasing every year in the past in China. It is a big problem to be solved about how to make the building energy consumption to be more reasonable. In order to obtain the feature of building energy consumption with its electrical instructions, a decision-tree algorithm is designed to mine the data of energy consumption of electricity. According to the analysis of the attribute quantity factors, comparing the factors classification information gain value, and analysis the proportion of power consumption factors, a typical energy consumption sample of a laboratory in university was modeled and analyzed in this paper. Experimental results show that the seasonal change factor has a great influence on the energy consumption of the building. And according to this conclusion, a suitable energy-efficiency device is planed.

Keywords: Energy consumption · Decision-tree · Information entropy · ID3 algorithm

1 Introduction

In the major energy-consuming industries of China, the building energy consumption makes a large proportion in the total energy of the country, which is increasing year by year. The proportion of building energy consumption is increasing at an annual rate of 1%. There are many factors influencing the energy consumption of buildings. The factors that effect building energy consumption are various and complicated. Based on accumulating raw data on building energy consumption, it studied every influence factors to describe the energy saving potential and energy saving value of buildings through detailed and in-depth study of the main factors. And finally get the implementation of energy-saving transformation programs and measures. A scientific and rational energy consumption analysis can not only guide the design direction of the new building for energy saving design, but also provide the theoretical basis for the relevant transformation program when the existing buildings are energy saving.

P. Krömer et al. (eds.), *Proceedings of the Fourth Euro-China Conference on Intelligent Data Analysis and Applications*, Advances in Intelligent Systems and Computing 682, DOI 10.1007/978-3-319-68527-4_29

The building energy conservation is a major move in the power industry to promote to complete the strategic goal of energy conservation and emissions reduction. To further strengthen the work of energy-saving emission reduction, what we need to do is to accurately grasp the energy consumption level of the building, scientifically analyze the operation of building energy, and deeply explore the energy use problems existing in the building [1–3].

Literature 1 is based on the investigation of 12 shopping malls in the main urban districts in Chongqing, and make some conclusions and judgments about the actual-value of building energy saving design in Chongqing shopping malls. Literature 2 In order to botain the status of energy consumption of central air conditioning in large commerical buildings of Chongqing, investgates and tests energy consumption in stores, hotels, and offices in Chongqing. Analyzes and assesses the energy consumption and management situation of large commerical buildings in Chongqing. Literature 3 according to the characteristics of the main factors influencing energy consumption, adopt the method of multivariate linear regression analysis, for different types of air conditioning system, establish the corresponding three energy consumption prediction model, through the F-test method test of significance of linear relationship of regression equation. The research results, not only has a guiding role for the energy-saving design of new buildings, energy saving renovation of existing buildings is at the same time for can offer reference and basis.

However, the focus is targeted for different type of buildings to devise suitable energy-efficiency plans. Taking a university as an example and obtain its energy consumption data, the energy consumption characteristics and influencing factors of building energy consumption are obtained by dealing with the energy consumption data. According to the multiple factors based on energy consumption data, and calculate their information gain,draw the most important influence factor, in order to guide the use of electricity, and provide suitable energy-efficiency plans.

2 The Principle of Decision Tree Algorithm

Decision tree algorithm has a wide range of applications in classification, prediction, rule extraction and other fields [4–6]. In the late 1970s and early 80s, the decision tree has been developed greatly in machine learning and data mining since the machine learning researcher Quinlan proposed the ID3 algorithm [7]. Then C4.5, proposed by Quinlan, has become a new supervised learning algorithm. In 1984, several statisticians proposed the CART classification algorithm. ID3 and CART algorithms are proposed at the same time, but similar methods are used to learn decision trees from training samples.

Taking Fujian University of Technology as an example, the energy consumption characteristics and inuencing factors of building energy consumption are obtained by dealing with the energy consumption data. The analysis result has certain reference significance to guide the use of electricity and energy conservation supervision.

The decision tree is a tree structure. Each leaf node corresponds to a classification, non-leaf nodes correspond to a certain attribute on the division, according to the sample in the attribute on the different values will be divided into several subsets. For the non-leaf node label are pure, the majority class at this node belongs to the class of samples. The core problem of the construction decision tree is how to select the appropriate attributes at each step to split the sample. For a classification problem, this is a top-down process that learns and constructs decision trees from known training samples.

3 Introduction and Basic Principle of ID3 Algorithm

The ID3 algorithm is to select the optimal test attribute through information entropy [8, 9]. It takes the attributes of the current sample with the maximum information gain value as the test attribute. The partition of the sample set is based on the value of the test attribute. How many different values of test attributes are used to classify the sample set into sub sample sets. At the same time, a new leaf node is generated on the decision tree corresponding to the sample set. According to the theory of information theory, the ID3 algorithm takes the uncertainty of the sample set as the criterion to measure the quality of the partition. It uses information gain values to measure uncertainty: the greater the information gain value, the less uncertainty. Therefore, the test attributes of each non leaf node are the attributes with the largest information gain. This allows you to get the purest resolution of the current situation, resulting in a smaller decision tree.

Let S be a collection of s data samples, assuming that class attributes have m different values: $C_i(\mathrm{i} = 1, 2, \ldots, \mathrm{m})$. Let S_i be the sample number in class C_i. For a given sample, its total entropy is expressed as Formula 1.

$$I(s_1, s_2, \ldots, s_m) = \sum_{i=1}^{m} P_i log_2(P_i) \tag{1}$$

In Formula 1, P_i is the probability that any sample belongs to C_i and can be estimated by $\frac{s_i}{s}$ in general. Set an attribute A with k different values $a_1, a_2, \ldots, a_k$, and use the attribute A to split the set S into k subsets $S_1, S_2, \ldots, S_k$, where S_j contains samples of the a_j value in the set S attribute A. If the attribute A is selected as the test attribute, these subsets are new leaf nodes that grow from the nodes of the set S. Let S_ij be the sample number of C_i in the subset S_j, then the entropy of the sample is divided into formula 2 according to attribute A.

$$\begin{aligned}
E(A) &= \sum_{j=1}^{k} \frac{s_{1j} + s_{2j} + \ldots + s_{mj}}{s} I(s_{1j} + s_{2j} + \ldots + s_{mj}) \\
I(s_{1j} + s_{2j} + \ldots + s_{mj}) &= -\sum_{i=1}^{m} P_{ij} log_2(P_{ij}) \\
P_{ij} &= \frac{s_{ij}}{s_{1j} + s_{2j} + \ldots + s_{mj}}
\end{aligned} \tag{2}$$

Finally, the information gain (Gain) of the sample set S is divided into formula 3 by attribute A.

$$Gain(A) = I(s_1, s_2, \ldots s_m) - E(A) \tag{3}$$

Obviously, the smaller the E(A), the greater the value of Gain(A), which shows that the greater the information provided by the test attribute A to the classification, the less uncertain the classification is after the choice of A. The k values of the attribute A correspond to the K subset or branch of the sample set S. By recursively invoking the above process (not including the selected attributes), other attributes are generated as the child nodes and branches of the node to generate the entire decision tree. As a typical decision tree learning algorithm, the ID3 decision tree algorithm takes the information gain as the criterion of attribute selection at all levels of nodes. In this way, the maximum class classification gain can be obtained for each non - leaf node test, and the entropy of the classified data set is minimized. Moreover, this method can make the average depth of the tree smaller, and can effectively improve the classification efficiency

4 Modeling and Analysis

This paper takes the energy consumption of a laboratory building in the University as the experimental data source. The data is selected from the energy consumption data collected from July 2016 to May 2017.

4.1 Energy Consumption Data Analysis

The power consumption model can be obtained by analyzing the data of daily electricity consumption, as shown in Fig. 1. The following four modes of power consumption can be analyzed from the Figure.

1. From July 2016 to the end of October 2016, the staff in order to reduce indoor temperature, in addition to the use of basic electricity, laboratory air conditioning equipment also long time running, resulting in relatively high consumption.
2. From November 2016 to mid January 2017, the temperature gradually became cool, leading to a periodic pattern of energy consumption in the laboratory.
3. Because from the middle of January 2017 to the middle of February 2017, it belongs to the winter vacation, and the energy consumption of the laboratory gradually tends to be steady and the fluctuation is small. Moreover, the electricity consumption is much smaller than the characteristics of the workday.
4. From the middle of February 2017 to the beginning of May, the energy consumption of the laboratory was the same as that the second models of power consumption that mentioned above, and remained regular for one year as the school opened normally.

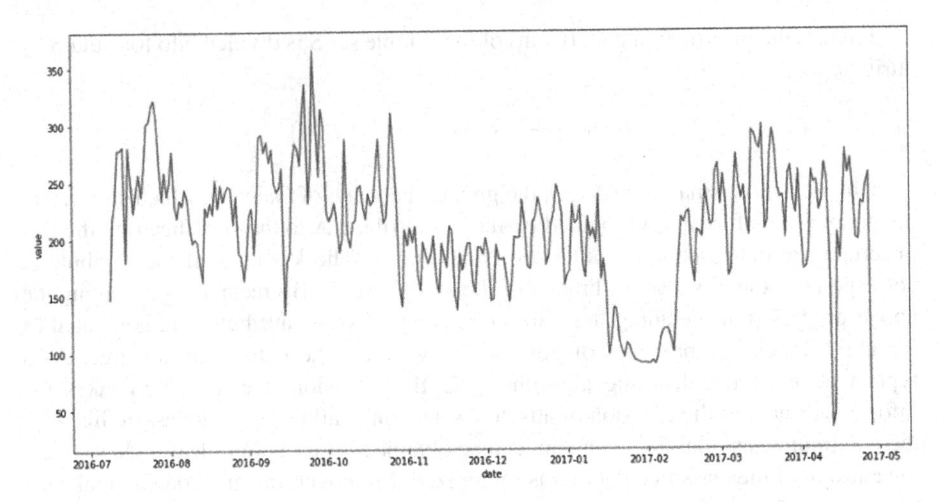

Fig. 1. Energy consumption of a laboratory in the University of Technology

4.2 Constructing Energy Consumption Decision-Tree Model

There are a number of different values in the data source, seasonal factor and holiday factor are two causations that influence the energy consumption. Since its own special geography location, the change of temperature caused by Fuzhou's seasonal change are not very evident, even the weather was very hot last through at least the end of October. Therefore, seasonal for changes of affecting energy consumption is devided into two parts, they are summer and winter. Since its own special nature of school's workers, holiday for changes of affecting energy another effect consumption is devided into two parts, they are holiday and non-holiday.

Electricity consumption is numeric, so it is necessary to analyze the properties. Dividing the 1/4 digit number and the 3/4 digit number as the demarcation points of the numerical value, numerical division less than 1/4 percentile to the category of "low energy consumption", more than 3/4 digit numerical division to the category of "high energy consumption", the numerical in between is numerical normal energy consumption. The power consumption decision tree model is established as shown in the Fig. 2:

It can be seen from the decision tree model that according to the level of power consumption, the total information entropy can be obtained that I(74,146,74) = 1.5034, For the seasonal properties, in the season as the "winter" category, there are 182 energy consumption values, as the "summer" category, there are 112 energy consumption values. The information entropy of seasonal attributes is obtained, I winter (27,88,67) = 1.446, I summer (47,58,7) = 1.2674. Thus, the information entropy of attribute as season is obtained by formula 4.

$$E(season) = \frac{182}{294} I(27, 88, 67) + \frac{112}{294} I(47, 58, 7) = 1.3780 \tag{4}$$

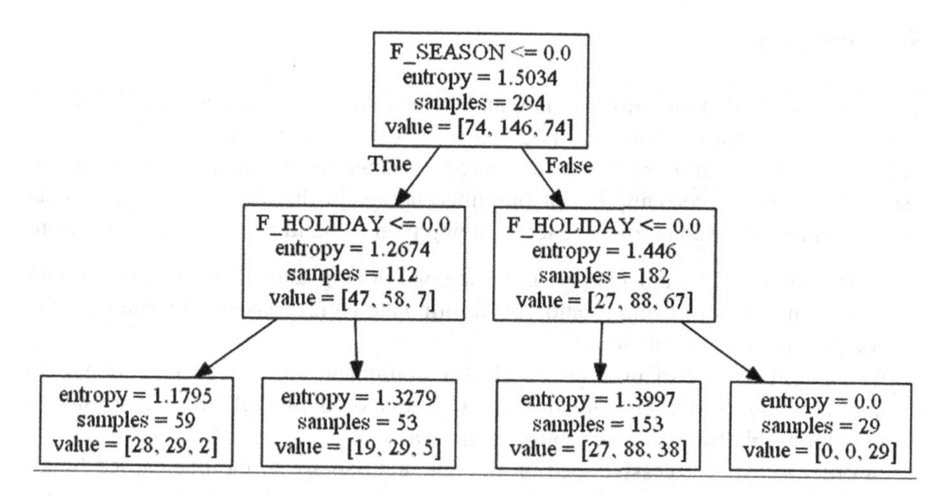

Fig. 2. Decision tree model of energy consumption

For the holiday properties, in the category of "Non holiday", there are 212 energy consumption values; in the category of "Holiday", there are 82 energy consumption values. The information entropy of holiday attributes is obtained, $I_n\, on\, holiday\,(55, 117, 40) = 1.4322.$, $I_h\, holiday\,(19, 29, 34) = 1.5458.$ Thus, The information entropy of attribute as holiday is obtained by formula 5.

$$E(holiday) = \frac{212}{294} I(55, 117, 40) + \frac{82}{294} I(19, 29, 34) = 1.4639 \quad (5)$$

The information gain value of the seasonal and vacation properties can be calculated by the formulas 6 and 7:

$$Gain(Season) = I(74, 146, 74) - E(Season) = 0.1254 \quad (6)$$

$$Gain(holiday) = I(74; 146; 74) - E(holiday) = 0.0395 \quad (7)$$

According to the calculation results of the information gain value, the information gain value of the seasonal attribute is the largest, it shows that in the test attribute, the season provides the most information for the classification and the most important feature, that is, the attribute has the greatest impact on the classification. Therefore, the influence of seasonal change on power consumption is the most important. Due to the specific seasonal characteristics of Fuzhou, the summer cycle is long and the temperature is high, so the laboratory will be open for a long time in the state of air-conditioning, air-conditioning and other refrigeration equipment consumption cannot be ignored. The winter cycle is short and the temperature is moderate, so no heating equipment is needed. The characteristics of the spring and autumn period are not obvious and the cycle is short, so it could be ignored.

5 Summary

The results show that the information gain between the seasonal transformation factor and the energy consumption of building is 0.1254, it is greater than the other factor, "vacation", and it can be seen that the seasonal change factor has an important influence on the energy consumption of building energy. In this regard, this paper puts forward three suggestions for the energy saving plan of building energy consumption:

1. In the summer of high temperature, air conditioning and other air conditioning equipment could be using rotation, And formulate regulations about turning off the equipment when people leave.
2. Centralized control of high power electric equipment such as air conditioner. In non-work hours, the air-conditioning will be closed through the energy-saving monitoring platform to avoid unnecessary waste.
3. In order to avoid excessive load on the line and damage to the line caused by air conditioning equipment and other high-power electrical equipment open at the same time, it is suggested that these high-power equipments should be reformed and controlled centrally.

The results of this study show the factors that influence the energy consumption of electricity, but for the different buildings, there are more influential factors. Therefore, modeling studies are needed for different buildings to tailor appropriate energy conservation plans for different buildings.

Acknowledgments. This work is supported by Fujian Science and Technology Department (No. 2014H0008), Fujian Transportation Department (No. 2015Y0008), Fujian Education Department (Nos. JK2014033, JA14209, JA1532), and Fujian University of Technology (Nos. GYZ13125, 61304199, GY-Z160064). Many thanks to the anonymous reviewers, whose insightful comments made this a better paper.

References

1. Lu, Y., Yan, Y.: Research on energy consumption and energy efficiency design for department stores in Chongqing. J. Chongqing Univ. **4**, 195–197 (2005)
2. Hongwei, W., Bai, X., Sun, C., Guo, L.: Analysis of energy consumption status and energy efficiency potential in large commercial buildings of Chongqing. Contam. Control Air-cond. Technol. **4**, 47–50 (2005)
3. Wan, S.: Research on Electrical Energy Consumption Model and Energy saving Control of Largescale Emporium Buildings in Xi'an. Xi'an University of Architecture and Technology (2016)
4. Lu, D.: Research and Application on the Data Mining Algorithm Basedon Decision Tree. Wuhan University of Technology (2008)
5. Guan, X.: Research on the Classifying Algorithm based on Decision Tree. Shanxi University (2006)
6. Tan, J., Wu, J.: Classification algorithm of rule based on decision tree. Comput. Eng. Des. **31**, 1017–1019 (2010). doi:10.16208/j.issn1000-7024.2010.05.023

7. Quinlan, J.R.: Induction of decision trees. Mach. Learn. **1**, 81–106 (1986). doi:10.1007/BF00116251
8. Zhang, X.: Research on the ID3 Algorithms of Decision Tree. Zhejiang University of Technology (2014)
9. Wang, X., Jiang, Y.: Analysis and improvement of ID3 decision tree algorithm. Comput. Eng. Des. **9**, 3069–3076 (2011). doi:10.16208/j.issn1000-7024.2011.09.069

The Optimal Reactive Power Compensation of Feeders by Using Fuzzy Method

Jing Tang[1,2(✉)], Wenxin Zhang[1,2], Jeng-Shyang Pan[1,2], Yen-Ming Tseng[1,2], Hui-Qiong Deng[2], and Ren-Wu Yan[2]

[1] Fujian Provincial Key Laboratory of Data Mining and Applications, Fujian University of Technology, Fuzhou 350108, China
tangjingdndx@163.com
[2] School of Information Science and Engineering, Fujian University of Technology, Fuzhou 350108, China

Abstract. The research object is aimed to the small power distribution underground system with 22.8 kV voltage level which consists of two feeders connected by a tie switch. According to the switches of the feeder adjacent to the formation of the segment is divided into the feeder section and each section has its own complex power with the load characteristics of the varying. Therefore, base on the changing of the complex power of the feeder outlet and the complex power of the segment load of the feeder to derive the all of power factor and the maximum power factor of each section of the feeder maximum power factor of each segments of the feeder line are deduced to find the compensation point of the best power factor. In order to improve the feeder output and feeder segments of the power factor. Due to the feeder segment by each adjacent two switch to investigate the load concentration, complex power, load changing trend for 24 h a day and the fuzzy theory derive the best in each segment of the feeder reactive power compensation point and compensation quantity. Based on result by the three-phase load flow program to reach the relative error of the power factor deduced from the fuzzy theory is discussed. The average relative error of the three-phase load current is 1.94%, but the three-phase load power consumption is large and the calculation time more than the fuzzy theory of reactive power compensation has a superior effect.

Keywords: Fuzzy theory · Reactive power compensation · Three phase load flow · Distribution underground system · Power factor

1 Introduction

Power delivery from the power plant to the substations and then to the factory, the family, and power transmission is line indispensable. To improve the quality of transmission which need to improve the feeder segment of the power factor so that line loss will be effectively improved so power Factor improvement is an important part of the energy companies to save energy, but also the user to pursue the high power factor of the good environment. From view of energy management [1–3], except the most effective management for energy applications must be considered about the power supply quality and other related issues. For the research of power quality [4, 5] such as

P. Krömer et al. (eds.), *Proceedings of the Fourth Euro-China Conference on Intelligent Data Analysis and Applications*, Advances in Intelligent Systems and Computing 682, DOI 10.1007/978-3-319-68527-4_30

the study of voltage curve diagnosis [6, 7], voltage surge and drop suddenly [8, 9] and voltage flicker [10, 11] issues. But for the distribution of small systems combination by feeders to investigate the reactive power compensation [12–15] literature which can improve the power factor and voltage profile made the quality of power supply more reliable.

The objective of the study was to select an underground feeder 22.8 kV for a small power distribution system consisting of two feeders, seven switches and eight load segments. Collection the feeder parameters and loadings of load segments to calculate by load flow program to achieve the magnitude and phase angle of voltage and current, real power and reactive power 24 h a day as the reference data. Using fuzzy theory to analyze and simulate the power system to complete the virtual power compensation of the distribution system to improve the power factor of the feeder outlet and the load segments. The real power and reactive power through the fuzzy theory to calculate whether the power factor value or within the allowable error, what compensate quantity that need and where the optimal point to compensate. Through the fuzzy transformation, fuzzy logic, inverse fuzzy transformation to calculate the reactive power compensation of solution. Furthermore, compare with the load flow analysis of the reactive power compensation of solution with the relative error.

2 Load Variation of Distribution Feeder

Figure 1 shows the distribution of small system research object that including two *i* and *j* feeders with a tie switch that base on the segment switch linkage divide into eight load segments to derive the load concentration, and according to the load characteristics of the 24 h a day. Figures 2 and 3 are the feeder *i* and *j* real power and reactive power curves. By the load type of feeder statistic that two feeders are commercial load-oriented.

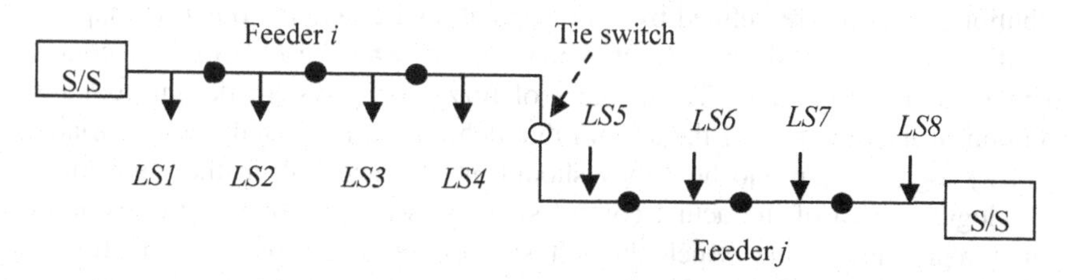

Tie switch =1, Switches = 6 , Load Segment = 8

Fig. 1. Segment load division of feeder

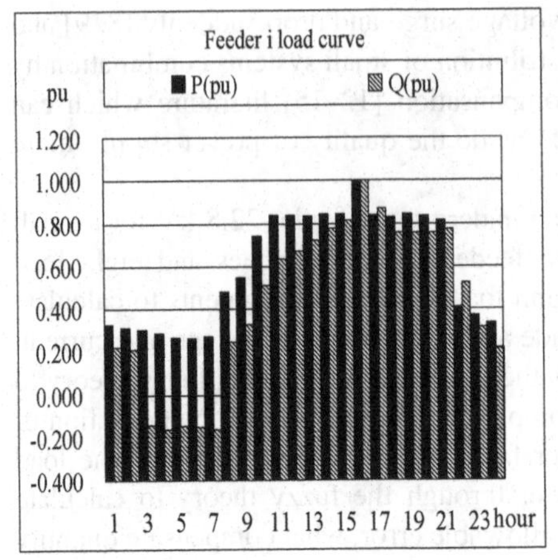

Fig. 2. Feeder *i* outlet daily load curve

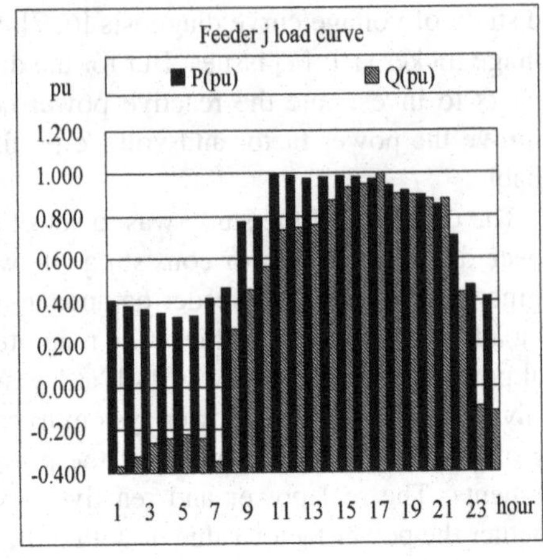

Fig. 3. Feeder *j* outlet daily load curve

3 Fuzzy System Architecture

3.1 System Architecture

It is using "fuzzy TECH 5.5" of fuzzy control software to process study. Fuzzy logic is an innovative technology; this technology enhances the traditional which is from the technical point of view to carry out the system design. The first step in the design of the fuzzy logic system is to define the structure of the system. To define the input and output variables of the fuzzy logic system and how the interaction should be in the input interface. Figure 4 contains the fuzzy values of the input values, and the left frame indicates the blurring method used. It lists the attribution function of the language variable of the predefined language variable by its name, and most of the design attribution functions are defined by way of point, and define the point or shape of the attribution function. Rule box as shown in Fig. 5 which each rule box contains a separate set of fuzzy rules. The process of fuzzy logic system design contains the definition of the structure of the system that defines the input and output variables of the fuzzy logic system and how they should interact and link in the input interface. Fuzzy logic system of the actual control strategy is definite of fuzzy rules and each column represents a rule which the left side of the line represents the fuzzy rule number. the first column has three buttons, respectively, "matrix" button, representing the matrix Rule editor, the other two "IF" and "THEN" button, is to provide the rules of the IF part and THEN part of the title, the use of these three keys can change the inference method, the button "DOS" on behalf of each rule weight.

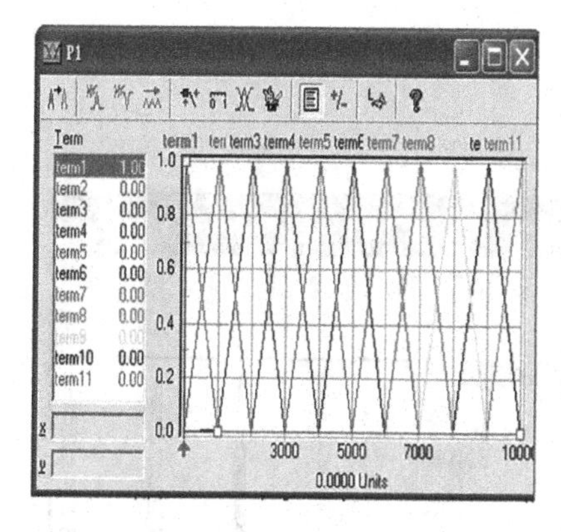

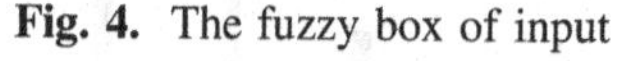

Fig. 4. The fuzzy box of input

Spreadsheet Rule Editor - RB1

#	IF in1	in2	in3	in4	THEN DoS	out1	THEN DoS	out2
1	low	low	low	low	0.00	very_low	0.00	very_low
2	low	low	low	low	0.00	low	0.00	low
3	low	low	low	low	0.00	medium	0.00	medium
4	low	low	low	low	0.00	high	0.00	high
5	low	low	low	low	0.00	very_high	0.00	very_high
6	low	low	low	medium	0.00	very_low	0.00	very_low
7	low	low	low	medium	0.00	low	0.00	low
8	low	low	low	medium	0.00	medium	0.00	medium
9	low	low	low	medium	0.00	high	0.00	high
10	low	low	low	medium	0.00	very_high	0.00	very_high
11	low	low	low	high	0.00	very_low	0.00	very_low
12	low	low	low	high	0.00	low	0.00	low
13	low	low	low	high	0.00	medium	0.00	medium

Fig. 5. Fuzzy logic rule box

4 Fuzzy Logic Applied to Power Factor Improvement

The 24-h real power P and reactive power Q input to the fuzzy logic program to calculate the PF1–PF8 power factor. From PF1–PF4 find out which points to be compensated, how much reactive power quantity to be compensated and how to improve the power factor. That PF5–PF8 has the same method to do. Input variables there are P1–P8 input real power and Q1–Q8 input reactive power sixteen variables and the output are PF1–PF8 eight power factor of load segments. Input variables there are P1–P8 input real power and Q1–Q8 input reactive power sixteen variables and the output are PF1–PF8 eight power factor of load segments.

Figure 6 simulation and verification of fuzzy logic system debug mode. Where AllPF1 is PF1 to PF4 before reactive power compensation the average power factor and AllPF2 is PF5 to PF8 before reactive power compensation the average power factor, var5 or var6 is indicate which point to be used to reactive power compensate, newQ1–newQ8 compensate how much reactive power need compensated? And newpf1–newpf8 are after reactive power compensation the power factor. n the var5, var6 window which the red arrow indicates the best power factor point PF6 (LM6 load segment) to be compensated as shown in Fig. 7.

5 Examples of Discussion

The 24-h actual feeder data and the load data input to the three-phase load flow and fuzzy logic to calculate the 24-h results of the power factors to make a comparison to assess whether the error within the allowable value as a basis. Table 1 is the feeder *i* and *j* daily 24 h the best reactive power compensation position and compensation quantity of the statistical table. From Table 1 the compensation location covers LM1, LM2 and LM4, only LM3 for the non-compensation location in feeder *i*. From 24:00 P.M. to 7:00 A.M. compensate for the leading power factor and the rest of the time to compensate for the lagging power factor of feeder *i* which the compensation in the range of −1289–2592 kVAR. At 9:00 A.M., feeder *i* need reactive power 2375 kVAR

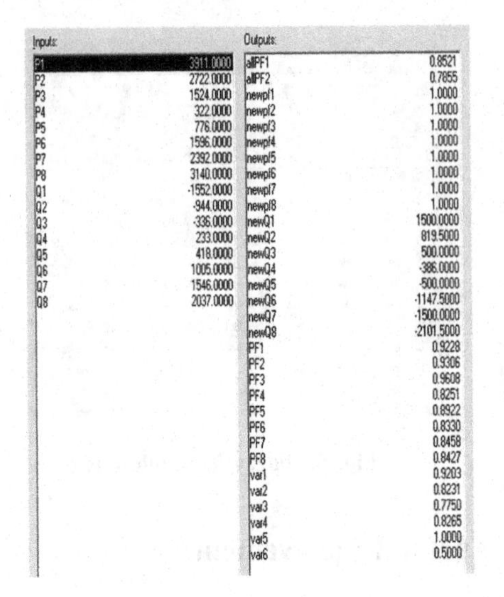

Fig. 6. Simulation and verification of fuzzy logic system debug mode

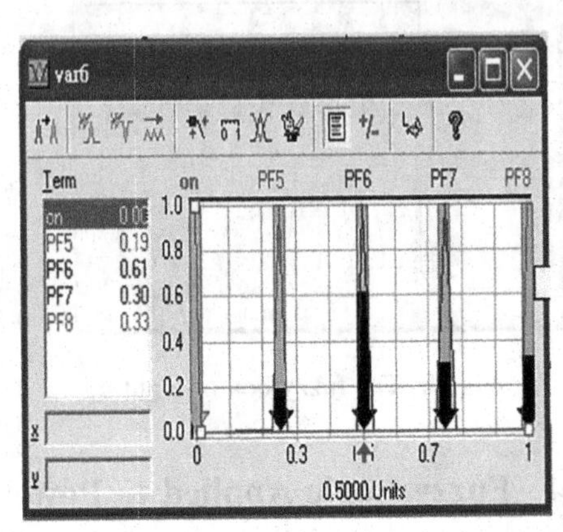

Fig. 7. The feeder in the PF6 point for the best compensation

for compensate, At 10:00 A.M., feeder i need reactive power 2592 kVAR for compensate made the curve to form two peaks. For 24 h a day, the absolute average quantity for compensation is 77.21 kVAR and the reactive power compensation curve shown in Fig. 8.

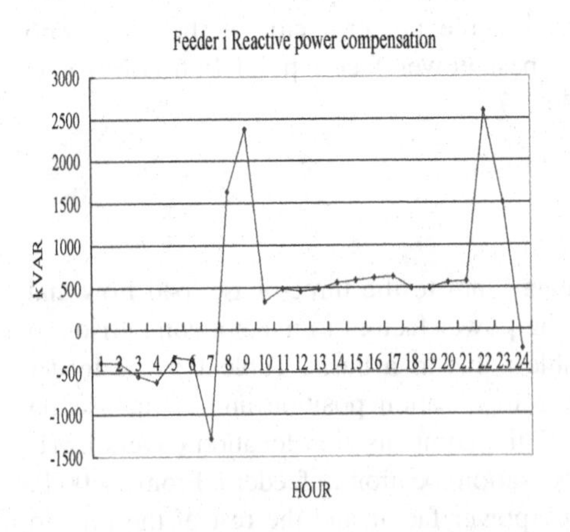

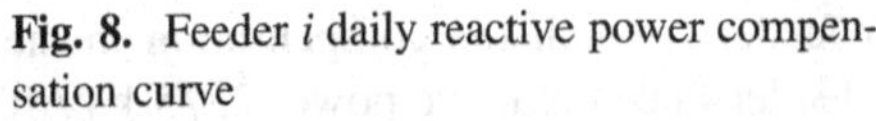
Fig. 8. Feeder i daily reactive power compensation curve

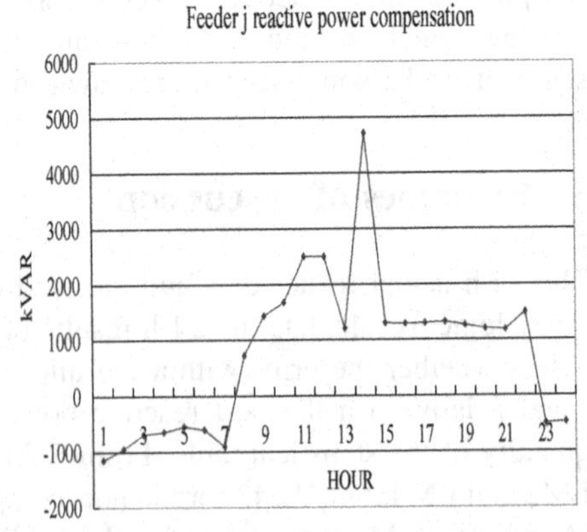

Fig. 9. Feeder j daily reactive power compensation curve

From Table 1 the compensation location covers LM4 and LM5, the other point of feeder j is the non-compensation location. From 23:00 P.M. to 7:00 A.M. compensate for the leading power factor and the rest of the time to compensate for the lagging power factor of feeder *j* which the compensation in the range of −1147–4719 kVAR. At 10:00 P.M., feeder *i* need reactive power 4719 kVAR for compensate made the curve to form one peaks. For 24 h a day, the absolute average quantity for compensation is 1318.38 kVAR and the reactive power compensation curve shown in Fig. 9.

Table 2 shows the 24 h a day average power factor of the feeder i average power factor by three phase load flow and fuzzy method for comparison. The relative error *REk* is defined as Eq. (1).

$$RE_k = \frac{PF_{tpl} - PF_{fz}}{PF_{tpl}} \times 100\% \tag{1}$$

Table 1. The optimal compensation locations and quantity for feeders *i* and *j*

Hour	Feeder *i*		Feeder *j*	
	Optimal location	Compensation quantity (kVAR)	Optimal location	Compensation quantity (kVAR)
1	LM4	−386	LM 6	−1147
2	LM 4	−396	LM 6	−947
3	LM 2	−549	LM 6	−694
4	LM 2	−622	LM 6	−647
5	LM 4	−325	LM 6	−552
6	LM 4	−355	LM 6	−606
7	LM 1	−1289	LM 6	−907
8	LM 1	1632	LM 6	722
9	LM 1	2375	LM 6	1441
10	LM 4	333	LM 6	1682
11	LM 4	488	LM 6	2500
12	LM 4	456	LM 6	2500
13	LM 4	482	LM 5	1203
14	LM 4	556	LM 8	4719
15	LM 4	583	LM 5	1308
16	LM 4	612	LM 5	1295
17	LM 4	626	LM 5	1307
18	LM 4	496	LM 5	1330
19	LM 4	493	LM 5	1266
20	LM 4	560	LM 5	1203
21	LM 4	572	LM 5	1185
22	LM 2	2592	LM 6	1501
23	LM 2	1500	LM 6	−500
24	LM 4	−231	LM 6	−479
average		77.21		1318.38

where

k: Representative is i or j.
PF_{tpl}: power factor investigate by Three-phase load flow
PF_{fz}: Power factor derive by fuzzy method

Form Table 2 which the maximum relative error of 3.97%, the minimum relative error of 0.36%, the average relative error of 1.86% in feeder *i* and the maximum relative error of 2.84%, the minimum relative error of 0.65%, the average error of 2.02% in feeder *j*. In general, the relative error is quite accurate that the three-phase load power consumption is large and the calculation time more than the fuzzy theory of reactive power compensation has a superior effect.

Table 2. Comparison of the average power factor of feeder *i* and *j* with fuzzy method and three phase load method

Hour	Feeder *i*			Feeder *j*		
	Three- phase load flow	Fuzzy method	*REi* (%)	Three- phase load flow	Fuzzy method	*REj* (%)
1	0.81	0.83	1.85%	0.85	0.83	1.89%
2	0.78	0.81	3.45%	0.86	0.87	0.93%
3	0.96	0.96	0.41%	0.90	0.92	2.34%
4	0.95	0.94	0.74%	0.89	0.90	1.01%
5	0.92	0.89	3.26%	0.90	0.92	2.34%
6	0.88	0.85	3.08%	0.89	0.91	2.13%
7	0.95	0.93	1.59%	0.88	0.89	1.14%
8	0.96	0.92	3.97%	0.93	0.92	0.65%
9	0.94	0.93	1.38%	0.93	0.91	1.62%
10	0.88	0.85	2.97%	0.90	0.88	1.79%
11	0.85	0.84	0.94%	0.89	0.87	2.14%
12	0.87	0.85	1.73%	0.88	0.86	2.60%
13	0.85	0.85	0.47%	0.86	0.84	2.33%
14	0.81	0.81	0.62%	0.85	0.83	1.89%
15	0.78	0.80	2.04%	0.84	0.82	2.84%
16	0.76	0.79	3.40%	0.84	0.82	2.84%
17	0.77	0.79	2.73%	0.83	0.80	3.03%
18	0.84	0.84	0.36%	0.80	0.78	2.74%
19	0.84	0.84	0.36%	0.81	0.79	2.23%
20	0.80	0.81	0.87%	0.82	0.80	1.96%
21	0.78	0.80	2.56%	0.81	0.79	2.11%
22	0.80	0.79	1.25%	0.90	0.88	1.90%
23	0.91	0.88	3.30%	0.99	0.97	2.02%
24	0.98	0.96	1.94%	0.98	0.96	1.94%
Average	0.86	0.86	1.86%	0.88	0.87	2.02%

Table 3 for the small distribution systems 24 h a day using the theory of fuzzy theory derived from the average power factor and three-phase load power flow calculation of the average power factor relative error comparison. The small distribution system covers two feeder i and j and let tie-switch fixed and immovable. The relative error is 1.58% at PM 5:00, the relative error is 0.36% at AM 4:00 and the average error is 1.94%, which is quite accurate. The average relative error curve is shown in Fig. 10.

Table 3. Relative error values for small hours of 24 h distribution

Hour	Feeder i	Feeder j	Average
1	1.85%	1.89%	1.87%
2	3.45%	0.93%	2.19%
3	0.41%	2.34%	1.38%
4	0.74%	1.01%	0.88%
5	3.26%	2.34%	2.80%
6	3.08%	2.13%	2.61%
7	1.59%	1.14%	1.37%
8	3.97%	0.65%	2.31%
9	1.38%	1.62%	1.50%
10	2.97%	1.79%	2.38%
11	0.94%	2.14%	1.54%
12	1.73%	2.60%	2.17%
13	0.47%	2.33%	1.40%
14	0.62%	1.89%	1.26%
15	2.04%	2.84%	2.44%
16	3.40%	2.84%	3.12%
17	2.73%	3.03%	2.88%
18	0.36%	2.74%	1.55%
19	0.36%	2.23%	1.30%
20	0.87%	1.96%	1.42%
21	2.56%	2.11%	2.34%
22	1.25%	1.90%	1.58%
23	3.30%	2.02%	2.66%
24	1.94%	1.94%	1.94%
average	1.86%	2.02%	1.94%

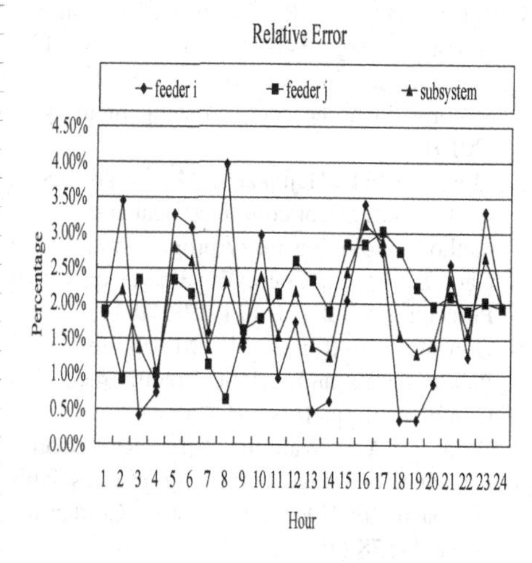

Fig. 10. Distribution of small system and feeder relative error curve

6 Conclusion

Reactive power is an indispensable in distribution system which made an impact on the system stability and reliable with the reactive power serious lack or excess. But reactive power in slight lack or excess will affect the electricity quality in supply side; especially for voltage and power factor and so on have a very important effect. In the system, real power is a continuous power and its flow can be derived by topology connection, but only production by the generator. But the reactive power in the system is discontinuous power but the flow can be derived by connecting the number of it is more difficult than real power that can be generators, capacitors and other components supply.

In order to improve the feeder output and feeder segments of the power factor. Due to the feeder segment by each adjacent two switch to investigate the load concentration, complex power, load changing trend for 24 h a day and the fuzzy theory derive the best in each segment of the feeder reactive power compensation point and compensation quantity. Based on result by the three-phase load flow program to reach the relative error of the power factor deduced from the fuzzy theory is discussed. The average relative error of the three-phase load current is 1.94%, but the three-phase load power consumption is large and the calculation time more than the fuzzy theory of reactive power compensation has a superior effect.

References

1. Zelazo, D., Dai, R., Mesbahi, M.: An energy management system for off-grid power systems. Energy Syst. **3**(2), 153–179 (2012)
2. Zhao, Z., Lee, W.C., Shin, Y., Song, K.B.: An optimal power scheduling method for demand response in home energy management system. IEEE Trans. Smart Grid **4**(3), 1391–1400 (2013)
3. Moradi, M.H., Hajinazari, M., Jamasb, S., Paripour, M.: An energy management system (EMS) strategy for combined heat and power (CHP) systems based on a hybrid optimization method employing fuzzy programming. Energy **49**(1), 86–101 (2013)
4. Gao, X., Li, L., Chen, W.: Power quality improvement for mircrogrid in islanded mode. Procedia Eng. **23**(5), 174–179 (2011)
5. González, P., Romero, E., Miñambres, V.M., Guerrero, M.A.: Grid-connected PV plants. Power quality and technical requirements. In: Electric Power Quality & Supply Reliability Conference, pp. 169–176 (2014)
6. Walls, J.A., Walton, A.J., Robertson, J.M., Crawford, T.M.: Interpretation of capacitance-voltage curves for process fault diagnosis: a machine-learning expert system approach. In: IEEE International Conference on Microelectronic Test Structures, vol. 41, pp. 174–178 (1988)
7. Wolny, S., Zdanowski, M.: Analysis of recovery voltage parameters of paper-oil insulation obtained from simulation investigations using the cole-cole model. IEEE Trans. Dielectr. Electr. Insul. **16**(6), 1676–1680 (2009)
8. Cheng, M.Z., Zhang, J.W., Cai, X., Cao, Y.F.: Influence of power system voltage drop on the DIFG shaft. Power Syst. Prot. Control **39**(20), 17–22 (2011)
9. Lahaçani, N.A., Mendil, B.: Modelling and simulation of the SVC for power system flow studies: electrical network in voltage drop. Leon. J. Sci. **7**(13), 153–170 (2008)
10. Chai, L., Sun, J.R.: Voltage flicker extraction based on wavelet analysis. Appl. Mech. Mater. **385–386**, 1389–1393 (2013)
11. Zhang, R., Liu, G.Q., Zhang, L.Y., Zhang, X.H.: The detection of voltage flicker based on complex wavelets. Appl. Mech. Mater. **48–49**, 1231–1234 (2011)
12. Li, F., Pilgrim, J.D., Dabeedin, C., Chebbo, A., Aggarwal, R.K.: Genetic algorithms for optimal reactive power compensation on the national grid system. Electr. Eng. **20**(1), 493–500 (2002)

13. Singh, M., Khadkikar, V., Chandra, A.: Grid synchronisation with harmonics and reactive power compensation capability of a permanent magnet synchronous generator-based variable speed wind energy conversion system. IET Power Electron. **4**(1), 122–130 (2011)
14. Majumder, R.: Reactive power compensation in single-phase operation of microgrid. IEEE Trans. Ind. Electron. **60**(4), 1403–1416 (2013)
15. Luo, A., Shuai, Z., Zhu, W., Shen, Z.J.: Combined system for harmonic suppression and reactive power compensation. IEEE Trans. Ind. Electron. **56**(2), 418–428 (2009)

Finite Element Analysis for No-Load and Rated Load of Low-Speed Permanent Magnet Hydro-Generator

Bing-Huang Chen[1,2,3(✉)]

[1] College of Electrical Engineering and Automation, Fuzhou University, Fuzhou 350118, China
chenbh@fjut.edu.cn
[2] School of Information Science and Engineering, Fujian University of Technology, Fuzhou 350118, China
[3] The Key Laboratory for Digital Equipment of Fujian Province, Fujian University of Technology, Fuzhou 350118, China

Abstract. Low-speed permanent magnet synchronous generator (PMSG) has the characteristics of high torque and low voltage. This study presents an optimized design about an outside rotor 10 kW low speed PMSG which is in the vertical vane wheel hydro-generator unit. Maxwell software is used to establish the finite element model of the low-speed PMSG, and the simulation analysis including three-phase flux waveforms, induced voltage waveforms, voltage waveforms, current waveforms, flux density distribution and torque characteristic curve is carried out under the no-load and rated load. Simulation results show that the low-speed PMSG, which has the skew stator core and fractional slot concentrated winding, has a series of advantages of low cogging torque, high efficiency, small output voltage harmonics and low waveform sine distortion rate, fully meeting the design requirements.

Keywords: Low speed · Operation · Finite element · Simulation · Permanent magnet

1 Introduction

In the last decade, develop clean and renewable energy has become the focus of the world. In the green renewable resources, water is a kind of important energy, the development and the use of water is of great significance [1].

The current way to develop water resources generally is the construction of hydropower station, using high efficient turbine to convert water. But there is some request for site selection, for example some scattered low head stream will not be able to use. Low-speed PMSG has high efficiency, low cost and good performance advantages. With the continuous development of power electronics and related technology, low-speed PMSG become more and more popular in wind power, hydropower, tidal power and other renewable energy power generation system [2, 3].

Considering that the system should be compact, this paper is aimed to build a low-speed PMSG that can be simply integrated with an ultra-low head hydroelectric

P. Krömer et al. (eds.), *Proceedings of the Fourth Euro-China Conference on Intelligent Data Analysis and Applications*, Advances in Intelligent Systems and Computing 682, DOI 10.1007/978-3-319-68527-4_31

power system with direct drive. For this purpose, the rotor of the low-speed PMSG and the vertical vane of waterwheel are connected directly by bolts [4]. So, the generator and the turbine are at the same speed. The design and performance of the generator are validated using laboratory experiment.

2 Structure and Electromagnetic Design of Low-Speed PMSG

2.1 Structure of Low-Speed PMSG

The low-speed PMSG is installed in the vertical vane waterwheel and the water wheel structure is shown in Fig. 1. The structure of the low-speed PMSG is shown in Fig. 2 (a). The rotor is designed which is out of the stator. The vertical vane of waterwheel is nested directly on the rotor. The constructive shape of generator is shown in Fig. 2(b).

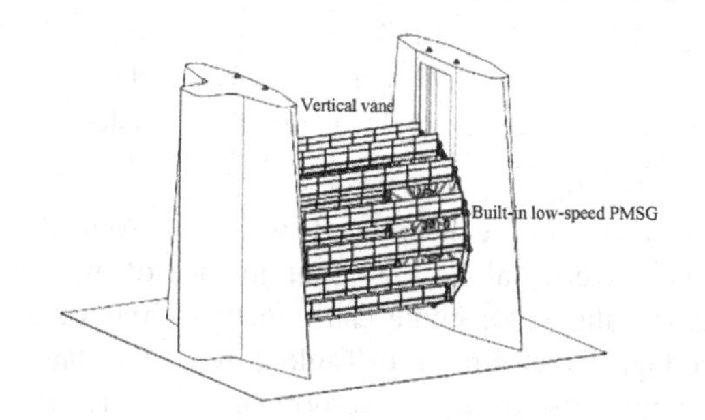

Fig. 1. Structure of the waterwheel

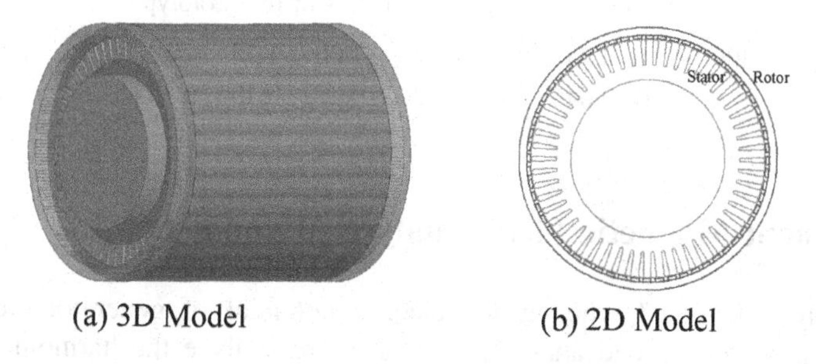

Fig. 2. Structure of the low-speed PMSG

The generator adopts chute design. The magnetic circuit configuration adopts radial direction, which the tiled magnet steel made of Nd-Fe-B pasted in the internal surface of the rotor. The stator winding is double fractional slot concentrated winding [5].

2.2 Design and Construction

Compared with the ordinary permanent magnet generator, the generator structure is especially suitable for water turbine because of its exterior rotor adopts multipolar structure. The magnetic circuit is shortened, which not only improve the utilization rate of the permanent magnetic material and reduced the yoke height. So, the generator manufacturing cost is reduced greatly [6].

The following Table 1 gives a summary of the main characteristics of the Low-speed PMSG.

Table 1. Main characteristics of the prototype

Characteristics	Value	Characteristics	Value
Exterior diameter of rotor	1000 mm	Interior diameter of rotor	940 mm
Exterior diameter of stator	904 mm	Interior diameter of stator	600 mm
Active length of electrical machine	950 mm	Air gap	3 mm
Rated power	10 kW	Output voltage	3 × 230 V
Rated speed	20 rpm	Number of slots	54
Voltage frequency	10 Hz	Number of poles	60

The stator adopts fraction concentrated winding. And the slot of stator is semi-closed trapezoidal groove along the circumference of the stator. The stator core adopts chute design, as the stator chute can effectively reduce distortion rate of the sinusoidal output voltage waveform. From Table 2, when the stator skew width 1 times that of the stator slot pitch, the voltage waveform sine distortion rate of the generator is minimal, and the generator output voltage waveform sine is the best [7].

Table 2. Structure parameters of the prototype

Skew width	0	0.25	0.50	0.75	1.00	1.25	1.50	1.75	2.00
THD%	1.910	0.645	0.409	0.297	0.222	0.409	0.357	1.835	0.988

3 Transient Magnetic Field Analysis Without Load

It is mainly considered whether the electric potential of generator can meet the requirements when no-load analysis, as well as to analyze the harmonic of no-load voltage and to verify the cogging torque of the generator, etc. The speed of the generator is set to the rated speed 20 RPM and load type is set to the current source, and the current value is set to 0 in transient field solver of Maxwell software. The generator running related data would be got by model simulation without load.

The Three-phase flux waveforms are shown in Fig. 3. The generator flux waveform is sine wave and flux linkage amplitude is 6.6959 Wb, which RMS is 4.7347 Wb.

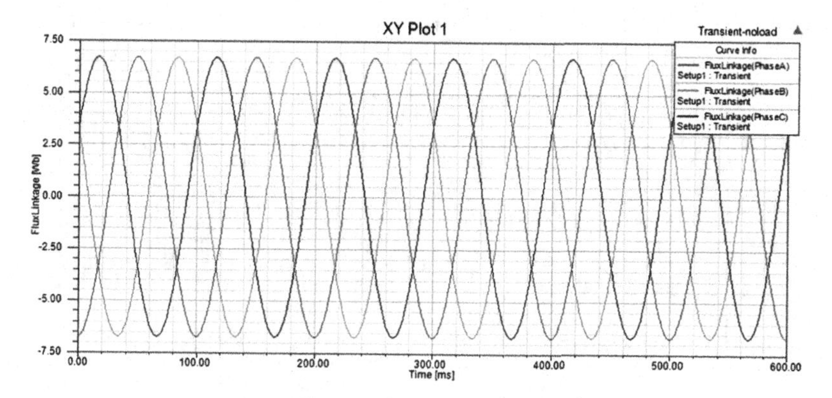

Fig. 3. Three-phase flux waveforms without load

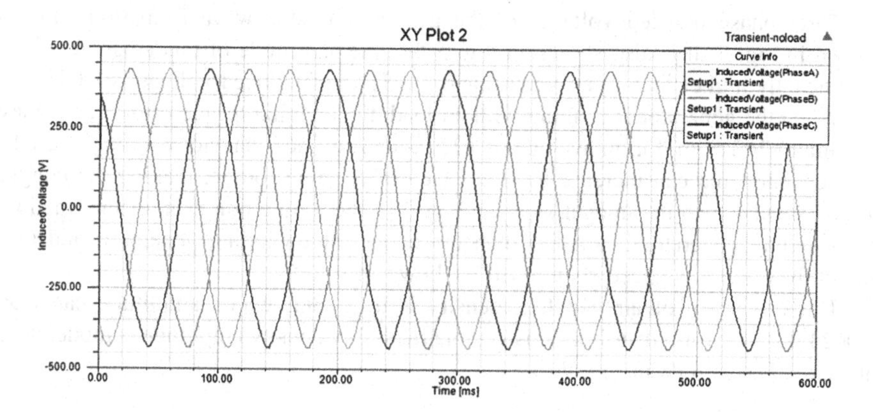

Fig. 4. Three-phase induced voltages waveform

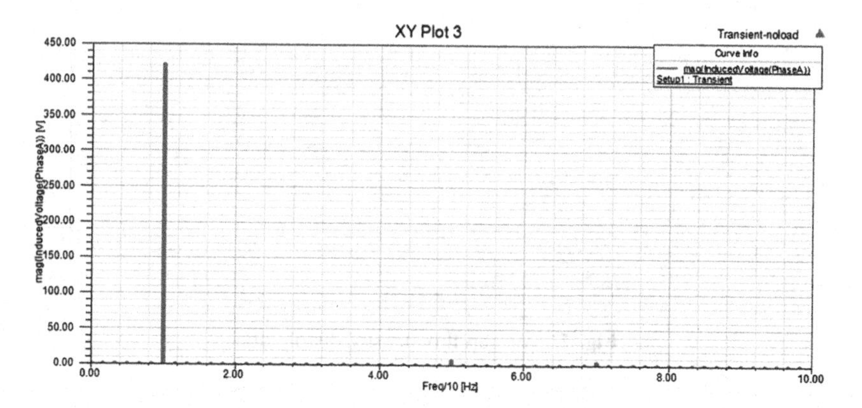

Fig. 5. Fourier analysis of three-phase induced voltages

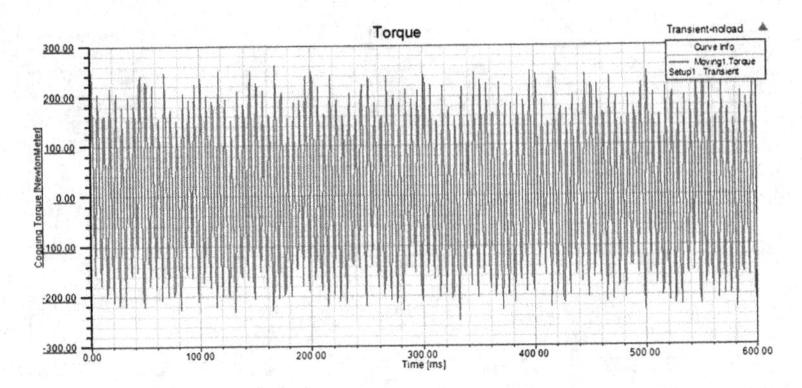

Fig. 6. The cogging torque waveform

Three-phase induced voltages of the generator is sine wave formation which is shown in Fig. 4. It is also shown in Fig. 5 that the amplitude of fundamental wave is significantly higher than other waves, and the THD of induced voltage is 0.04%.

When the generator is running without load, the cogging torque ripple can achieve maximum 262 N·m, for an average of 192 N·m. The fractional slot winding is used in the generator that can increase the frequency of cogging torque, making the cogging torque smaller than the integer slot winding [8]. The torque pulsation of the fractional slot winding generator is smaller than the integer slot winding generator, and more smoothly when low speed running. It is shown in Fig. 6.

The waveform of air gap flux density is trapezoid, which amplitude can reach 1.0624 T. Waveform distribution is symmetrical, flux density amplitude is moderate. In this case, it is shown in Fig. 7.

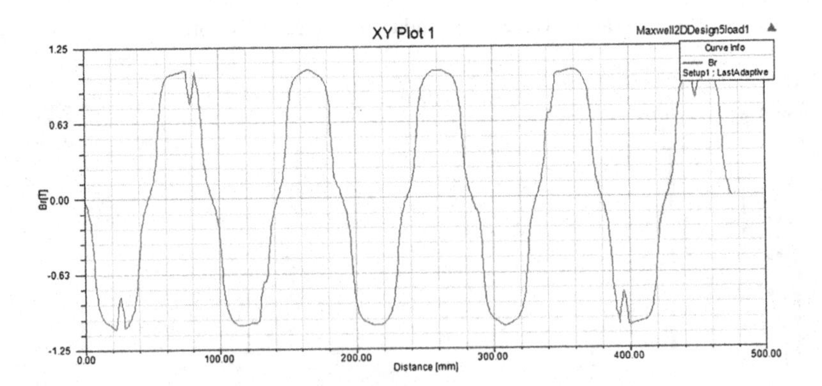

Fig. 7. Air gap flux density distribution

4 The Runtime Transient Magnetic Field Analysis with Rated Load

It is mainly considered in analysis of Low-speed PMSG whether the rated voltage and the output power can meet the requirements, as well as to analyze the harmonic of voltage and to verify the torque of the generator, etc. The speed of the generator is set to the rated speed 20 RPM and load type is set to the voltage source, and the current value is set to 0 in transient field solver of Maxwell software. The generator running relevant data would be got by model simulation with rated load through setting the value of resistance and inductance.

Figure 8 shows that flux linkage amplitude of the generator under the rated load is 6.0388 Wb, which is down by 10.21% over the flux amplitude without load. The decrease of the flux values is mainly due to the magnetic field generated by load current in the generator. The direction of the magnetic fields generated by the current and the permanent magnet is instead, which leading to the flux value of generator reduced.

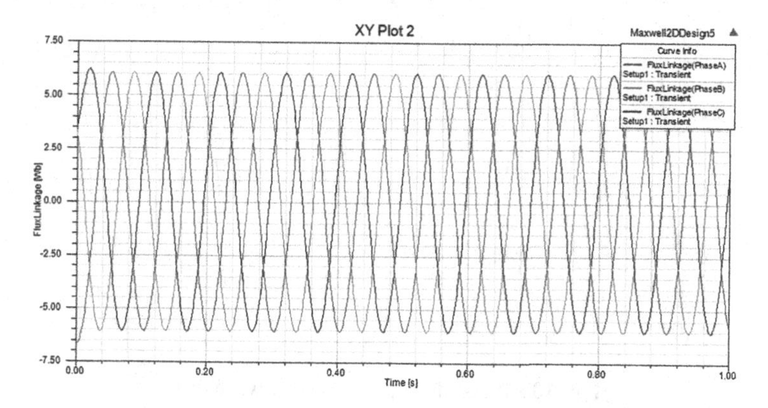

Fig. 8. Three-phase flux waveforms with rated load

The amplitude of rated phase voltage and phase current is 187.794 V and 35.9757 A respectively from Figs. 9 and 10. According to the power calculation formula, the output power of the generator can reach 10.134 kW, meeting the requirements of design.

$$P = 3U_N I_N \cos\varphi \tag{1}$$

The rated torque of the generator is smooth, less volatile, torque amplitude can reach 9.4468 kN·m, which output torque is large (see Fig. 11).

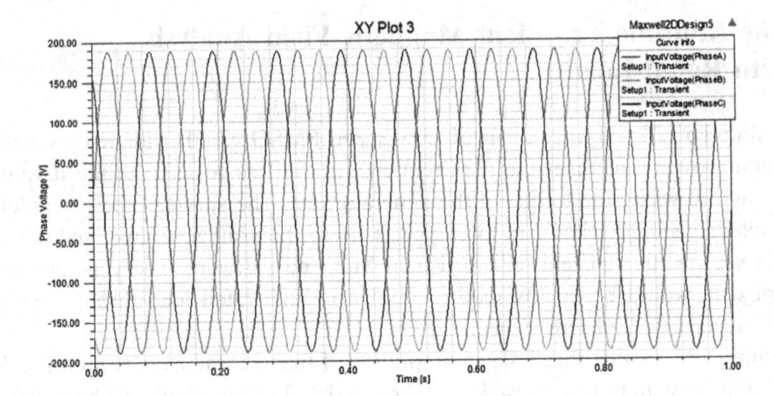

Fig. 9. Rated three-phase voltage waveforms

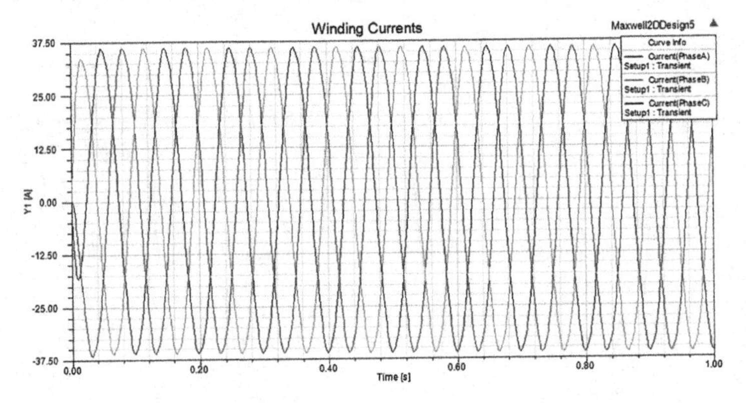

Fig. 10. Rated three-phase current waveforms

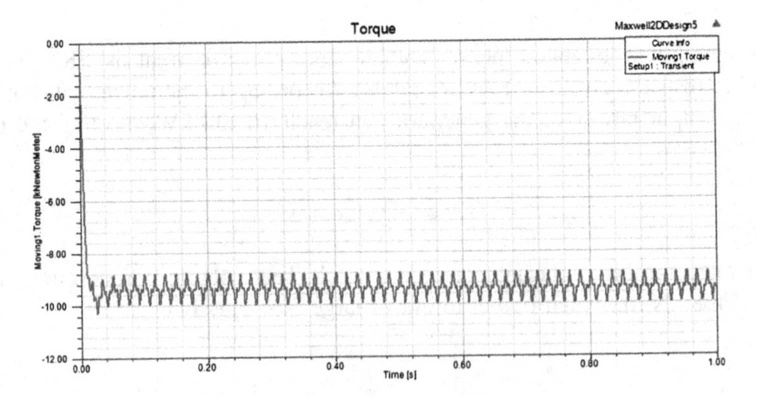

Fig. 11. Rated torque waveform

5 Conclusion

The finite element analysis of a 10 kW low-speed PMSG has been discussed in this paper, which is based on the Maxwell software. The simulation results show that low-speed PMSG has lower cogging torque and high efficiency, the output voltage waveform harmonic component of the generator is small and the waveform sine distortion rate is low. It can satisfy the demands of the turbine matching [9].

Acknowledgement. This work was partially supported by National Science and Technology Support Project under grant 2015BAF24B01, and also supported by Fujian province science and technology innovation platform under grant 2014H2002, Middle-aged and young teachers' scientific research foundation of Fujian province under grant JA15345, Fuzhou science and technology plan project under grant 2016-G-57.

References

1. Williamson, S.J., Stark, B.H., Booker, J.D.: Performance of a low-head pico-hydro Turgo turbine. Appl. Energy **102**, 1114–1126 (2013)
2. Adhau, S.P., Moharil, R.M., Adhau, P.G.: Mini-hydro power generation on existing irrigation projects: case study of Indian sites. Renew. Sustain. Energy Rev. **16**(7), 4785–4795 (2012)
3. Irasari, P., Sutikno, P., Widiyanto, P.: Performance measurement of a compact generator - hydro turbine system. Int. J. Elect. Comput. Eng. (IJECE) **5**(6), 1252–1261 (2015)
4. Chen, B., Lei, S.: A vertical vane waterwheel based on ultra-low head hydroelectric power. China Rural Water Hydropower **21**(11), 60–61 (2012)
5. Wang, B., Li, S.: Low speed PM synchronous motor with approximate number of poles and slots. MicroMotors **26**(40), 87–88 (2007)
6. Wen, J., Song, J.: Performance analysis of external-rotor permanent magnet synchronous generator based on ANSOFT. Explosion-Proof Electric Mach. **45**(1), 5–8 (2013)
7. Gong, J.: The influence of skew width & non-uniform air-gap on PMSM's performance. Large Electric Mach. Hydraul. Turbine **13**(1), 17–20 (2008)
8. Ni, Y., Li, W., Lang, X.: Research of cogging torque and performance of PM generators. MicroMotors **46**(4), 187–190 (2013)
9. Liu, Z.: Optimum design and simulation of low-speed permanent magnet hydro-generator. Eng. J. Wuhan Univ. **48**(4), 502–506 (2015)

Design and Implementation of Intelligent Outlet System Based on Android and Wifi

Lisang Liu[1,2](✉), Chengwen Lian[1], Ying Ma[1,2], Dongwei He[1,2], Jianxing Li[1,2], and Tianjian Li[1]

[1] Fujian University of Technology, Fuzhou, China
liulisang@fjut.edu.cn
[2] Key Laboratory of Digital Equipment, Fuzhou, Fujian Province, China

Abstract. With the deepening of the concept of Internet of things and the development of science and technology, smart home has developed by leaps and bounds. Intelligent wireless outlet is an essential part of smart home. In this paper, a wireless outlet system is designed based on Wifi technology and Android platform, using SCM and ESP8266 module to build the hardware system. The cloud server is adopted to transmit data between host computer and slave computer, while the smart phone based on Android can collect the data and control the outlet nodes through the Wifi communication mode so as to realize the monitoring of the home appliance. The system is implemented successfully and can reform the traditional home without complicated wiring. Stable and reliable as tested, the system has great practical value.

Keywords: Intelligent outlet · Android · Wifi · Smart home

1 Introduction of Smart Home

Internet of things (IoT) is a novel type of network, which combines the internet, wireless technology and process technology of micro electrical-mechanical. The objects, people and animals in a certain case would be assigned a unique identifier that allows them to transmit data over the network without direct interaction between human and computers [1]. As the development of IoT, more and more attention has been paid to the industrial applications of IoT. Smart home, which is one of the applications, is gradually permeating our daily life. Aim to enhance convenience, safety, comfort and artistry of house, smart home conduct an efficient management system to realize an energy saving residential environment using wiring technology, security technology, network communication technology and automatic control technology [2]. Smart home not only can provide people with convenient and comfortable family life, but also can change the home environment from the original passive structure into active structure, and realize the omni-directional information interaction. Actually, it is the intelligent module added to the electrical equipment that makes our lives more comfortable.

P. Krömer et al. (eds.), *Proceedings of the Fourth Euro-China Conference on Intelligent Data Analysis and Applications*, Advances in Intelligent Systems and Computing 682, DOI 10.1007/978-3-319-68527-4_32

1.1 Research Status of Smart Home

The concept of smart home was first proposed in 1989, when the United Technologies Corporation (UTC) applied it to a building in Hartford city of Connecticut. A prototype as it was, though, it brought endless temptation that Japan, Korea, the European Union and other countries followed the USA to put forward the solution of smart home [3]. The smart home got rapid development in China since 1990 and it was improved as the development of communication.

At present, various vendors have launched their own smart home system, but they are too costly to be accepted. Compared with the whole set of system transformation, it is easier and cheaper to carry out the intelligent transformation of the outlet [4]. It is expected to control the electrical appliances and equipment far away from home by simply operating on the mobile phone. Imaging that one can enjoy cool air, a hot bath and fresh coffee as soon as the home arrival only because he/she turns on the air condition, the water heater and the coffee machines before going back home! Therefore, an intelligent outlet control system, using the smart phone as the mobile control terminal, came into being.

1.2 Research Status of Intelligent Outlet System

As an emerging electrical appliance of smart home, intelligent outlet is connected with the power socket of various household appliances. It enables users to realize the switch and timing remote control of the connected electric appliance. According to statistics of York central air-conditioning, the demand of the intelligent outlet is over that of intelligent security and is accounted for more than half of total demand in the use of a variety of intelligent home furnishing. The outlet is one of the most frequently used equipment in all kinds of family that its performance and market prospect is essential to its future development.

According to the "Industry Research and Investment Strategy Research Report of intelligent outlet during 2013–2018 years", the intelligent outlets are generally divided into four categories [5]: energy metering type, timing type, leakage protection type, remote control type. Many well-known manufacturers have introduced their own wireless switches or intelligent outlet, such as Schneider, ABB. Take ABB products for instance, a single wireless switch or wireless socket costs $150 while the price of two-way switch ranges from $175 to $305. Even an ordinary house, with two bedrooms and one living room, requires at least 30 switches and 40 outlets, that is, at least $5000. The Wemo switch, one of the first generation intelligent outlet produced by Belkin, was sold at $49.99. With performance of remote control and timing control though, its size was large, which is similar to Broadlink switch made by BroadLink Corporation. The Millet outlet is a well-designed advanced product, however, its functionality and security is not fully considered. Further study is necessary on the facilitate interaction, security and stability. It is found that there are mainly three deficiencies according to the relevant performance survey of intelligent outlet:

(1) The security issues like short circuit and high temperature are concerned about, the technical program security is not considered.

(2) The agent servers are adopted by most vendors, that is to say, control commands are sent from mobile phone to server at first, then passed to intelligent outlet by server directly. This method of control not only increases the pressure on the server, but also causes a control delay when the server is full load.
(3) Only simple functions such as on-off and timing control can be completed, the status of server and hardware of intelligent outlet cannot be inquired. Thus it is difficult to maintain the whole system.

As mentioned above, this paper designed an intelligent outlet control system consists of hardware, server and smart phone to achieve remote management of smart home by wifi technology and android platform.

2 Key Technologies of Intelligent Outlet

The wireless communication and Android technology are the key points.

2.1 Wireless Communication Technology

The main wireless communication technology in smart home are:

- Infrared communication technology;
- Radio-frequency technique less than 1 GHz;
- Bluetooth technology;
- ZigBee communication technology;
- Z-Wave technology;
- WiFi technology.

The Bluetooth, ZigBee and Wifi are the most commonly used technologies [5], the differences among which are listed in Table 1.

Piyare and Tazil [6] showed a home automation system based on Bluetooth, which connected to the device and smart phone. However, only short transmission distance is

Table 1. Comparison of the three wireless communication technologies.

Protocol	Bluetooth	ZigBee	Wifi
Transmission speed	1 MB/s	20 MB/s 40 MB/s 250 MB/s	11 MB/s 54 MB/s
Transmission distance	10 m	10–100 m	>100 m
Network topology	Point-to-point	Point-to-point, Star topology, Mesh topology	Point to hub
Band	2.4 GHz	2.4 GHz	2.4 GHz
Power consumption	Low	Very low	Low
Reliability	High	Low	High
Interactivity	Poor	Poor	Good
Construction convenience	Good	Good	Good
Confidentiality	Good	Poor	Good

supported and only the devices within the range can be controlled. Researchers tried to utilize the gateway to improve interoperability of network. Xu [7] introduced the ZigBee technology and its application in intelligent outlet. Although ZigBee has the advantages of low power consumption and low cost, the ZigBee-to-Wifi transfer function need to be add to the gateway so as to transmit data from end-and-end.

Unlike other communication technology, Wifi does not need gateway conversion to access the internet. The mobile phone and other handheld terminals can access the internet by Wifi with higher transmission speed, non-wiring and scalability. As a result, Wifi is widely spread to millions of families around the world. More and More devices are equipped with Wifi function and wireless module becomes the standard configuration for notebooks, panel computer and smart phone. There is no doubt that smart home based on Wifi technology has great potential and bright market prospect.

2.2 Android Technology

The four components of Android applications are Activity, Service, Broadcast and Content Provider, where Activity is the most basic module that provides a visual interface for user. Each application is composed of several Activity, each Activity contains buttons, list, display boxes and other components.

In this paper, a low-cost intelligent outlet system is designed based on Wifi and Android technology, with the performance of remote control, timing and query. It is safe, space-saving and scalable. The five sections are as follows:

Section 1: introduce the background and the research status of smart home and intelligent outlet, including merits and deficiencies. Section 2: show the key technologies of intelligent outlet. Section 3: propose the intelligent outlet system, design the hardware system and present the program flow chart. Section 4: test the outlet system and discuss the result. Finally, a conclusion is given.

3 The Proposed Intelligent Outlet Architecture in Smart Home

3.1 The Architecture of Smart Home System Based on Wifi Technology

The smart home system is composed of intelligent outlet nodes, gateway, wireless router, remote server, control terminals, 3G/4G communication network and internet, (see Fig. 1).

The principle of operation are as follows:

- Configure the Service Set Identifier (SSID) and cipher code of route via handheld terminals.
- Configure the IP address of the cloud server via handheld terminals.
- Power on the intelligent outlets nodes to connect to cloud server.
- Connect handheld terminals to the server to send command orders.
- The server send the command order to the intelligent outlet in the house.
- The intelligent outlet executes the command.

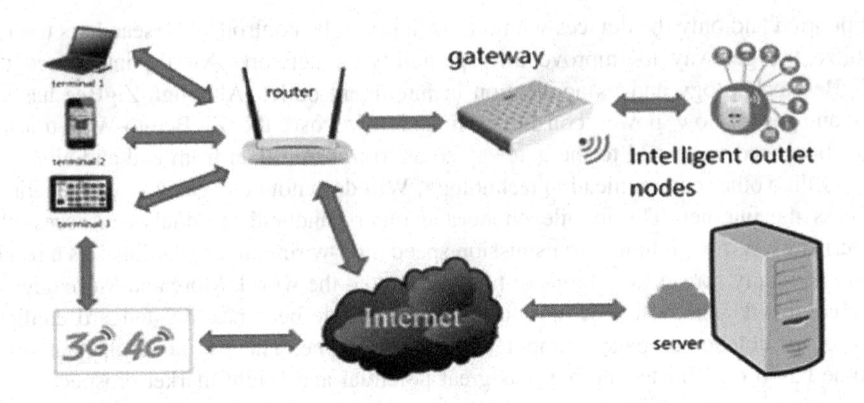

Fig. 1. The architecture of smart home based on Wifi

3.2 Hardware Design

The proposed intelligent outlet system includes power module, MCU control module, relay control circuit, Wifi module, display module and detection module. The hardware structure of the intelligent outlet is shown in Fig. 2.

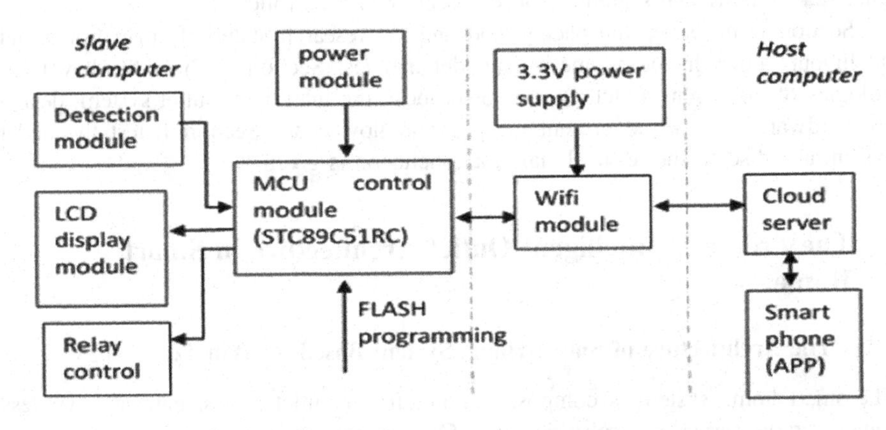

Fig. 2. Hardware structure of the intelligent outlet

According to the Fig. 2, the paper carries on the design simulation in Keil. The hardware schematic diagram includes the switching power supply circuit, the monolithic integrated circuit main control module, the Wifi module, the transformer module, the relay and the peripheral circuit and so on (see Fig. 3).

The power supply module contains two parts, the high-voltage side and the low-voltage side, which should be isolated when wiring to ensure the security. The Wifi module adopts ESP8266 chip, which transmit data to mobile phone based on ASCII in serial mode. There are 4 relays in relay module, which adopts

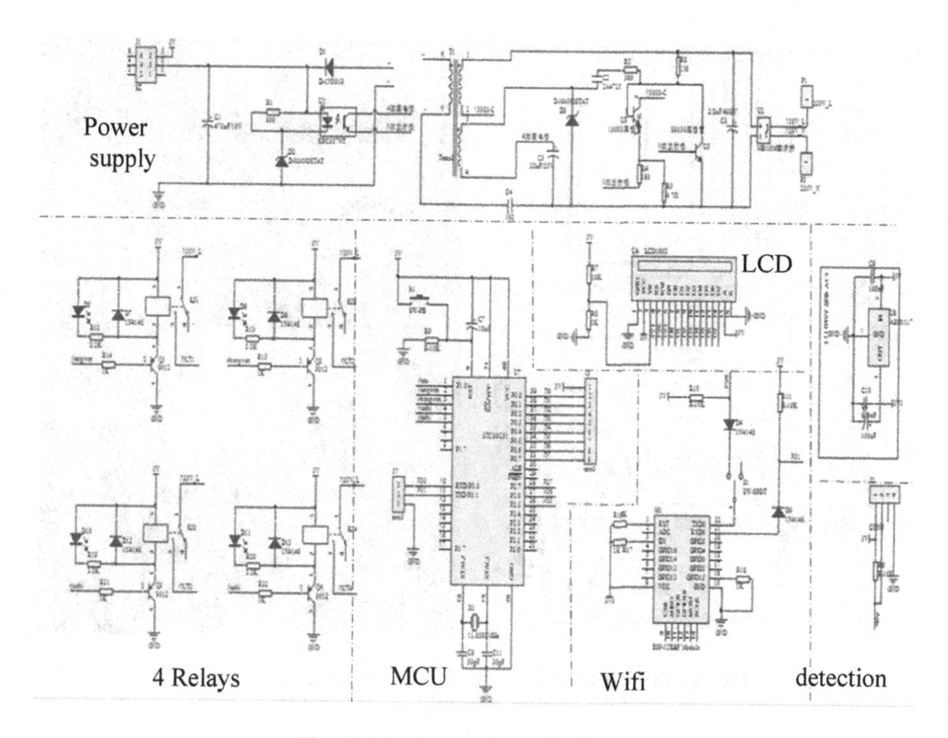

Fig. 3. Hardware schematic diagram

HK4100F-DC5V-SHG with 6 pins. The detection module adopts DHT11 as the temperature and humidity sensor, which can send 40 bits of data at one time through one data wire. The display module adopts LCD1602 liquid crystal display.

The Printed Circuit Board (PCB) is designed after simulation and the practical material object is shown in Fig. 4.

3.3 Software Programming

After choosing Wifi as the communication protocol of the intelligent outlet control system, it is essential to design a applicable software system to fulfill the control function. The software design of the proposed system mainly includes the design of single-chip program, the design of App of the mobile phone and the design of server programming. The programming of slave computer includes ESP8266 Wifi communication module, LCD1602 display module, DHT11 temperature and humidity detection module, relay process and interrupt servicing. The flow chart is shown in Fig. 5.

The application (App) is developed on the base of Eclipse exploitation environment and is supported by Android 4.0. Eclipse is an integrated cross-platform exploitation environment, which was first used for Java project development. The main user activity profiles can be written in the MainActivity.java file under the SRC directory. The framework of App project directory is shown in Fig. 6.

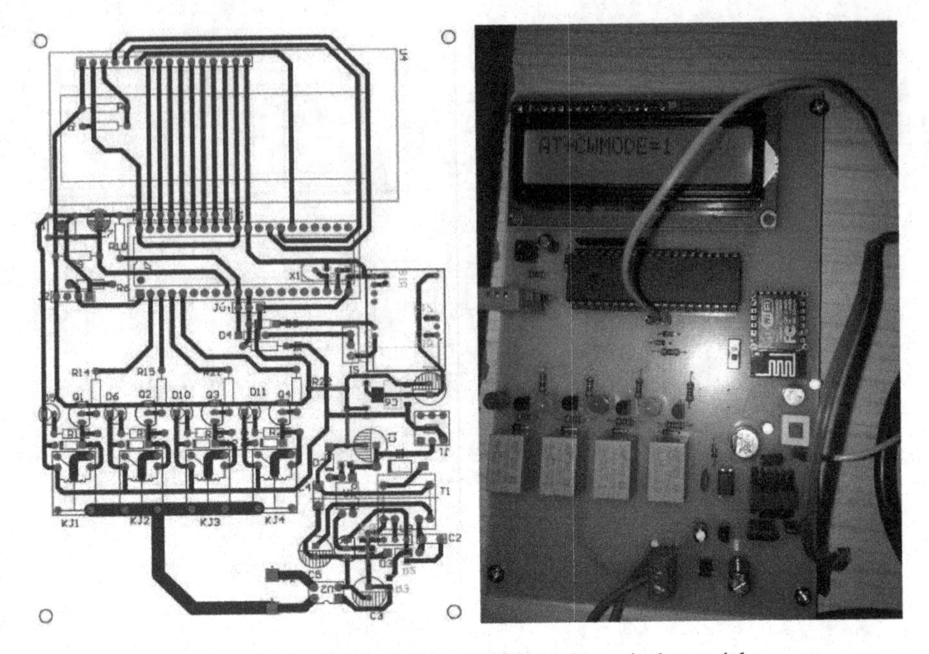

Fig. 4. Printed Circuit Board (PCB) and practical material

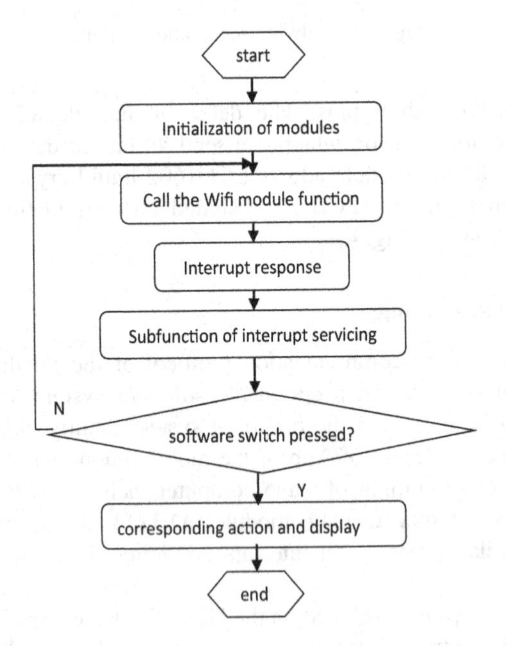

Fig. 5. Main flow chart

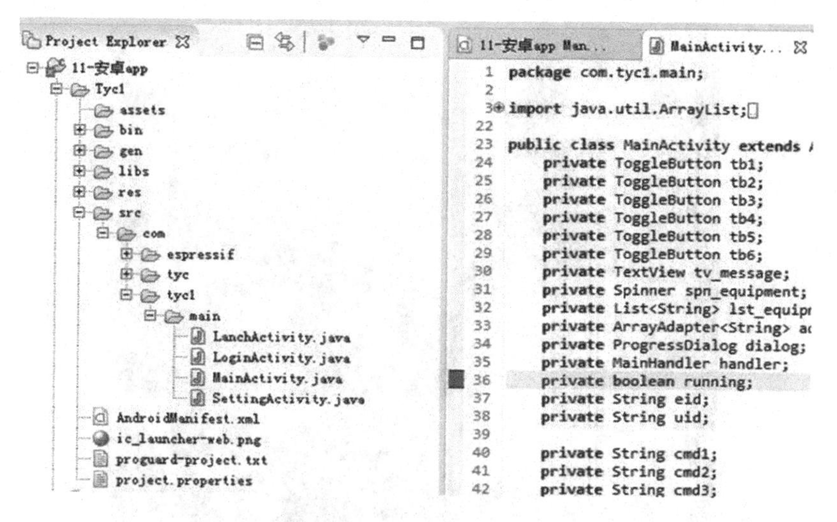

Fig. 6. App project directory

The cloud server Elastic Compute Service (ECS) offered by Ali Corporation is adopted to collect data and command. The Android App is deployed on Ali cloud server and then transmitted by the server to the Wifi module, which can establish TCP links with the server through the TCP address assigned by the server [8].

4 Results and Discussion

In the process of building the hardware platform of the system, we debug each module to eliminate the faults. In the process of software programming, we also correct the bug. Finally, the overall test is carried out to check whether the performance specifications meet the requirements. The debugging practical material is shown in Fig. 7. The communication between host machine and slave machine is established and works well. The three buttons "LED_ON" in smart phone from left to right are correspond to light 1 to 3 in the board. To make it clearer, light 3 is connected to the socket and fluorescent lamp in the socket is on. Meanwhile the "tp" and "hm" in the phone represent temperature and humidity of the current environment. These tests show that the host machine can query the data of slave machine and remote control the intelligent outlet.

The system not only can serve as an intelligent light control system, but also can expand other functions in smart home such as smart curtains, smart cameras, smart water heater, etc. For example, the remote video surveillance can be implemented by changing the socket into a camera and modifying the corresponding software program.

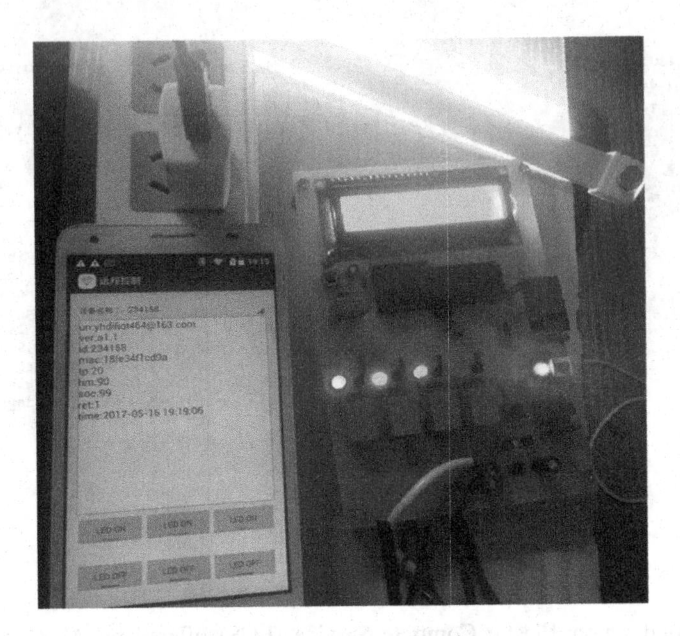

Fig. 7. Debugging and testing.

5 Conclusion

An intelligent outlet control system is designed and implemented based on Android and Wifi technology in this paper. It can be applied to smart home to meet people's primary requirements for home intelligence. The system can monitor the environment data, remote control and query by mobile phone, and it has advantages of free placement, low power consumption, low cost and scalability. However, the App interface and appearance design are still need further improvement to make it more attractive.

Acknowledgement. In this paper, the research was supported by Initial Scientific Research Fund of Fujian University of Technology and Pre-research Project of Fujian University of Technology.

References

1. Sun, Q., Liu, J., Li, S., et al.: Internet of things: summarize on concepts, architecture and key technology problem. J. Beijing Univ. Posts Telecommun. **33**(3), 1–9 (2010)
2. Zhang X.: The Design of Smart Outlet based on Android Phone. Zhejiang University (2014)
3. Gross, M.B.: Smart house and home automation technologies. IEEE Spectr. **7**(2), 26–30 (2012). University of Washington, Seattle WAUSA
4. Zhang, Z.Y.: Formal reasoning system based on fuzzy propositional modal logic. J. Softw. IEEE Trans. Appl. Supercond. **22**(1), 336–339 (2012)
5. Zhang, X., Sun, Z., Zhu, C.: Design of wireless smart outlet based on bluetooth and smart phone. Ind. Control Comput. **26**(11), 42–44 (2013)

6. Piyare, R., Tazil, M.: Bluetooth based home automation system using cell phone. In: 15th IEEE International Symposium on Consumer Electronic, pp. 192–195. IEEE Press, Washington D.C. (2011)
7. Xu, W., Jiang, Y., Wang, B.: The application of ZigBee technology in smart outlet design. Telecommun. Electric Power Syst. **32**(3), 78–81 (2011)
8. WIFI Module ESP8266 for IoT. http://en.itupdates.info/news/WIFI-module-ESP8266-for-IoT/

Intelligent Methods for Optimization and Learning

Design and Performance Simulation of Direct Drive Hub Motor Based on Improved Genetic Algorithm

Zhong-Shu Liu[1,2](✉)

[1] School of Information Science and Engineering, Fujian University of Technology, Fuzhou 350118, China
lzs@fjut.edu.cn
[2] The Key Laboratory for Automotive Electronics and Electric Drive of Fujian Province, Fujian University of Technology, Fuzhou 350118, China

Abstract. By using improved genetic algorithm (IGA), a direct drive hub motor with high efficiency and low-cost was designed. According to the characteristics of the hub motor, a mathematical model for optimize of direct drive hub motor was established, and optimized calculation was extended, and finally a prototype was developed. The simulation results show that the novel direct-drive hub motor can meet the requirements of great torque when low speed and high speed when constant power, so it is very suitable for the vehicle.

Keywords: Improve · Genetic algorithm · Direct drive · Hub motor · Simulation

1 Introduction

Due to the shortage of energy and environmental pollution, the development of electric vehicle has become the focus of national governments. The motor drive system is the core component of electric vehicle, and its driving mode can be grouped into two categories: centralized motor drive and hub motor drive. Motor drive is a new way of electric vehicle drive, it changes the power transmission hardware connection into wheel connection, saves the mechanical manipulation of shifting device which the traditional cars required, all of above makes the car drive structure be greatly simplified, each of the electrical control system of electric wheel can be easy to implement and braking energy feedback be easy to achieve. There are two driving modes of hub motor, namely, direct drive and wheel side reducer. In this paper, the direct drive outer rotor hub motor is studied, and its structure is shown in Fig. 1, it is an outer rotor three-phase permanent magnet synchronous motor (PMSM) with great torque when low speed and high speed when constant power, meanwhile, it does not need retarding mechanism, which makes the drive system be simple and efficiency be further improved [1].

The hub motor designed in this paper has advantages of high power density, high efficiency, simple structure, great starting torque and strong ability for overload. Optimum design of hub motor is a process to change engineer problem into optimal

P. Krömer et al. (eds.), *Proceedings of the Fourth Euro-China Conference on Intelligent Data Analysis and Applications*, Advances in Intelligent Systems and Computing 682, DOI 10.1007/978-3-319-68527-4_33

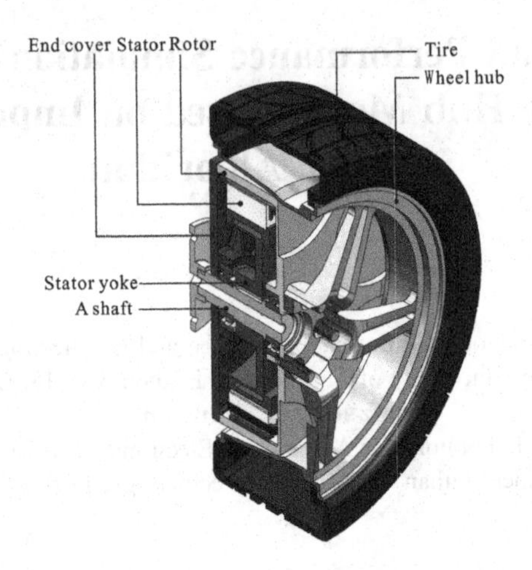

Fig. 1. Structure of direct drive motor

problem, it is a complex problem of nonlinear, constrained, discrete and the traditional optimization algorithm is difficult to find the global optimal solution of the hub motor. In this paper, based on the study of permanent magnet synchronous motor, according to the characteristics of the direct drive hub motor [2], an three-phase permanent magnet synchronous motor (PMSM) used as direct drive motor was developed by using improved genetic algorithm (IGA).

2 Improved Genetic Algorithm

2.1 Improved Genetic Algorithm

The basic idea of improved genetic algorithm is using adaptive crossover operator and mutation operator of genetic algorithm to optimize the initial scheme, thus the optimized solutions can be used as the initial scheme of pattern search method, finally, a global optimal solution with highest efficiency is obtained.

Due to the accuracy and efficiency of crossover rate and mutation rate of genetic algorithm depending on the fitness of population individual, when the fitness is bigger, in order to ensure that reduce the probability of genetic damage, an individual will use the smaller crossover rate and mutation rate [3], while fitness is small, in order to ensure its searching area, the individual will use the bigger crossover rate and mutation rate, therefore, the crossover operator p_c and mutation operator p_v of improved genetic algorithm (IGA) are shown as follows

$$p_c = \begin{cases} p_{c1}, f' > f'_{avg} \\ \frac{p_{c1}-p_{c2}}{2}, \cos\left[\frac{f-f_{avg}}{f_{\min}-f_{avg}}\pi\right] + \frac{p_{c1}+p_{c2}}{2}, f' \leq f'_{avg} \end{cases} \tag{1}$$

$$p_v = \begin{cases} p_{v1}, f' > f'_{avg} \\ \frac{p_{v1}-p_{v2}}{2}, \cos\left[\frac{f-f_{avg}}{f_{\min}-f_{avg}}\pi\right] + \frac{p_{v1}+p_{v2}}{2}, f' \leq f'_{avg} \end{cases} \tag{2}$$

In formula (1) and (2), p_{c1} and $p_{v1} \in (0, 1)$, $f_{\min}$ is the minimum fitness of the group; f_{avg} is the average fitness for each generation group; f' is the smaller fitness of the two individuals crossing; f'_{avg} is the average fitness of the variation individuals. After improving the crossover and mutation rates of improved genetic algorithm (IGA), the calculation flow chart is shown in Fig. 2.

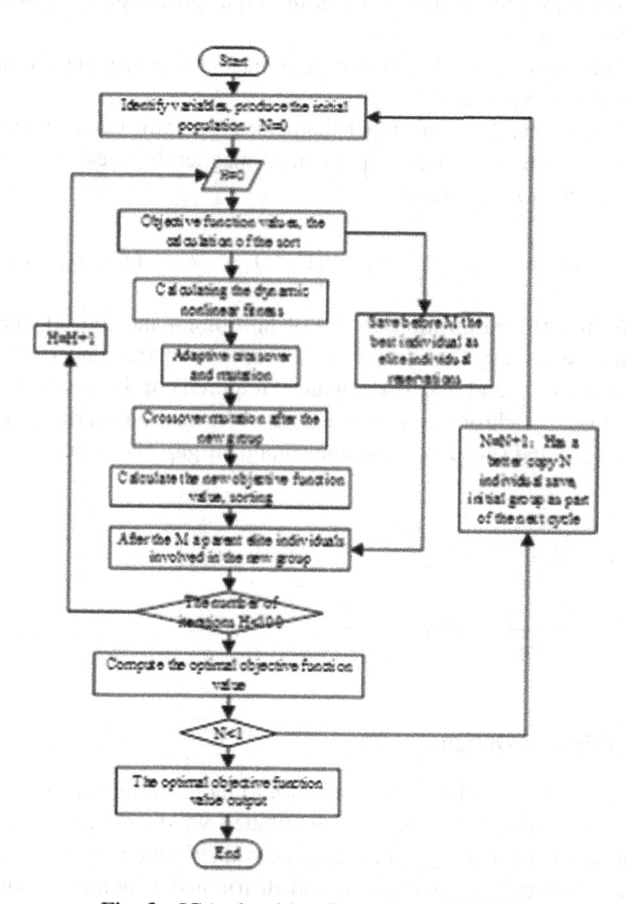

Fig. 2. IGA algorithm flow chart of PMSM

2.2 Overall Electromagnetic Structure Design

To select objective function and optimization variable is very important in optimization design of direct drive hub motor. To comprehensively consider the performance and cost of hub motor, the difficulty of processing, the structure arrangement and other factors, the objective function of the hub motor is defined as the cost of a three-phase permanent magnet synchronous motor corresponding to per unit efficiency. That is:

$$f(x) = \min \frac{F(x)}{\eta} = \min \frac{(m_{cu}t_{cu} + m_{Fe}t_{Fe} + m_m t_m)}{P_1 - \sum p} P_1 \tag{3}$$

In above, m_{cu} is the price for the copper materials; m_{Fe} is the price of steel materials; m_m is the price of nd-fe-b materials; t_{cu} is the total weight of copper materials used; t_{Fe} is the total weight of the steel materials used for; t_m is the total weight of nd-fe-b permanent magnet materials. P_1 is the input power of the motor; $\sum p$ is the total loss of the motor [4].

The following parameters of the permanent magnet synchronous motor are selected as the optimal design variables:

Rotor outer diameter D_2; rotor inner diameter D_{i2}; stator outer diameter D_1; stator inner diameter D_{i1}; length of core l_{eff}; permanent magnet thickness h_m and the width of the permanent magnet b_m, namely:

$$X = \begin{bmatrix} x_1 & x_2 & x_3 & x_4 & x_5 & x_6 & x_7 \end{bmatrix}^T = \begin{bmatrix} D_2 & D_{i2} & D_1 & D_{i1} & l_{eff} & h_m & b_m \end{bmatrix}^T \tag{4}$$

The constraints of optimization design of hub motor include boundary constraint and performance constraint, the constraint conditions in this paper according to the engineering simulation and the performance requirement in design of three-phase permanent magnet synchronous motor are as follows: the starting current I_{st}. the starting torque T_{st}; rated speed n_N; heat load H and air gap magnetic induction intensity B_δ. That is:

$$\begin{aligned} g(1) &= \frac{I_{st} - I_{st0}}{I_{st0}} \le 0; g(2) = \frac{T_{st} - T_{st0}}{T_{st0}} \le 0; g(3) = \frac{n_N - n_{N0}}{n_{N0}} \le 0; \\ g(4) &= \frac{H - H_0}{H_0} \le 0; g(5) = \frac{B_\delta - B_{\delta 0}}{B_{\delta 0}} \le 0 \end{aligned} \tag{5}$$

2.3 Stator Winding Design

Stator winding design is an important part of hub motor design, which can be divided into two types: centralized winding and distributed winding. The objective of design for stator winding is to make gap magnetic field distribution close to sine wave. The main difference between the centralized and distributed windings of the stator is the winding pitch, centralized winding pitch is 1, which means two effective edges of each coil of the motor across a tooth [5], and distributed winding pitch is greater than 1, which means the two effective edges of each coil across the slot number greater than 1. The end part of the centralized winding is small and the winding process is simple, which makes the hub motor have a better electromagnetic performance and a lower copper loss and higher efficiency [6]. The hub motor stator winding designed in this paper have a centralized winding, which is shown in Fig. 3.

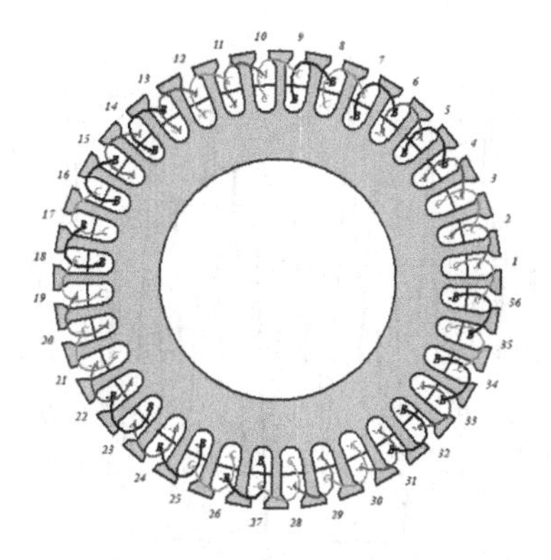

Fig. 3. Centralized winding wiring diagram

2.4 Measures to Reduce the Cogging Torque

The cogging torque of the motor is an endemism in permanent magnetic synchronous motor (PMSM), it is a reluctance torque caused by the interaction between the stator core slot and permanent magnet in rotor, which will cause vibration and noise in the motor, and affect the speed control system of low speed performance and positioning accuracy of position control system. At present, there are various measures to reduce the cogging torque of permanent magnet motor, such as magnetic pole deviation, magnetic pole eccentric, and slot [7].

In this paper, the cogging torque is reduced by two measures as following:

2.4.1 Inner Stator Chute

This method is to adjust the phase relationship of the unit tooth groove torque at the superposition, which can make each other cancel out. In theory, the stator groove tilting a stator pitch relative to the rotor could eliminate the cogging torque completely. After taking all the factors into consideration, we adopted the chute stator iron with 1.5 times pitch in the prototype.

2.4.2 Optimizing z/2p

In above, z means the stator slot number and 2p means the poles of rotor. By using least common multiple (LCM) of the slot number and the poles of rotor to calculate the minimum order number of cogging torque. It is known that the greater the number of the cycle of fundamental wave of the cogging torque [8], the smaller the amplitude. So we should choose the least common multiple (LCM) as larger as possible, and through the magnitude of least common multiple (LCM) to calculate minimum order of cogging torque [8].

The analysis result shows that in order to reduce the cogging torque, we should choose the LCM as large as possible In view of the manufacturing process and cost factors, the z/2p ratio of the motor in this paper is chosen to be 36/20 and it equals 180. So the minimum order of cogging torque equals $K = LCM/Z = 180/36 = 5$, which means all cogging torque less than 5 order will be eliminated. The cogging torque curve is shown in Fig. 4.

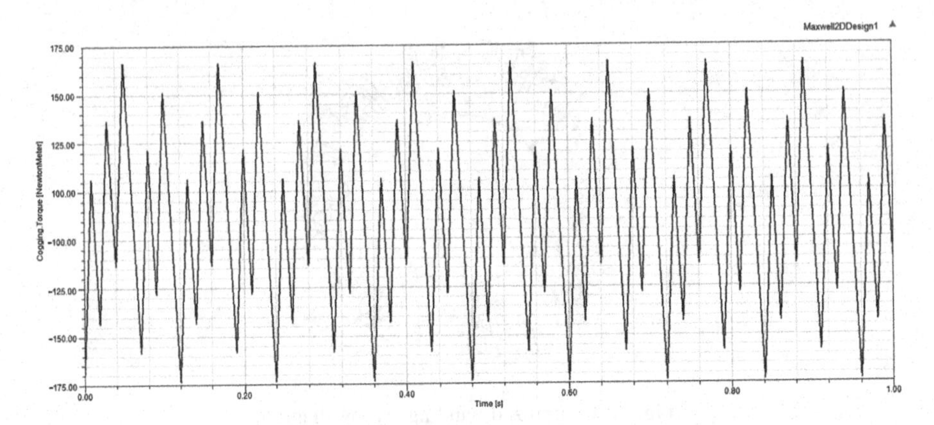

Fig. 4. Cogging torque of hub motor

2.5 Optimum Design Result

Based on theory above and method of the improved genetic algorithm, optimize the hub motor electromagnetic structure, the result is as follows: rated voltage 255 v, rated power 45 kw, poles of rotor 20, stator slots 36, double-layer centralized stator winding, Y connection. The structural parameters are shown in Table 1.

Table 1. Structural parameters of hub motor

Parameter	Value	Parameter	Value
Outer diameter of the stator	270/mm	Outer diameter of the rotor	303/mm
Inner diameter of the stator	190/mm	Inner diameter of the rotor	283/mm
The length of the core	55/mm	The air gap length	1.5/mm
The thickness of the magnet	5/mm	Slot number	36
The width of the magnet	30/mm	Pole	20

3 Finite Element Analysis of Magnetic Field

3.1 Grid Subdivision

Now we will analyze the hub motor by using two-dimensional finite element method, which is a numerical calculation method based on discretization. Figure 5 is the grid subdivision of hub motor designed in this paper. It can be seen from Fig. 5 that the grid density of stator and rotor would be minimum while the grid density of air gap would be maximum. So using different grid density subdivision we can save computing resources and ensure the accuracy of the calculation [9].

3.2 Analysis of the Hub Motor Internal Electromagnetic Field

Using Ansoft Maxwell we can analyze the flux density distribution of cloud of generator shown in Fig. 6(a) when it is in no-load. It can be seen from the diagram, the flux

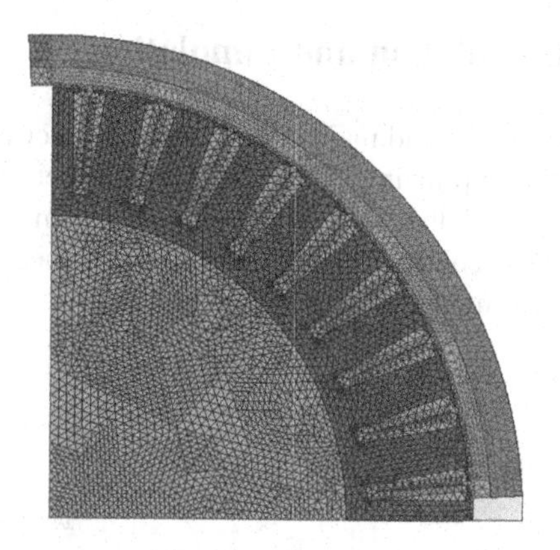

Fig. 5. Grid division of hub motor

density value of stator yoke of motor is maximum which has reached 2.71 T, we can get the flux density distribution of cloud of generator shown in Fig. 6(b) when it is in load. Compared with Fig. 6(a), we found that when the motor is in load, the iron core saturation degree decrease relative to the case in no-load, motor stator yoke of flux density value decrease to 2.58 T, at the same time, all parts of the stator iron core flux density amplitude is decreasing, the center line of the air-gap magnetic field of electric skewed compared with no-load, the reason is that the armature reaction weakens the air gap field and causes gap magnetic field distortion.

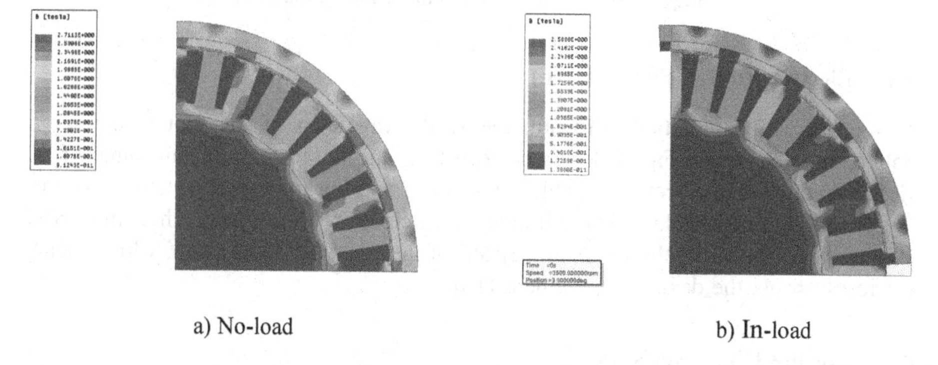

a) No-load b) In-load

Fig. 6. Flux density cloud diagram of hub motor

We could make a conclusion from above that the magnetic flux near the stator and rotor yoke and the air gap are saturated relatively, and the field distribution of motor is reasonable. It does not exist local magnetic oversaturation and the prototype meets the design requirements [10].

4 Performance Calculation and Simulation

Using Simplorer software we build main circuit and control circuit model of hub motor as following, when setting up main circuit model we can select related components in the model library [11], and then use wire connects them according to the actual structure of the circuit. The control circuit model is established by using state machine. Figure 7 is the simulation model of hub motor system.

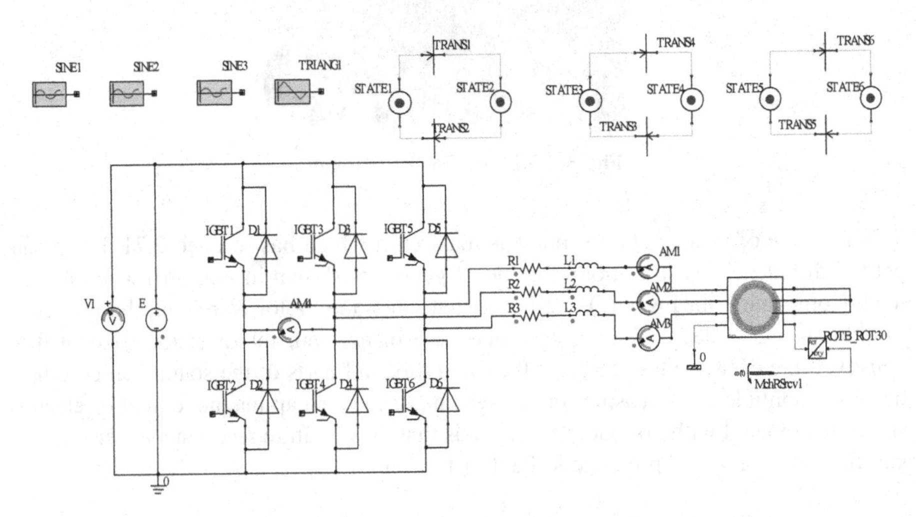

Fig. 7. Hub motor system circuit model diagram

4.1 Electric Motive Force (EMF)

Using Maxwell simulation software, the EMF waveform of the motor running at 3500 rpm is shown in Fig. 8. It can be seen from the figure that the sinusoidal of the EMF waveform is perfect [12], which indicating that the improved genetic algorithm design has a remarkable effect. Simulation calculation shows that when the phase voltage peak is 160 V, the maximum speed of the motor is 3500 r/min, which means the result meets the design requirement [13].

4.2 Torque Characteristics

The torque characteristic of the hub motor is shown in the Fig. 9. From the characteristic it can be seen that the motor has great starting torque and small torque ripple, so it is very suitable for the vehicle [14].

4.3 Input Voltage and Input Current Characteristics

The input voltage and input current waveform of generator when it is in-load cases are shown in Fig. 10(a) and (b) respectively. As it can be seen from the diagram, they are all sine wave without distortion, the peak value of them are 375 V and 200 A

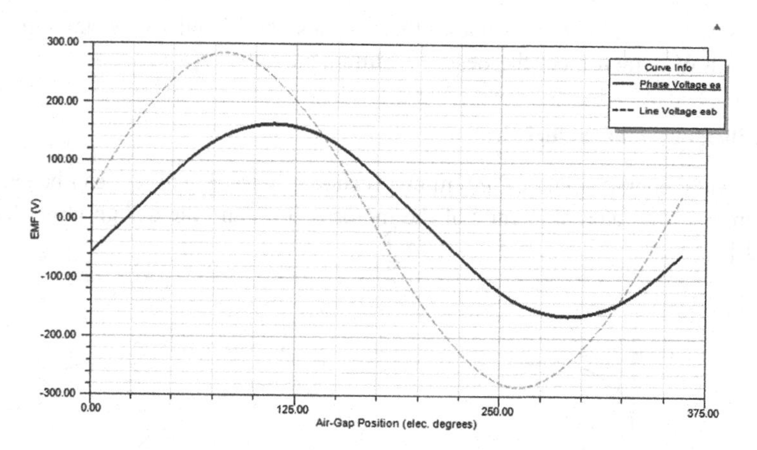

Fig. 8. EMF waveform diagram of hub motor

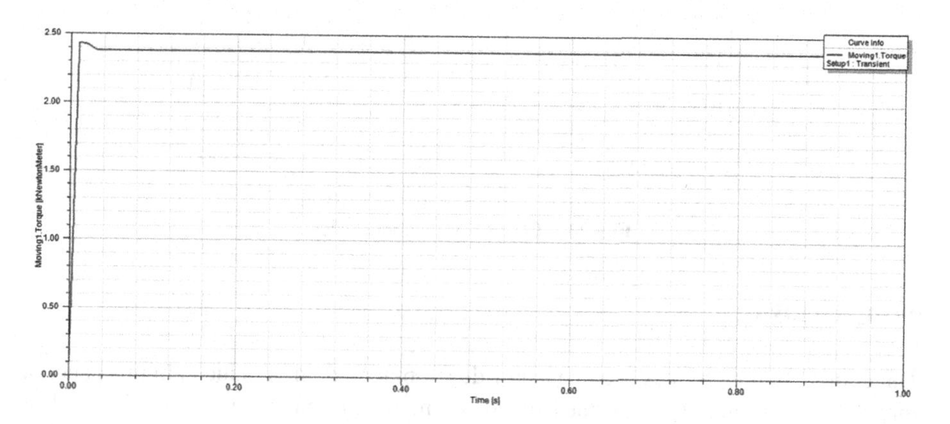

Fig. 9. Torque characteristics

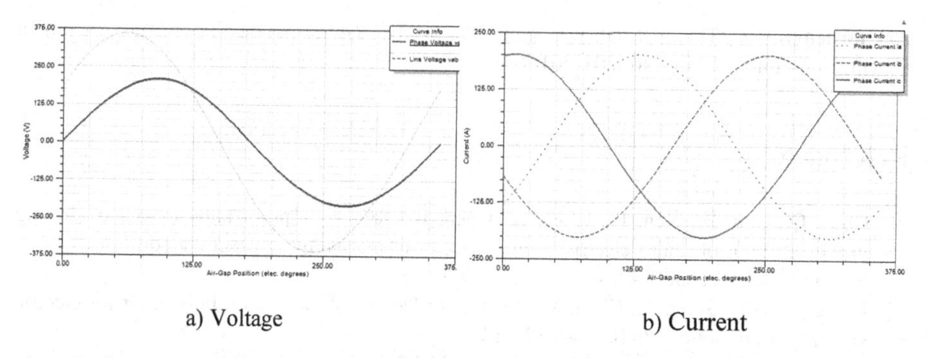

a) Voltage b) Current

Fig. 10. Input voltage and input current characteristics

respectively, and the effective value of them equals 262 V and 131 V, the input power equals 46 KW, which meet the design requirement fully.

4.4 Efficiency Characteristics

The efficiency characteristic of hub motor is shown in the Fig. 11. It can be seen from the figure that the rated efficiency of the motor is 96% and the design target has been reached [15].

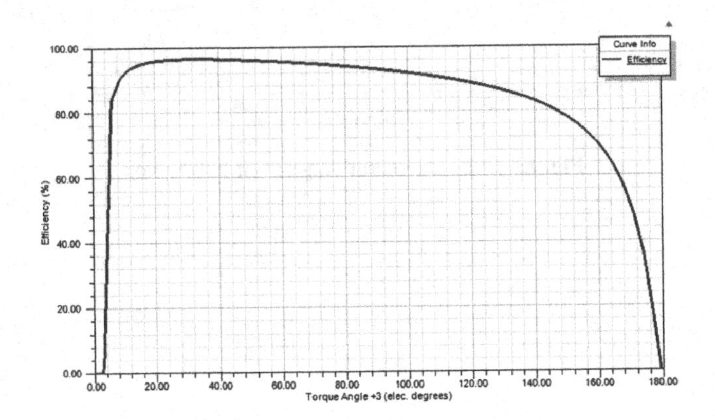

Fig. 11. Efficiency characteristics

5 Conclusion

Direct drive hub motor has become a new development trend of vehicle. In this paper, a direct drive hub motor is designed based on improved genetic algorithm (IGA). The simulation results show that the electric motor has the characteristics of great torque when low speed, fast response, low torque ripple, high efficiency and low-cost. The above results show that design of prototype is reasonable.

Acknowledgment. This research was supported by the Natural Science Foundation of Fujian Province of China. (Grant No. 2017J01667).

References

1. Chen, Q., Shu, H., Ren, K., Chen, L., Chen, B., Xie, A.: Optimization design of driving in-wheel motor of micro-electric vehicle based on improvement genetic algorithm. J. Cent. S. Univ. (Sci. Technol) **43**(8), 3013–3017 (2012)
2. Lei, L., Hu, Y., Song, Z., Chen, Y., Zhang, H.: Design of outer rotor hub motor for electric vehicle. Micromotors **49**(10), 6–9 (2016)
3. Li, S., An, R., Xu, Z., Dong, W.: Coordinated LVRT of IG and PMSG in hybrid wind farm. Electr. Power Autom. **35**(2), 21–27 (2015)

4. Lu, D., Ouyang, M., Gu, J., Li, J.: Field oriented control of permanent magnet brushless hub motor in electric vehicle. Electr. Mach. Control **16**(11), 76–83 (2016)
5. Chen, G., Tang, N., Li, D.: Optimization design of PMSG based on improved genetic algorithm. Explosion-Proof Electr. Mach. **46**(1), 5–9 (2011)
6. Wang, X., Gao, P.: Analysis of 3-D temperature of in-wheel motor with inner-oil improved cooling for electric vehicle. Electr. Mach. Control **20**(3), 36–41 (2012)
7. Shen, X., Zhao, Q., Xu, P.: Control research on maximum power of the direct-drive permanent magnet synchronous rotor. J. Electr. Power **29**(1), 28–31 (2014)
8. Gao, J.: Research on design technology and application of direct-drive permanent magnet rotor with wind turbine. Ph.D, Hunan University, Changsha, pp. 36–55 (2013)
9. Kun, Y., Xiu, Y.: Modeling optimization and simulation of grid-connected direct-driven permanent magnet wind power system based on small signal stability analysis. East China Electr. Power **42**(10), 2028–2033 (2014)
10. Zhai, Y.: Research on vector control of PMSG in low temperature waste heat power generation system. Ph.D., Shenyang University of Technology, Shenyang, pp. 17–30 (2015)
11. Feng, H., Huo, Z., Hu, X.: Constant power control for wind turbine based on finite-time method. Renew. Energy Resour. **33**(12), 1831–1834 (2015)
12. Shen, J., Miao, D.: Variable Speed permanent magnet synchronous generator systems and control strategies. Trans. China Electrotech. Soc. **28**(3), 2–8 (2013)
13. An, Q., Li, S., Lizhi, S.: Research on novel open-end wind permanent magnet synchronous motor vector control systems. Proc. CSEE **35**(22), 5891–5898 (2015)
14. Sun, M.: Design and flux-weaking control for in-wheel –motor of electric vehicle, pp. 36–55 Ph.D., Harbin Institute of Tecnology, Harbin (2009)
15. Weng, M., Li, Q., Cao, M.: Parameters calculation and structure design for wheel hub permanent magnet brushless direct current motor of drive method. Electr. Mach. Control Appl. **42**(7), 12–15 (2015)

QUATRE Algorithm with Sort Strategy for Global Optimization in Comparison with DE and PSO Variants

Jeng-Shyang Pan[1], Zhenyu Meng[2(✉)], Shu-Chuan Chu[3], and John F. Roddick[3]

[1] Fujian Provincial Key Lab of Big Data Mining and Applications, Fujian University of Technology, Fuzhou, China

[2] Department of Computer Science and Technology, Harbin Institute of Technology Shenzhen Graduate School, Shenzhen, China

mzy1314@gmail.com

[3] School of Computer Science, Engineering and Mathematics, Flinders University, Adelaide, Australia

Abstract. Optimization algorithm in swarm intelligence is getting more and more prevalent both in theoretical field and in real-world applications. Many nature-inspired algorithms in this domain have been proposed and employed in different applications. In this paper, a new QUATRE algorithm with sort strategy is proposed for global optimization. QUATRE algorithm is a simple but powerful stochastic optimization algorithm proposed in 2016 and it tackles the representational/positional bias existing in DE structure. Here a sort strategy is used for the enhancement of the canonical QUATRE algorithm. This advancement is verified on CEC2013 test suite for real-parameter optimization and also is contrasted with several state-of-the-art algorithms including Particle Swarm Optimization (PSO) variants, Differential Evolution (DE) variants on COCO framework under BBOB2009 benchmarks. Experiment results show that the proposed QUATRE algorithm with sort strategy is competitive with the contrasted algorithms.

Keywords: Differential evolution · Particle swarm optimization · QUATRE · Real parameter optimization · Swarm intelligence

1 Introduction

Optimization algorithm aims at tackling different optimization demands arising in our daily lives, and the standard approach to tackle these problems often begins by designing an objective function that can model the problems' objectives while incorporating any constraints [1]. Consequently, these optimization algorithms can be divided into different categories according to the property that they possess or that the target objective functions possess. These

P. Krömer et al. (eds.), *Proceedings of the Fourth Euro-China Conference on Intelligent Data Analysis and Applications*, Advances in Intelligent Systems and Computing 682, DOI 10.1007/978-3-319-68527-4_34

categories are continuous optimization, discrete optimization, stochastic optimization, numerical optimization and combinatorial optimization, etc. Computational Intelligence (CI) gives computational methodologies and approaches to address and tackle complex real-world optimization problems, especially for these that models/objective functions are of noise and uncertainty. Evolutionary Computation (EC) is a subset of CI, and it presents the methodologies with evolutionary thought. Optimization algorithms in EC are often equipped with a meta-heuristic or stochastic optimization characteristic to solve black box optimization problems, and they belong to the family of trial and error problem solvers and distinguished by the use of a population of candidate solutions[1].

Particle Swarm Optimization (PSO) [2] and Differential Evolution (DE) [1] are two popular and powerful algorithms in EC, and the variants of them often secure the front ranks of Conference on Evolutionary Computation (CEC) Competitions. There were still many other optimization algorithms proposed in the recent few decades such as Ant Colony Optimization [3], Artificial Bee Colony Algorithm [4], Ebb-tide-fish algorithm [5,6,23], Monkey King Evolution [13], and QUATRE algorithm [14,21,22] etc. These algorithms are all powerful ones, and have advantages and disadvantages of their own. In order to avoid reinventing the wheel, a new proposed algorithm should performs better than the state-of-the-art variant of a popular algorithm rather than the popular algorithm itself. Here for these comparison, we mainly focus on two branches, PSO variants and DE variants.

The canonical PSO was introduced by Kennedy and Eberhart in 1995 [2], particles in the next generation are influenced by their elder generation's global best location and the ancestor's history best location. This evolution scheme is different from the ones that utilize crossover, mutation, and selection operations inspired by Charles Darwin's evolution theory. Velocity in PSO is used to generate much more diversity of the population, and the proper value of velocity makes a good balance between the particle exploitation and exploration. As PSO algorithm is simple and powerful, many researches have learned about the technique and proposed many variants [7–9]. DE was firstly introduced in 1995 [1], and it was also arguably one of the most powerful operator for optimization. Price et al. [10] introduced a total of 10 working strategies including 5 mutation strategies and 2 crossover strategies, and it also gave a general convention of different DE schemes, DE/x/y/z. Also, many researchers proposed many DE variants to enhance the canonical DE algorithm [10–12,17]. As we know from literature and evolutionary computation competitions, state-of-the-art DE variant usually has an overall better performance than state-of-the-art PSO variant. Moreover, there are still weakness, representation/positional bias for a higher dimensional perspective of view, in DE structure, and QUATRE structure is proposed to tackle such inborn weakness in DE structure [18].

The rest of the paper is organized as follows. The QUATRE algorithm with sort strategy is given in Sect. 2, Sect. 3 presents the experimental analysis of QUATRE algorithm with sort strategy under CEC2013 test suite for real para-

[1] https://en.wikipedia.org/wiki/Evolutionary_computation.

meter optimization. Moreover, COCO framework under BBOB2009 benchmarks are also used to contrast some state-of-the-art PSO and DE variants with the proposed S-QUATRE algorithm in this section. Section 4 gives the final conclusion.

2 QUATRE Algorithm with Sort Strategy (S-QUATRE)

QUATRE is abbreviation of QUasi-Affine TRansformation Evolution, and that the reason we name the algorithm QUATRE algorithm is because individuals in QUATRE algorithm evolve in a quasi-affine transformation form. The detailed equation of this evolution is given in Eq. 1. $\widehat{X}$ denotes the population matrix with ps individuals, $\widehat{X} = [X_{1,G}, X_{2,G}, ..., X_{ps,G}]^T$. $X_{i,G}, i \in \{1, 2, ..., ps\}$ denotes the ith row vector in matrix $\widehat{X}$ of the Gth generation, $X_{i,G} = [x_{i1}, x_{i2}, ..., x_{iD}]$, D is the dimension number of the objective function. $\widehat{B}$ denotes the donor matrix in which the ith row vector is $B_{i,G}$. "$\bigotimes$" denotes component-wise multiplication of the elements in each matrix, and it is the same operation as ".*" in Matlab Software.

$$\widehat{X} \leftarrow \overline{M} \bigotimes \widehat{X} + M \bigotimes \widehat{B} \tag{1}$$

M is the evolution matrix, and it is transformed from an initial matrix M_{ini}. M_{ini} is initialized by a lower triangular matrix with the elements equaling to ones. Equation 2 gives an example to show the initialization of M_{ini} and the transformation from M_{ini} to M when population size ps equaling to dimension number D. "$\sim$" denotes the transformation, and there are two steps for the transformation, the first step is to randomly permute the elements in each D-dimensional row vector in M_{ini}, and the second step is to randomly permute the row vectors with the elements of each row vector unchanged, then we can get M.

$$M_{ini} = \begin{bmatrix} 1 & & & \\ 1 & 1 & & \\ & & \cdots & \\ 1 & 1 & \cdots & 1 \end{bmatrix} \sim \begin{bmatrix} 1 & 1 & & \\ & & \cdots & \\ 1 & 1 & \cdots & 1 \\ & & 1 & \end{bmatrix} = M \tag{2}$$

$\overline{M}$ is binary inverse operation of M. The corresponding values of non-zero elements in M are zeros in matrix $\overline{M}$, while the corresponding values of zero elements in M are ones in $\overline{M}$. Equation 3 shows a example of binary inverse operation.

$$M = \begin{bmatrix} 1 & & & \\ 1 & 1 & & \\ & & \cdots & \\ 1 & 1 & \cdots & 1 \end{bmatrix}, \overline{M} = \begin{bmatrix} 0 & 1 & \cdots & 1 \\ 0 & 0 & \cdots & 1 \\ & & \cdots & 1 \\ 0 & 0 & \cdots & 0 \end{bmatrix} \tag{3}$$

Usually, the population size is larger than the dimension number of the objective function, matrix M_{ini} needs to be extended according to population size (ps).

For example, when $ps = 2D$, M_{ini} is initialized by piling two lower triangle matrix together shown in Eq. 4. Generally, when $ps\%D = k$, the first k rows of the $D \times D$ lower triangular matrix are added in the rear of matrix M_{ini}, and M is adaptive changed according to the "$\sim$" operation.

$$M_{ini} = \begin{bmatrix} 1 & & & \\ 1 & 1 & & \\ & & \cdots & \\ 1 & 1 & \cdots & 1 \\ 1 & & & \\ 1 & 1 & & \\ & & \cdots & \\ 1 & 1 & \cdots & 1 \end{bmatrix} \sim \begin{bmatrix} 1 & 1 & & \\ & & \cdots & \\ 1 & 1 & \cdots & 1 \\ & & & 1 \\ & & \cdots & \\ 1 & \cdots & 1 & \\ & 1 & & \\ 1 & & \cdots & 1 \end{bmatrix} = M \tag{4}$$

There are several different calculation schemes of $B_{i,G}$ in QUATRE algorithm [18,19], and they are listed in Table 1. $X_{gbest,G}$ denotes the global best individual in the Gth generation of the ps individuals in $\widehat{X}$, $X_{r_i,G}, i \in \{0,1,2,3,4\}$ denotes a randomly selected individual from the population. There are mainly three selection scheme of $X_{r_i,G}$ in DE structure [10], and they are "Random selection with restriction", "Permutation selection", and "Random offset selection". We employ a fourth selection, a new permutation scheme proposed in [20], in the implementation of the QUATRE structure. The indices of $X_{r_i,G}, i \in \{0,1,2,3,4\}$ in the $\widehat{X}_{i,G}$ is generated by random permutation of ps.

Table 1. The six schemes of matrix B calculation in QUATRE algorithm.

No.	QUATRE variants	Equation
1	$QUATRE/best/1$	$B_{i,G} = X_{gbest,G} + F \cdot (X_{r_1,G} - X_{r_2,G})$
2	$QUATRE/rand/1$	$B_{i,G} = X_{r_0,G} + F \cdot (X_{r_1,G} - X_{r_2,G})$
3	$QUATRE/target/1$	$B_{i,G} = X_{i,G} + F \cdot (X_{r_1,G} - X_{r_2,G})$
4	$QUATRE/target-to-best/1$	$B_{i,G} = X_{i,G} + F \cdot (X_{gbest,G} - X_{i,G}) + F \cdot (X_{r_1,G} - X_{r_2,G})$
5	$QUATRE/best/2$	$B_{i,G} = X_{gbest,G} + F \cdot (X_{r_1,G} - X_{r_2,G}) + F \cdot (X_{r_3,G} - X_{r_4,G})$
6	$QUATRE/rand/2$	$B_{i,G} = X_{r_0,G} + F \cdot (X_{r_1,G} - X_{r_2,G}) + F \cdot (X_{r_3,G} - X_{r_4,G})$
7	$QUATRE/target/2$	$B_{i,G} = X_{i,G} + F \cdot (X_{r_1,G} - X_{r_2,G}) + F \cdot (X_{r_3,G} - X_{r_4,G})$

For the S-QUATRE algorithm, a sort strategy is employed to enhance the performance of QUATRE algorithm. The ps individuals in the population $\widehat{X} = [X_1, X_2, ..., X_{ps}]^T$ are firstly sorted after initialization according to the fitness values of the objective function (target function/benchmark function). Then the individuals in the sorted sequence are divided into two parts, the first part contains $\frac{ps}{2}$ (ps is supposed to be an even number.) individuals that obtained better fitness values, and the second part contains the last $\frac{ps}{2}$ individuals obtained worse fitness values. Figure 1 gives an illustration of sort strategy in S-QUATRE algorithm. The better half individuals are reserved into next generation and the worse half individuals need to evolve according to QUATRE algorithm. After the evolution step, selection is made between the original worse individual and the corresponding evolved individuals. The winner individuals are also reserved into next generation. The pseudo code of S-QUATRE algorithm is given in Algorithm 1.

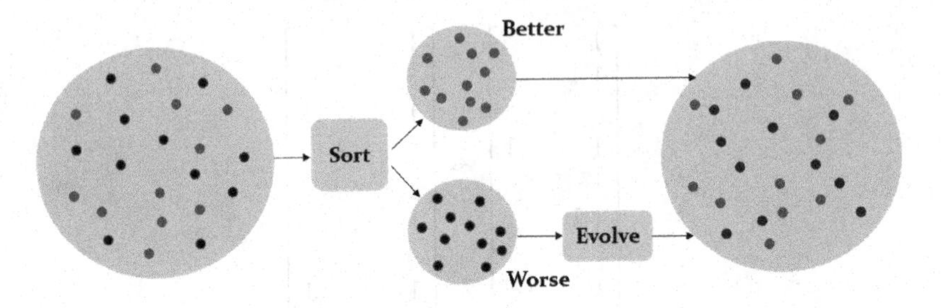

Fig. 1. Illustration of sort scheme in S-QUATRE algorithm.

3 Experiment Analysis of S-QUATRE Algorithm

3.1 Test Environment Description

We employ some commonly used black-box test suite, including BBOB2009 [15] and CEC2013 [16], to verify the proposed S-QUATRE algorithm. For a stochastic optimization algorithm, there usually are two different measurement criteria, one is fix-cost (evolution terminates when reaching fixed number of function evaluation, and CEC2013 test suites suggest to use fix-cost measurement), and the other is fix-target (evolution terminates when reaching a target value, BBOB2009 recommend this measurement). For the CEC2013 benchmark functions, 28 benchmark functions can be divided into 3 parts, uni-modal functions (f_1–f_5), basic multi-modal functions (f_6–f_{20}) and composition functions (f_{21}–f_{28}). All the functions are shifted to the same global best location, $O\{o_1, o_2, ..., o_d\}$. For this test suites, 51 runs are conducted, and the best, mean and standard deviation of these runs are selected and calculated for comparison. For the COCO framework with BBOB2009 testbeds, 24 noiseless functions can be divided into 5 parts, separable functions (f_1–f_5), functions with low or moderate conditioning (f_6–f_9), functions with high condition (f_{10}–f_{14}), multi-modal functions with adequate global structure (f_{15}–f_{19}), and multi-modal functions with weak global structure (f_{20}–f_{24}). All functions are scalable to dimensions with the search domain in each dimension set to $[-5, 5]$, and most BBOB2009 functions have their global optimum in the range $[-4, 4]$ of each dimension. All the experiments are conducted on a PC with Intel(R) Core(TM)2 Duo CPU T6670@ 2.2 Hz on RedHat Linux Enterprise Edition 5.5 Operating System, and all the algorithms are implemented in Matlab 2011b Unix version.

Algorithm 1. Pseudo code of S-QUATRE Algorithm

Initialization:

Initialize the searching space V, dimension D, and the benchmark function $f(X)$.

Iteration:

1: **while** $exeTime < MaxIteration \| !stopCriterion$ **do**
2: Generate evolution matrix M of S-QUATRE according to Eq. 4.
3: **if** $exeTime = 1$ **then**
4: Initialize the individual matrix $\widehat{X}$, and calculate fitness values of all individuals.
5: **end if**
6: **if** $exeTime > 1$ **then**
7: Sort individuals in the population and then label them into better and worse groups.
8: Individuals evolution according to Eq. 1, and only the worse group individuals in $\widehat{X}$ are evolved.
9: **for** all individuals in worse group **do**
10: **if** $f(X_{i,G})$ optimal than $f(X_{i,G+1})$ **then**
11: $X_{i,G+1} \leftarrow X_{i,G}$.
12: Update fitness value of the individuals.
13: **end if**
14: **end for**
15: **end if**
16: find the global best individual and label it $X_{gbest,G}$.
17: **end while**

Output:

The global optima $X_{gbest,G}$, $f(X_{gbest,G})$.

3.2 Experiment Comparisons Between S-QUATRE and Other Algorithms

In order to evaluate the performance of QUATRE algorithm and S-QUATRE algorithm, contrast is made between these two algorithms first over CEC2013 test suite for real-parameter optimization. The parameter settings of the two algorithms are $ps = 100$, $FC = 0.7$, $nfe = ps \times D \times 1000$ (nfe denotes the number of function evaluation). 51 runs are conducted to make statistical analysis and the dimension is set to $D = 10$ for all these benchmark functions. The comparison result is shown in Table 2. We can see from Table 2 that QUATRE algorithm can find the global optima on function f_1–f_6, f_9, and f_{11}, S-QUATRE algorithm can find the global optima on function f_1–f_6 and f_9. For the best result of the total 51 runs, QUATRE algorithm can find two more results than S-QUATRE, but S-QUATRE has a better average performance than QUATRE algorithm on these testbeds. We also contrast the new S-QUATRE algorithm with several recently proposed state-of-the-art algorithms including iwPSO [7], ccPSO [8], SLPSO [9], canonical DE [1], ODE [17], and jDE [12]. The recommend parameter settings of all these contrasted algorithms are listed in Table 3.

Table 2. Comparison between QUATRE algorithm and the S-QUATRE algorithm under CEC2013 test suite. Best value (minimum), mean and standard deviation of 51-run fitness error comparisons of the two algorithms with the same number of function evaluations. The best results of the comparisons are emphasized in **BOLDFACE** and the tier results are highlighted in *ITALIC* fonts.

10D	QUATRE algorithm			S-QUATRE algorithm		
Setting:	$FC = 0.7, ps = 100, nfe = ps \times D \times 1000$			$FC = 0.7, ps = 100, nfe = ps \times D \times 1000$		
No.	Best	Mean	Std	Best	Mean	Std
f_1	*0*	*0*	*0*	*0*	*0*	*0*
f_2	*0*	**2.2291E−14**	**6.8285E−14**	*0*	9.0949*E*−14	1.1428*E*−13
f_3	*0*	4.5959*E*−03	1.7094*E*−02	*0*	**6.1331E−09**	**2.7428E−08**
f_4	*0*	**0**	**0**	*0*	9.0949*E*−14	1.3602*E*−13
f_5	*0*	*0*	*0*	*0*	*0*	*0*
f_6	*0*	4.4064E+00	4.8315E+00	*0*	**2.9437E+00**	**4.6134E+00**
f_7	**2.6603E−11**	7.0377*E*−01	2.1134E+00	1.2135*E*−05	**3.3668E−01**	**1.3667E+00**
f_8	2.0227E+01	2.0457E+01	8.6529*E*−02	**2.0128E+01**	**2.0448E+01**	**1.1127E−01**
f_9	*0*	2.0279E+00	1.4519E+00	*0*	**1.7545E+00**	**1.3019E+00**
f_{10}	4.9170*E*−02	**1.0887E−01**	**8.8611E−02**	**3.9502E−02**	2.2177*E*−01	1.6925*E*−01
f_{11}	**0**	3.1214E+00	1.8022E+00	9.9496E+00	**3.0346E+00**	**1.5970E+00**
f_{12}	4.7202E+00	**1.3973E+00**	**6.7581E+00**	**2.9849E+00**	1.5729E+00	7.5676E+00
f_{13}	9.9496*E*−01	1.9075E+01	8.6885E+00	**9.5938E+00**	**1.8396E+01**	**8.4462E+00**
f_{14}	3.8531E+00	1.2560E+02	1.1706E+02	**2.7075E+00**	**1.1560E+02**	**8.8505E+01**
f_{15}	**2.9998E+02**	**1.0043E+03**	**3.2006E+02**	5.9533E+02	1.2182E+03	2.0490E+02
f_{16}	**6.1354E−01**	**1.2167E+00**	**3.5948E−01**	8.4330*E*−01	1.3312E+00	3.1680*E*−01
f_{17}	**3.9636E−01**	**8.7361E+00**	**4.3007E+00**	7.3216*E*−01	1.0226E+01	3.7966E+00
f_{18}	1.2956E+01	**3.1576E+01**	**8.1329E+00**	**2.3460E+00**	3.8516E+01	6.4190E+00
f_{19}	2.8013*E*−01	5.9791*E*−01	1.3885*E*−01	**5.0128E−01**	**6.9164E−01**	**1.7944E−01**
f_{20}	**1.0848E+00**	**1.8075E+00**	**5.0166E−01**	1.7177E+00	2.9443E+00	6.0567*E*−01
f_{21}	**1.0000E+02**	3.7665E+02	7.37814E+01	2.0000E+02	**3.6016E+02**	**8.2158E+01**
f_{22}	3.1368E+01	9.0642E+01	7.4036E+01	**3.2767E+01**	**1.8733E+02**	**1.0161E+02**
f_{23}	**1.5338E+02**	**9.8230E+02**	**3.1535E+02**	4.6476E+02	1.1177E+03	3.2404E+02
f_{24}	**1.1186E+02**	2.0435E+02	1.3891E+01	2.0000E+02	**2.0383E+02**	**4.7791E+01**
f_{25}	2.0000E+02	2.0553E+02	5.2419E+00	**1.1209E+02**	**1.9323E+02**	**3.1779E+01**
f_{26}	**1.0398E+02**	**1.1269E+02**	**4.9148E+00**	1.0602E+02	1.7974E+02	5.0107E+01
f_{27}	*3.0000E+02*	**3.2536E +02**	**8.9250E+01**	*3.0000E+02*	3.4895E+02	8.7042E+01
f_{28}	*1.0000E+02*	2.9608E+02	2.8006E+01	*1.0000E+02*	**2.8000E+02**	**6.1559E+01**
Win	10	12		8	14	
Lose	8	14		10	12	
Draw	10	2		10	2	

The comparison results are shown in Fig. 2, the empirical cumulative distribution functions of the expected running length 10D and 20D optimization over these contrasted algorithms are figured and we can see that S-QUATRE has an overall better performance than other algorithms.

Table 3. Recommended parameter settings of all these contrasted algorithms.

Algorithms	Parameters settings
iwPSO	$c_1 = c_2 = 2.0, iw = 0.5, vel = rnd$
ccPSO	$c_1 = c_2 = 2.05, iw = 1, K = 0.7298, v = rnd$
SLPSO	$M = 100, c_3 = D/M * 0.01$
DE	$Cr = 0.1, F = 0.5$
ODE	$Cr = 0.1, F = 0.5, Jr = 0.37$
jDE	$Cr = 0.1, Fl = 0.1, Fu = 0.9, Tf = 0.1, Tcr = 0.1$
S-QUATRE	$FC = 0.7$

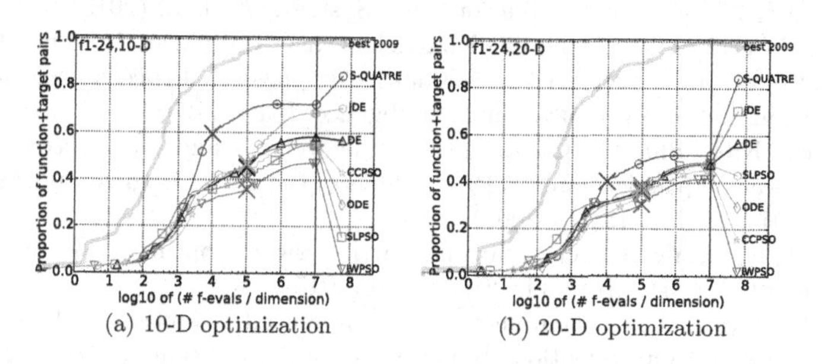

(a) 10-D optimization (b) 20-D optimization

Fig. 2. Algorithm verification on COCO framework under BBOB2009 testbeds over 24 functions.

4 Conclusion

In this paper, we discussed a new QUATRE algorithm with sort strategy for optimization problems. Sort strategy was incorporated into the former QUATRE algorithm to improve the overall performance of it. The new algorithm was verified on CEC2013 test suite for real-parameter optimization and COCO framework under BBOB2009 benchmarks. Several state-of-the-art PSO and DE variants were contrasted herein the paper. Conducted experiments show that the proposed S-QUATRE algorithm had better performance than all the compared ones. Moreover, the sort strategy enhanced the stability of the former QUATRE algorithm and this made the overall performance improved on many optimization benchmarks.

Acknowledgement. This work is partially funded by National Natural Science Foundation of China (61371178) and Shenzhen Innovation and Entrepreneurship Project (GRCK20160826105935160).

References

1. Storn, R., Price, K.: Differential evolution - a simple and efficient heuristic for global optimization over continuous spaces. J. Global Optim. **11**(4), 341–359 (1997)

2. Kennedy, J., Eberhart, R.: Particle swarm optimization. In: Proceedings of IEEE International Conference on Neural Networks, vol. 4, pp. 1942–1948. IEEE (1995)
3. Dorigo, M., Maniezzo, V., Colorni, A.: Ant system: optimization by a colony of cooperating agents. IEEE Trans. Syst. Man Cybern. Part B Cybern. **26**(1), 29–41 (1996)
4. Karaboga, D., Basturk, B.: A powerful and efficient algorithm for numerical function optimization: artificial bee colony (ABC) algorithm. J. Global Optim. **39**(3), 459–471 (2007)
5. Meng, Z., Pan, J.-S.: A simple and accurate global optimizer for continuous spaces optimization. In: Genetic and Evolutionary Computing, pp. 121–129. Springer International Publishing, Cham (2015)
6. Meng, Z., Pan, J.S., Alelaiwi, A.: A new meta-heuristic ebb-tide-fish-inspired algorithm for traffic navigation. Telecommun. Syst. **62**(2), 1–13 (2016)
7. Shi, Y., Eberhart, R.: A modified particle swarm optimizer. In: The 1998 IEEE International Conference on Evolutionary Computation Proceedings, 1998. IEEE World Congress on Computational Intelligence. IEEE (1998)
8. Eberhart, R.C., Shi, Y.: Comparing inertia weights and constriction factors in particle swarm optimization. In: Proceedings of the 2000 Congress on Evolutionary Computation, vol. 1. IEEE (2000)
9. Cheng, R., Jin, Y.: A social learning particle swarm optimization algorithm for scalable optimization. Inf. Sci. **291**, 43–60 (2015)
10. Price, K., Storn, R.M., Lampinen, J.A.: Differential Evolution: A Practical Approach to Global Optimization. Springer Science & Business Media, Heidelberg (2006)
11. Liu, J., Lampinen, J.: A fuzzy adaptive differential evolution algorithm. Soft. Comput. **9**(6), 448–462 (2005)
12. Brest, J., et al.: Self-adapting control parameters in differential evolution: a comparative study on numerical benchmark problems. IEEE Trans. Evol. Comput. **10**(6), 646–657 (2006)
13. Meng, Z., Pan, J.-S.: Monkey king evolution: a new memetic evolutionary algorithm and its application in vehicle fuel consumption optimization. Knowl.-Based Syst. **97**, 144–157 (2016)
14. Meng, Z., Pan, J.S., Xu, H.: QUasi-Affine TRansformation Evolutionary (QUATRE) algorithm: a cooperative swarm based algorithm for global optimization. Knowl.-Based Syst. **109**, 104–121 (2016)
15. Hansen, N., Finck, S., Ros, R., Auger, A.: Real-Parameter Black-Box Optimization Benchmarking 2009: Noiseless Functions Definitions. Research report RR-6829 (2009)
16. Liang, J.J., et al.: Problem definitions and evaluation criteria for the CEC 2013 special session on real-parameter optimization. Computational Intelligence Laboratory, Zhengzhou University, Zhengzhou, China and Nanyang Technological University, Singapore, Technical report 201212 (2013)
17. Rahnamayan, S.: Opposition-based differential evolution. IEEE Trans. Evol. Comput. **12**(1), 64–79 (2008)
18. Meng, Z., Pan, J.S.: QUasi-affine TRansformation Evolutionary (QUATRE) algorithm: a parameter-reduced differential evolution algorithm for optimization problems. In: IEEE Congress on Evolutionary Computation (CEC), pp. 4082–4089. IEEE (2016)

19. Pan, J.-S., Meng, Z., Xu, H., Li, X.: QUasi-Affine TRansformation Evolution (QUATRE) algorithm: a new simple and accurate structure for global optimization. In: International Conference on Industrial, Engineering and Other Applications of Applied Intelligent Systems, pp. 657–667. Springer International Publishing, Heidelberg (2016)
20. Pan, J.-S., Meng, Z., Xu, H., Li, X.: A matrix-based implementation of DE algorithm: the compensation and deficiency. In: International Conference on Industrial, Engineering and Other Applications of Applied Intelligent Systems, pp. 72–81. Springer, Cham (2017)
21. Meng, Z., Pan, J.S.: A Competitive QUasi-Affine TRansformation Evolutionary (C-QUATRE) algorithm for global optimization. In: 2016 IEEE International Conference on Systems, Man, and Cybernetics (SMC), pp. 001644–001649. IEEE (2016)
22. Meng, Z., Pan, J.S.: QUasi-affine TRansformation Evolutionary (QUATRE) algorithm: the framework analysis for global optimization and application in hand gesture segmentation. In: 2016 IEEE 13th International Conference on Signal Processing (ICSP), pp. 1832–1837. IEEE (2016)
23. Pan, J.S., Meng, Z., Chu, S.C., et al.: Monkey King Evolution: an enhanced ebb-tide-fish algorithm for global optimization and its application in vehicle navigation under wireless sensor network environment. Telecommun. Syst. **65**(3), 351–364 (2017)

The QUasi-Affine TRansformation Evolution (QUATRE) Algorithm: An Overview

Zhenyu Meng[1(✉)], Jeng-Shyang Pan[1,2], and Xiaoqing Li[3]

[1] Department of Computer Science and Technology,
Harbin Institute of Technology Shenzhen Graduate School, Shenzhen, China
mzy1314@gmail.com
[2] Fujian Provincial Key Lab of Big Data Mining and Applications,
Fujian University of Technology, Fuzhou, China
[3] Shenzhen Institute of Advanced Technology,
Chinese Academy of Sciences, Shenzhen, China

Abstract. QUasi-Affine TRansformation Evolution (QUATRE) algorithm is a new simple but powerful stochastic optimization algorithm proposed recently. The QUATRE algorithm aims to tackle the representational/positional bias inborn with DE algorithm and secures an overall better performance on commonly used Conference of Evolutionary Computation (CEC) benchmark functions. Recently, several QUATRE variants have been already proposed since its inception in 2016 and performed very well on many benchmark functions. In this paper, we mainly have a brief overview of all these proposed QUATRE variants first and then make simple contrasts between these QUATRE variants and several state-of-the-art DE variants under CEC2013 test suites for real-parameter single objective optimization benchmark functions. Experiment results show that the movement trajectory of individuals in the QUATRE structure is much more efficient than DE structure on most of the tested benchmark functions.

Keywords: Benchmark function · Differential evolution · Global optimization · QUATRE algorithm

1 Introduction

Recently, many heuristic and meta-heuristic optimization algorithms have been proposed and asserted to be good ones with the inspiration of the nature since 1990s, and some may be far-fetched to serve as an inspiration of these algorithm while some are excellent inspirations with which the associated algorithms become excellent branches of evolutionary computation. A new good nature-inspired algorithm, avoiding reinventing the wheels, usually conquers the weaknesses of an already existing algorithm/model and secures an overall better performance on a large domain of benchmark functions or real-world optimization

P. Krömer et al. (eds.), *Proceedings of the Fourth Euro-China Conference on Intelligent Data Analysis and Applications*, Advances in Intelligent Systems and Computing 682, DOI 10.1007/978-3-319-68527-4_35

applications. QUATRE algorithm [2,9–11] is such an algorithm which tackles the representational/positional bias of Differential Evolution (DE) [5].

DE is a very popular and powerful optimization algorithm which was proposed by Storn [1] in 1995. The original DE is developed from a former Genetic Anneal algorithm, a hybrid algorithm combing Genetic Algorithm (GA) [3] and Simulated Annealing (SA) [4]. DE tackles the weakness of GA on continuous parameter optimization and achieves a big success. There are three arguments in the general convention of DE, "DE/x/y/z". x denotes the base vector which is employed in mutation vector generation, y denotes the number of difference pairs in the generation equation of mutation vector, z denotes the crossover scheme. There are mainly two crossover schemes, one is "binomial crossover" (bin) and the other is "exponential crossover" (exp). From DE literature [1,5,6] we know that "bin" crossover performs better on continuous optimization problems rather than combinatorial optimization problems, and "bin" crossover is also the default crossover scheme in most of DE variants. An example is given in Eq. 1 to show the mutation generation scheme in canonical DE, "DE/rand/1/bin".

$$DE/rand/1/bin: \quad V_{i,G} = X_{r_0,G} + F * (X_{r_1,G} - X_{r_2,G}) \tag{1}$$

In Eq. 1, there are three vectors, $X_{r_0,G}$, $X_{r_1,G}$, $X_{r_2,G}$, to be selected from the individual population in order to calculate vector $V_{i,G}$. Usually, there are three different selection categories, "random selection with/without restriction", "permutation selection" and "random offset selection", the detailed information can be found in [6,7]. The canonical DE employs random selection with restriction to select $X_{r_0,G}$, $X_{r_1,G}$ and $X_{r_2,G}$. After the generation of donor/mutant vector, the crossover operation is proceeding according to Fig. 1.

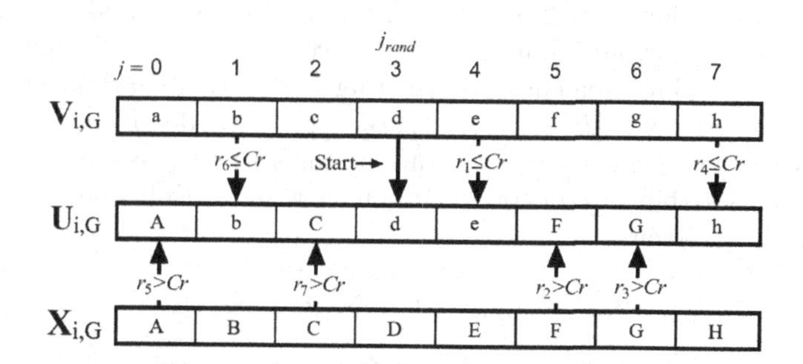

Fig. 1. Illustration of binomial crossover for an 8-D example.

The binomial crossover employed in Fig. 1 can be considered as an extension from N-point crossover. Moreover, the exponential crossover can be considered as a combination of 1-point and 2-point crossover [6]. Usually, the dependence on parameters separation for the optimization performance is known as "representational/positional bias". The binomial/uniform crossover is considered to be able to eliminate this bias as all the D parameters are selected or not by independent random trials. From a high dimensional perspective of view, it is not true. The key point to eliminate "representational/positional bias" is equalization the selection probability of all trial vectors, and form this view, there still are representational bias in binomial crossover. That' why we proposed QUATRE structure to tack such bias for optimization problems.

The rest of the paper is organized as follows. In Sect. 2, we mainly have a brief overview of several QUATRE variants proposed in literature. In Sect. 3, all these QUATRE variants are verified under CEC2013 test suites for real-parameter optimization. Moreover, several state-of-the-art DE variants are also contrasted in this section. Finally, we give the conclusion in Sect. 4.

2 QUATRE Variants Mentioned in Literature

2.1 The Canonical QUATRE Algorithm

The canonical QUATRE algorithm is proposed by Meng [9] in 2016, individuals in QUATRE structure evolve according to Eq. 2. $\widehat{X}$ denotes the individual population matrix which consist of ps different individuals with the location of ith individual to be the ith row vector of the matrix, $\widehat{X} = [X_{1,G}, X_{2,G}, \ldots, X_{i,G}, \ldots, X_{ps,G}]^T$. $X_{i,G} = [x_{i,1}, x_{i,2}, \ldots, x_{i,D}]$ denotes the location of the ith individual in the Gth generation, and the location of each individual is a candidate solution of a specific D-dimension optimization problem.

$$\widehat{X} \leftarrow \overline{M} \bigotimes \widehat{X} + M \bigotimes \widehat{B} \tag{2}$$

M is the evolution matrix, which is in the same dimension as matrix $\widehat{X}$, and $\overline{M}$ is the associated evolution matrix of M. Every element in M is either 0 or 1, and the corresponding element in $\overline{M}$ is either 1 or 0, the binary inverse value of the element in M. $\widehat{B} = [B_{1,G}, B_{2,G}, \ldots, B_{ps,G}]^T$ is the donor matrix, and $B_{i,G}$ can be calculated by Eq. 3. This calculation scheme of $B_{i,G}$ is in correspondence with "DE/1/rand/bin". $\bigotimes$ denotes component-wise multiply operation, same as ".*" in Matlab software.

$$B_{i,G} = X_{r_0,G} + F \cdot (X_{r_1,G} - X_{r_2,G}) \tag{3}$$

Evolution matrix M is transformed from a piled lower triangular matrix M_{ini}. When $ps = j \cdot D + k$, M_{ini} is initialized by piling j D-dimensional lower triangular matrices, and then the first k-rows of a lower triangular matrix is added in the rear of the piled matrix. Equation 4 gives an example when ps equals to $j \cdot D + 2$. "$\sim$" operation consists two consequent steps. The first step is to randomly permute every parameter in each vector of M_{ini}, and the second step is to randomly permute the row vectors with all the D parameters in the vectors unchanged.

$$M_{ini} = \begin{bmatrix} 1 \\ 1\,1 \\ \quad\cdots \\ 1\,1\cdots 1 \\ 1 \\ 1\,1 \\ \quad\cdots \\ 1\,1\cdots 1 \\ \vdots \\ 1 \\ 1\,1 \end{bmatrix} \sim \begin{bmatrix} 1\;1 \\ \quad\cdots \\ 1\;1\;\cdots 1 \\ \qquad 1 \\ \quad\cdots \\ 1\cdots\;1 \\ \quad 1 \\ 1\quad\;\cdots 1 \\ \vdots \\ \qquad 1\;1 \\ 1\quad\;\cdots 1 \end{bmatrix} = M \tag{4}$$

The main difference between the canonical QUATRE algorithm and DE algorithm lies in the evolution matrix M. M in DE structure can be initialized according to Eq. 5. M_{ini} of DE algorithm in QUATRE structure can be written in Eq. 6, and matrix One is a piled D-dimensional identity matrix, same piling strategy as Eq. 4. "|" in Eq. 5 denotes the binary "OR" operation. The aim of "$\sim$" operation is to equalize the selection probability of s parameters (s is the number of 1-elements in a row vector, $s \in \{1, 2, \ldots, D\}$) in a D-dimensional donor vector generated from a population with ps individuals. The matrix M_{ini} initialization strategy as well as "$\sim$" operation in QUATRE structure implements the equal selection probability that s parameters in the trial vector are inherited form the donor vector, and a deeper analysis can be found in [7,8]. Paper [2] also gives an extension of six schemes for the calculation of $B_{i,G}$, and these different $B_{i,G}$ generation schemes usually perform different on different optimization problems, which is similar to the DE variants.

$$M = M_{ini}|One \tag{5}$$

$$M_{ini} = \begin{bmatrix} rnd_{11} \le Cr & rnd_{12} \le Cr & \cdots & rnd_{1D} \le Cr \\ rnd_{21} \le Cr & rnd_{22} \le Cr & \cdots & rnd_{2D} \le Cr \\ \vdots & \vdots & \vdots & \vdots \\ rnd_{i1} \le Cr & rnd_{i2} \le Cr & rnd_{ij} \le Cr & rnd_{iD} \le Cr \\ \vdots & \vdots & \vdots & \vdots \\ rnd_{ps,1} \le Cr & rnd_{ps,2} \le Cr & \cdots & rnd_{ps,D} \le Cr \end{bmatrix} \tag{6}$$

$$One_{ini} = \begin{bmatrix} 1\,0 \cdots 0 \\ 0\,1 \cdots 0 \\ \vdots\;\vdots \cdots \vdots \\ 0\,1 \cdots 0 \end{bmatrix} \sim \begin{bmatrix} 0\;1\;\cdots 0 \\ 0 \cdots\;0\;1 \\ \vdots\;\vdots\;\cdots \vdots \\ 0 \cdots\;1\;0 \end{bmatrix} = One \tag{7}$$

2.2 Competitive QUATRE (C-QUATRE) Algorithm

A competitive QUATRE algorithm was proposed in [12] to enhance the overall performance of QUATRE algorithm by employing a pair-wise competition mechanism. We suppose the population size is an even number, and the whole population is divided into two groups with the same group size, $\frac{ps}{2}$ (ps is the population size). During the competition, each individual in the first group, the winner group, is competed against an individual from the other group, the loser group. All the individuals are competed once and the winner individuals are reserved into next generation while the loser individuals need to evolve, generate new offspring, and select the better into the next generation. Figure 2 illustrates the pair-wise competition mechanism in C-QUATRE algorithm.

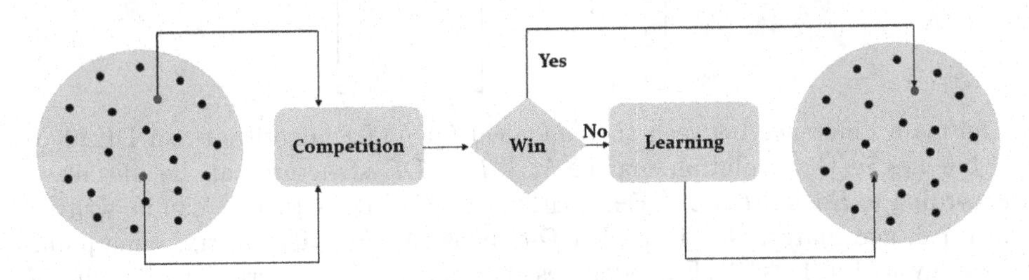

Fig. 2. Pairwise competition mechanism and individuals' evolution in C-QUATRE algorithm.

In the implementation of C-QUATRE algorithm, the ps individuals are distribution randomly into the search domain. Then the first half $\frac{ps}{2}$ individuals are classified into the winner group while the left half $\frac{ps}{2}$ individuals into the loser group. The ith individuals of the two groups are pair-wise competed and the winner individual and loser individual are placed into the ith location of winner and loser groups respectively. The individuals evolve according to the former QUATRE algorithm, and only the loser group individuals evolve and produce next generation individuals. Selection operation is consequently done between the loser individuals and their offsprings. The winner group individuals and the selected winner individuals merge into a new population, and then a permutation of all the ps individuals is done to change the individuals' sequence in the population. Finally, a new evolution circle begins and the pseudo code of C-QUATRE algorithm is given in Algorithm 1.

2.3 QUATRE Algorithm with Sort Strategy (S-QUATRE)

A QUATRE algorithm with sort strategy was proposed in [13] to improve the overall performance of the former QUATRE algorithm. The main idea is that the whole population is sorted first after initialization, and then the population is divided into the better and the worse groups. During the evolution, only the individuals in the worse group evolve according the QUATRE evolution equation with the better group individuals unchanged. Selection is consequently made between the former worse individuals and evolved ones. Then the next generation individuals is decided and a new circle of evolution begins. Figure 3 illustrate the evolution mechanism of S-QUATRE algorithm.

Algorithm 1. Pseudo code of the C-QUATRE Algorithm

Initialization:

Initialize the searching space V, dimension D, and the benchmark function $f(X)$.

Iteration:

1: **while** $exeTime < MaxIteration | !stopCriterion$ **do**
2: Generate evolution matrix M_{ini} and then calculate M according to the example shown in Eq. 4.
3: **if** $exeTime = 1$ **then**
4: Initialize M_{ini} and then calculate M by $\sim$ operation. Generate individual matrix $\widehat{X}$ by random distribution, calculate the fitness values of each individual.
5: **end if**
6: **if** $exeTime > 1$ **then**
7: Individuals are randomly divided into two groups.
8: Pairwise competition.
9: Evolution using Eq. 2, only loser individual coordinates in X are evolved.
10: **for** all loser individuals **do**
11: **if** $f(X_{i,G})$ is optimal than $f(X_{i,G+1})$ **then**
12: $X_{i,G+1} \leftarrow X_{i,G}$.
13: Update fitness value of the individual.
14: **end if**
15: **end for**
16: **end if**
17: label the global best individual $X_{gbest,G}$.
18: **end while**

Output:

The global optima $X_{gbest,G}$, $f(X_{gbest,G})$.

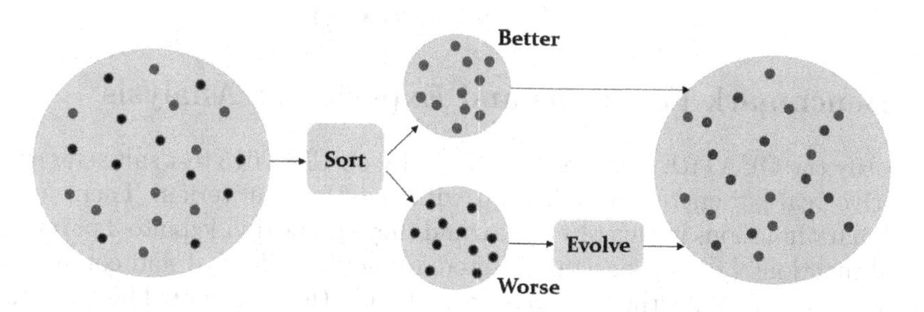

Fig. 3. Illustration of sort scheme in S-QUATRE algorithm.

2.4 QUATRE with External Archive (QUATRE-EAR)

A new powerful QUATRE variant with external archive (QUATRE-EAR) is proposed in [8]. The QUATRE-EAR algorithm presented a new mutation strategy for the calculation of $B_{i,G}$. The new mutation strategy employed an external archive to record inferior solution during the evolution process. The detailed equation of the calculation of $B_{i,G}$ is given in Eq. 8. $X^{p}_{best,G}$ denotes randomly selected top $100p\%$ individuals from the population, and $\widehat{X}_{r_2,G}$ denotes the union of external archive restored inferior individuals and the current population individuals. A time stamp mechanism is also employed in the external archive, and too old inferior solutions in the archive is discarded and this avoids the misleading evolution direction of the population.

$$\begin{aligned} B_{i,G} &= X_{i,G} + F \cdot (X^{p}_{best,G} - X_{i,G}) + F \cdot (X_{r_1,G} - \widehat{X}_{r_2,G}) \\ B &= [B_{1,G}, B_{2,G}, \ldots, B_{i,G}, \ldots, B_{ps,G}]^T, i \in \{1, 2, \ldots, ps\} \end{aligned} \tag{8}$$

Moreover, an self-adaptive initialization of evolution matrix M_{ini} in each generation is advanced and a new adaptive scheme of control parameter F is also presented in the QUATRE-EAR algorithm to implement a better choice of the control parameter during the evolution process. The initialization scheme of M_{ini} in each generation is changed according to Eq. 9, and the adaptive scheme of control parameter F is changed according to Eq. 10.

$$\begin{cases} P(U_{i,G}^{A=k}) = \begin{cases} \dfrac{ns_k^2}{\sum_{j=1}^{k}(ns_j) \times (ns_k + nf_k)}, & if\ ns_k > 0, \\ \epsilon, & otherwise. \end{cases} \\ N(k) = \dfrac{P(U_{i,G}^{A=k})}{\sum_{j=1}^{D} P(U_{i,G}^{A=j})} \times ps. \end{cases} \tag{9}$$

$$\begin{cases} w_{F_i} = \dfrac{\Delta f_i}{\sum_{F_i \in S_F} \Delta f_i} \\ \mu_F = \dfrac{\sum_{F_i \in S_F} w_{F_i} \cdot F_i^2}{\sum_{F_i \in S_F} w_{F_i} \cdot F_i} \end{cases} \tag{10}$$

$$\Delta f_i = f(U_{i,G}) - f(X_{i,G}) \tag{11}$$

3 Benchmark Functions and Experiment Analysis

We verify the QUATRE variants [2,9–13,17,18] on CEC2013 test suites for single objective real-parameter optimization with fixed cost criterion. There are 28 benchmark functions in this test suites, and they can be divided into 3 parts, uni-modal functions (f_1–f_5), basic multi-modal functions (f_6–f_{20}) and composition functions (f_{21}–f_{28}). All the functions are shifted to the same global best location, $O\{o_1, o_2, \cdots, o_d\}$. For this test suites, 51 runs are conducted, and the best, mean and standard deviation of these runs are selected and calculated for comparison. A total $10000 \times D$ number of function evaluations is used as the maximum fixed cost for each run. All the experiments are conducted on a PC with Intel(R) Core(TM)2 Duo CPU T6670@ 2.2 Hz on RedHat Linux Enterprise Edition 5.5 Operating System, and all the algorithms are implemented in Matlab 2011b Unix version. Several state-of-the-art DE variants are also contrasted here with our proposed QUATRE variants, these DE variants are DE [5], jDE [14], JADE [15], SHADE [16].

The parameter setting of all these algorithms are listed in Table 1. In consideration of page limitation, and the optimization accuracy of QUATRE, C-QUATRE, S-QUATRE, and QUATRE-EAR have been already presented in the associated papers, we do not present the best, average, and standard deviation of 51 runs here. We only figure the convergence curve to illustrate the convergence speed of these contrasted algorithms on several randomly selected benchmark functions, f_3, f_8, f_{14}, f_{26}. The convergence speed comparison is given in Fig. 4, and we can see that QUATRE structure have and overall better performance than DE structure and QUATRE-EAR algorithm is much more competitive than the contrasted state-of-the-art algorithms.

Table 1. Recommended Parameter settings of all these contrasted algorithms.

Algorithms	Parameters initial settings
DE/best/1	$F = 0.5, Cr = 0.1$
DE/best/1	$F = 0.5, Cr = 0.9$
jDE	$F = 0.5, Cr = 0.9, \tau_F = \tau_{Cr} = 0.1, F_l = 0.1, F_u = 0.9$
SHADE	$\mu_F = 0.5, F \sim C(\mu_F, 0.1), \mu_{Cr} = 0.5, Cr \sim N(\mu_{Cr}, 0.1), p = 0.2, H = 100$
QUATRE	$F = 0.7$
CQUATRE	$F = 0.7$
SQUATRE	$F = 0.7$
QUATRE-EAR	$\mu_F = 0.5, F_i \sim C(\mu_F, 0.2), \mu_{Cr} = 0.5, Cr \sim N(\mu_{Cr}, 0.1), c = 0.8, p = 0.1, a = 1.7, T_0 = 70$

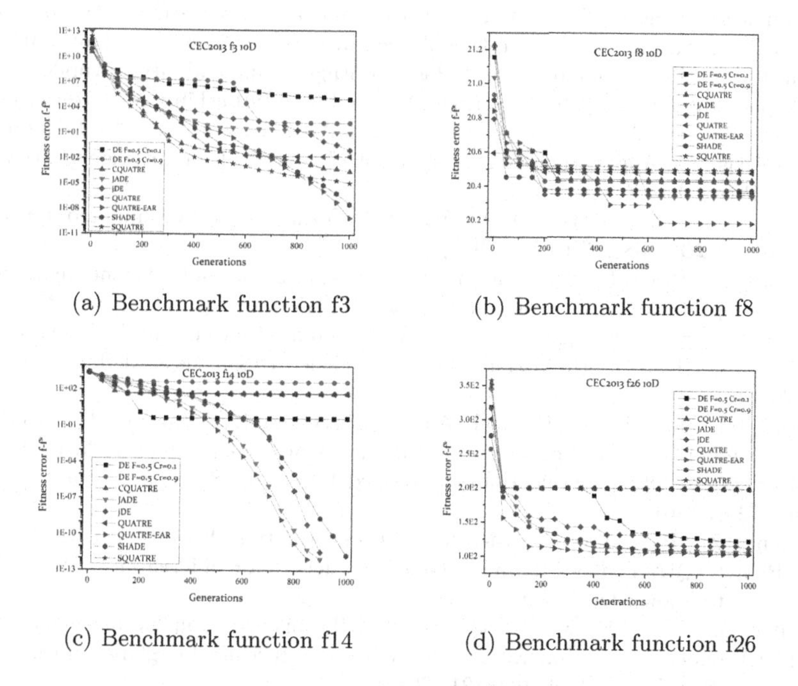

(a) Benchmark function f3

(b) Benchmark function f8

(c) Benchmark function f14

(d) Benchmark function f26

Fig. 4. Median values of the 51 runs for each algorithm is selected to analysis the convergence speed on a certain benchmark function. Four benchmark functions are randomly selected from the total 28 functions and the 3 groups, and they are f_3, f_8, f_{14} and f_{26}, and convergence speed of these algorithms are contrasted on the four functions.

4 Conclusion

In this paper, we had a brief overview of several recently proposed QUATRE variants. All these variants are verified on CEC2013 test suite for real-parameter single objective optimization benchmarks. Several state-of-the-art DE variants are also contrasted with these QUATRE variants in the paper. The QUATRE structure, from the experiment results, had an overall better performance on these benchmark functions and QUATRE-EAR algorithm secured the best performance in all these comparisons.

Acknowledgement. This work is partially funded by Shenzhen Innovation and Entrepreneurship Project (GRCK20160826105935160) and National Natural Science Foundation of China (61371178).

References

1. Storn, R., Price, K.: Differential evolution–a simple and efficient adaptive scheme for global optimization over continuous spaces. International Computer Science Institute, Berkeley, CA, Technical report, TR-95-012 (1995)
2. Pan, J.S., Meng, Z., Xu, H., et al.: QUasi-Affine TRansformation Evolution (QUATRE) algorithm: a new simple and accurate structure for global optimization. In: International Conference on Industrial, Engineering and Other Applications of Applied Intelligent Systems. Springer International Publishing, pp. 657–667 (2016)
3. Holland, J.H.: Adaptation in Natural and Artificial Systems: An Introductory Analysis with Applications to Biology, Control, and Artificial Intelligence. U Michigan Press, Ann Arbor (1975)
4. Kirkpatrick, S., Gelatt, C.D., Vecchi, M.P.: Optimization by simulated annealing. Science **220**(4598), 671–680 (1983)
5. Storn, R., Price, K.: Differential evolution - a simple and efficient heuristic for global optimization over continuous spaces. J. Glob. Optim. **11**(4), 341–359 (1997)
6. Price, K., Storn, R.M., Lampinen, J.A.: Differential Evolution: A Practical Approach to Global Optimization. Springer Science & Business Media, Heidelberg (2006)
7. Pan, J.S., Meng, Z., Xu, H., et al.: A matrix-based implementation of DE algorithm: the compensation and deciency. In: International Conference on Industrial, Engineering and Other Applications of Applied Intelligent Systems. Springer, Cham, pp. 72-81 (2017)
8. Meng, Z., Pan, J.S.: QUasi-Affine TRansformation Evolution with External ARchive (QUATRE-EAR): an enhanced structure for differential evolution, submitted to knowledge-based systems
9. Meng, Z., Pan, J.S., Xu, H.: QUasi-Affine TRansformation Evolutionary (QUATRE) algorithm: a cooperative swarm based algorithm for global optimization. Knowl.-Based Syst. **109**, 104–121 (2016)
10. Meng, Z., Pan, J.S., QUasi-Affine TRansformation Evolutionary (QUATRE) algorithm: a parameter-reduced differential evolution algorithm for optimization problems. In: IEEE Congress on Evolutionary Computation (CEC), pp. 4082–4089. IEEE (2016)

11. Meng, Z., Pan, J.S.: Monkey king evolution: a new memetic evolutionary algorithm and its application in vehicle fuel consumption optimization. Knowl.-Based Syst. **97**, 144–157 (2016)
12. Meng, Z., Pan, J.S.: A Competitive QUasi-Affine TRansformation Evolutionary (C-QUATRE) algorithm for global optimization. In: 2016 IEEE International Conference on Systems, Man, and Cybernetics (SMC). IEEE (2016)
13. Pan, J.-S., Meng, Z., Chu, S., Roddick, J.F.: QUATRE algorithm with sort strategy for global optimization in comparison with DE and PSO variants. In: Proceedings of the Fourth Euro-China Conference on Intelligent Data Analysis and Applications, Advances in Intelligent Systems and Computing 682. doi:10.1007/978-3-319-68527-4_34
14. Brest, J., et al.: Self-adapting control parameters in differential evolution: a comparative study on numerical benchmark problems. IEEE Trans. Evol. Comput. **10**(6), 646–657 (2006)
15. Zhang, J., Sanderson, A.C.: JADE: adaptive differential evolution with optional external archive. IEEE Trans. Evol. Comput. **13**(5), 945–958 (2009)
16. Tanabe, R., Fukunaga, A.: Success-history based parameter adaptation for differential evolution. In: IEEE Congress on Evolutionary Computation, Cancun, pp. 71–78 (2013)
17. Meng, Z., Pan, J.S.: QUasi-Affine TRansformation Evolutionary (QUATRE) algorithm: the framework analysis for global optimization and application in hand gesture segmentation. In: 2016 IEEE 13th International Conference on Signal Processing (ICSP), pp. 1832–1837. IEEE (2016)
18. Pan, J.S., Meng, Z., Chu, S.C., et al.: Monkey king evolution: an enhanced ebb-tide-fish algorithm for global optimization and its application in vehicle navigation under wireless sensor network environment. Telecommun. Syst. **65**(3), 351–364 (2017)

A Hybrid Central Force Optimization Algorithm for Optimizing Ontology Alignment

Xingsi Xue[1,2(✉)], Shijian Liu[1,2], and Jinshui Wang[1,2]

[1] College of Information Science and Engineering, Fujian University of Technology, No. 3 Xueyuan Road, University Town, Minhou, Fuzhou City 350118, Fujian Province, China
jack8375@gmail.com

[2] Fujian Provincial Key Laboratory of Big Data Mining and Applications, Fujian University of Technology, No. 3 Xueyuan Road, University Town, Minhou, Fuzhou City 350118, Fujian Province, China

Abstract. Ontology is regarded as an effective solution to data heterogeneity on the semantic web. However, different ontology engineers might use different ways to define the concept, which causes the ontology heterogeneity problem and raises the heterogeneous problem to a higher level. Ontology matching technology, which is able to identify the same concepts in two heterogeneous ontologies, is recognized as a ground solution to tackle the ontology heterogeneity problem. Since different ontology matchers do not necessarily find the same correct correspondences, usually several competing matchers are applied to the same pair of entities in order to increase evidence towards a potential match or mismatch. How to select, combine and tune various ontology matchers to obtain the high quality ontology alignment becomes a crucial challenges in ontology matching domain. Recently, swarm intelligent algorithms are appearing as a suitable methodology to face this challenge, but the slow convergence and premature convergence are two main shortcomings that makes them incapable of effectively searching the optimal solution for large scale and complex ontology matching problems. To improve the ontology alignment's quality, our work investigates a new emergent class of swarm intelligent algorithm, named Central Force Optimization (CFO) algorithm. To balance CFO's exploration and exploitation, we proposes a Hybrid CFO (HCFO) by introducing the local search strategy into CFO's evolving process, and utilize HCFO to automatically select, combine and tune various ontology matchers to optimize the ontology alignment. The experimental results show that our approach can significantly improve the ontology alignment's quality of existing swarm intelligent algorithm based ontology matching technologies.

Keywords: Hybrid central force optimization algorithm · Ontology matching · Ontology heterogeneity problem

P. Krömer et al. (eds.), *Proceedings of the Fourth Euro-China Conference on Intelligent Data Analysis and Applications*, Advances in Intelligent Systems and Computing 682, DOI 10.1007/978-3-319-68527-4_36

1 Introduction

Ontology is regarded as an effective solution to data heterogeneity on the semantic web. However, different ontology engineers might use different ways to define the concept, which causes the ontology heterogeneity problem and raises the heterogeneous problem to a higher level. Ontology matching technology, which is able to identify the same concepts in two heterogeneous ontologies, is recognized as a ground solution to tackle the ontology heterogeneity problem. Since different ontology matchers do not necessarily find the same correct correspondences, usually several competing matchers are applied to the same pair of entities in order to increase evidence towards a potential match or mismatch [11]. How to select, combine and tune various ontology matchers to obtain the high quality ontology alignment becomes a crucial challenges in ontology matching domain [13]. Recently, swarm intelligent algorithms are appearing as a suitable methodology to face this challenge. GOAL [8] is the first ontology matching system that utilizes Evolutionary Algorithm (EA) to determine the weight configuration for a weighted average aggregation of several matchers by considering a reference alignment. GAOM [15] also utilizes EA, where each chromosome represents an alignment of two ontologies and is evaluated by a fitness function. To improve the performance of EA based ontology matching technology, Acampora et al. [1] employ a Memetic Algorithm (MA), which introduces the local search strategy into EA's evolving process, to solve the alignment problem. MapPSO [2] exploits a Particle Swarm Optimization (PSO) algorithm to directly determine the optimal ontology alignment. In particular, it uses an evaluation method based on multiple objectives following a priori approach for the ontology matching problem, i.e. aggregating all the objectives through a given weighted function.

However, the slow convergence and premature convergence are two main shortcomings of the swarm intelligent algorithms, which makes them incapable of effectively searching the optimal solution for large scale and complex ontology matching problems. To improve the ontology alignment's quality, our work investigates an emergent class of swarm intelligent algorithm, named Central Force Optimization (CFO) algorithm, to solve the problem of ontology matching. Particularly, CFO is a population-based self-adaptive search optimization algorithm that was proposed by Formato [4] in 2007. CFO demonstrates and leverages two characteristics that make it unique when compared to other methodologies, i.e. a basis in Newton's law of gravity and a deterministic nature. Classic CFO has a very strong exploration ability, but, the numerical experiments show that stagnation is a drawback with CFO, i.e. CFO often stops proceeding towards the global optima even though the population has not converged to local optima or any other point [3]. To balance CFO's exploration and exploitation, we proposes a Hybrid CFO (HCFO) by introducing the local search strategy into CFO's evolving process, and utilize HCFO to automatically select, combine and tune various ontology matchers to optimize the ontology alignment.

The rest of the paper is organized as follows: Sect. 2 describes the ontology matching problem; Sect. 3 presents the HCFO based ontology matching technology; Sect. 4 presents the experimental studies and analysis; finally, Sect. 5 draws the conclusions.

2 Ontology Matching Problem

In this work, an ontology is defined as three-tuple (C, P, I) [17], where C, P, I are respectively referred to the set of classes, properties and instances. In addition, an ontology alignment A between two ontologies is a correspondence set and each correspondence inside is a 3-tuples $(e, e', =, confidence)$, where e and e' are respectively the entities of two ontology, the relation of the correspondence is the equivalence $(=)$, and $confidence$ is the trust extent of the equivalence holds between e and e'.

In this work, supposing in the golden alignment, one entity in source ontology is matched with only one entity in target ontology and vice versa, the optimal model of ontology matching problem is defined as follows:

$$\begin{cases} max \quad f(X) \\ s.t. \quad X = (x_1, x_2, \cdots, x_n)^T \\ \qquad \sum_{i=1}^{n} x_i = 1 \\ \qquad x_i \in [0, 1], i = 1, 2, \cdots, n \end{cases} \tag{1}$$

where n is the number of ontology matchers, $x_i, i = 1, 2, \cdots, n$ is the i-th ontology matcher's aggregating weight, and function $f()$ evaluates the quality of final alignment which is determined by X. In particular, function $f()$ is defined as follows:

$$f(X) = \frac{1}{2} \times (\frac{|C_{O_1-Match}| + |C_{O_2-Match}|}{|C_{O_2}| + |C_{O_2}|} + \frac{|C_{O_1-Match}| + |C_{O_2-Match}|}{2 \cdot |Corr_{O_1-O_2}|}) \tag{2}$$

where:

- O_1 and O_2 are two ontologies to be matched, respectively;
- $C_{O_1-Match}$ and $C_{O_2-Match}$ are the set of matched entities of ontologies O_1 and O_2 respectively;
- C_{O_1} and C_{O_2} are the set of all entities of ontologies O_1 and O_2 respectively;
- $Corr_{O_1-O_2}$ is the set of correspondences in a resulting alignment from ontology O_1 to ontology O_2.

3 Hybrid Central Force Optimization Algorithm Based Ontology Matching

CFO is a heuristic swarm intelligent global optimization algorithm, which simulate body surrounds the track which the heavenly body with Newton's gravity law. Particularly, Newton's universal law of gravitation describes the gravitational attraction of two masses m_1 and m_2. The magnitude of the force of attraction is as follows:

$$F = \gamma \frac{m_1 m_2}{r^2} \tag{3}$$

where r is the distance between m_1 and m_2, coefficient γ is the "gravitational constant". Because the force acts along the line connecting the masses' centers, gravity is a central force. Each mass is accelerated towards the other, and the vector acceleration experienced by mass m_1 due to mass m_2 is determined by the following formula:

$$\mathbf{a} = -\gamma \frac{m_2 \hat{r}}{r^2} \tag{4}$$

where $\hat{r}$ is a unit vector, which points towards mass m_1 from mass m_2 along the line joining the two masses. The position vector of a particle subject to constant acceleration during the interval t to $t + \Delta t$ is given by the following formula:

$$\mathbf{R}(t + \Delta t) = \mathbf{R}_0 + \mathbf{V}_0 \Delta t + \frac{1}{2} \mathbf{a} \Delta t^2 \tag{5}$$

The mass's position at time $t + \Delta t$ is $\mathbf{R}(t + \Delta t)$, where $\mathbf{R}_0$ and $\mathbf{V}_0$ are the position and velocity vectors at time t, respectively.

On this basis, in the N_d dimensional search space with N_p probes $x^p = (x_1^p, x_2^p, \cdots, x_{N_d}^p)$, $p = 1, 2, \cdots, N_p$, and in generation t, x^p moves according to the following formulas:

$$\mathbf{x}_{t+1}^p = \mathbf{x}_t^p + \mathbf{V}_t^p \Delta t + \frac{1}{2} \mathbf{A}_t^p (\Delta t)^2 \tag{6}$$

$$\mathbf{A}^p = G \sum_{k=1, pk \geq p}^{N_p} U(\mathbf{f}_{t-1}^k - \mathbf{f}_{t-1}^p)(\mathbf{f}_{t-1}^k - \mathbf{f}_{t-1}^p)^{\alpha} (\mathbf{x}_{t-1}^k - \mathbf{x}_{t-1}^p) \|\mathbf{x}_{t-1}^k - \mathbf{x}_{t-1}^p\|^{-\beta} \tag{7}$$

$$\mathbf{V}_{t-1}^p = \frac{\mathbf{x}_{t-1}^p - \mathbf{x}_{t-2}^p}{\Delta t} \tag{8}$$

where $\mathbf{x}_t^p$, $\mathbf{A}_t^p$ and $\mathbf{V}_t^p$ respectively represent x^p's position vector, acceleration vector and velocity vector in generation t, $\mathbf{f}_t^p$ is x^p's objective function value, U represents a piece-wise function $U(z) = \begin{cases} 1, & z \geq 0 \\ 0, & \text{otherwise} \end{cases}$, and G, α and β are three parameters for determining the gravitational acceleration with mass and distance. As the CFO algorithm progresses, it is possible that some probes will "fly" to points outside the defined decision space. In order to avoid searching in unallowable regions, the probe should be returned to the decision space when this happens, and the details of this procedure please see also [4].

In this paper, in order to better tradeoff the exploitation and exploration and improve CFO's performance, we propose to introduce the local search strategy into CFO, i.e. Hybrid CFO. HCFO may be considered as a synergy between population-based approaches and separate local improvement methods. The main advantage of HCFO is that the space of the possible solutions is reduced to the subspace of local optima, which ensure HCFO not only converges to high-quality solutions, but also searches vast solution spaces more efficiently than conventional CFO. These characteristics make HCFO apparently suitable to be applied to a complex process such as ontology matching. The basic steps of CFO are as follows: (1) compute initial probe positions, the corresponding fitnesses, and assign initial accelerations; (2) successively compute each probe's new position based on previously computed accelerations; (3) verify that each probe is located inside the decision space, making corrections as required; (4) update the fitness at each new probe position and update the elite solution; (5) when elite solution keeps unchanged for ϵ generations, execute the local search process; (6) compute accelerations for the next time step based on the new positions; and (7) loop over all time steps.

3.1 Chromosome Encoding Mechanism

In this work, we use the binary encoding where the parameter set is encoded in a solution. Since the weighted average approach is used to combine various similarity measures where the sum of the weights is equal to 1, our encoding mechanism indirectly represents them through the numbers in the interval [0,1]. Assuming that n is the number of weights required, the number set can be represented as $\{c_1, c_2, \ldots, c_n\}$, and the chromosome decoding is carried out by dividing the numbers with their sum, then we get the parameter set $\{\frac{c_1}{\sum_{i=1}^{n} c_i}, \frac{c_2}{\sum_{i=1}^{n} c_i}, \ldots, \frac{c_n}{\sum_{i=1}^{n} c_i}\}$.

3.2 Fitness Function and Elitism

In this work, fitness function is the objective function that evaluate the quality of the alignment obtained, see also Eq. (2). In addition, we regard the individual with the highest fitness value as the elite solution of current generation. When the algorithm terminates, the elite will be returned as the output.

Local Search Process. In general, the local search strategies perform iterative search for optimum solution in the neighborhood of a candidate. In order to trade off between the local search and the global search, the local search process in our work is designed as follows:

- the local search is applied when elite solution keeps unchanged for ϵ generations;
- the local search is executed after computing each probe's new position;
- the local search is applied to the elite solution;
- the local search method adopted in this paper is the hill climbing algorithm.

In particular, the hill climbing algorithm is a local search iterative method. During iterations, the algorithm attempts to find a better solution by randomly permutating the current one's position. If the permutation improves the current solution, then the new solution becomes the current one. The search is repeated until no further improvements can be found or after a maximum number of iterations. After the local search, we try to update the elite by comparing the best individual obtained in local search with the original elite.

4 Experimental Studies and Analysis

4.1 Testing Cases and Experimental Configuration

In the experiment, we utilize the bibliographic track of OAEI 2016[1] to test our approach's performance. Table 1 shows a brief description on the benchmark of OAEI 2016.

Table 1. Brief description on benchmark. 1XX, 2XX and 3XX stands for the test case whose ID beginning with the prefix digit 1, 2 and 3, respectively

ID	Description
1XX	The ontologies under alignment are the same or the first one is the OWL Lite restriction of the second one
2XX	The ontologies under alignment have different lexical, linguistic or structure features
3XX	The ontologies under alignment are real world cases

In general, the basic ontology matcher used in these matching systems can be divided into four broad categories [16]: (1) syntactic-based matcher [14], which computes a string distance or edit distance between the ontology entities; (2) linguistic-based matcher [10], which calculates the similarity between ontology entities by considering linguistic relations such as synonymy, hypernym, and so on; (3) structure-based matcher [9], which utilizes the specialization relation, e.g. the subsumption relation, in ontologies to calculate the similarity between ontology entities; (4) instance-based matcher [16], which exploits the similarity between classified instances to discover the correspondences between the concepts in ontologies. Therefore, in our work, we utilize the above four basic similarity matchers.

HCFO use the control parameters, which represent a trade-off setting obtained in empirical way to achieve the highest average alignment quality on all test cases:

- Numerical accuracy = 0.01;
- Population size $N_p = 200$;
- $G = 3$;
- $\alpha = 2$;
- $\beta = 2$;
- $\Delta t = 1$;
- Elite solution keeps unchanged threshold $\epsilon = 30$;
- Local Search intensity = 20 iterations;
- Maximum generation = 3000.

[1] http://oaei.ontologymatching.org/2016/benchmarks/index.html.

4.2 Results and Analysis

All the experimental results in the tables are the average values over ten independent runs. Specifically, Tables 2 and 3 show the statistical comparison among three EA based ontology matching approaches and our approach. we carry out the statistical comparison on the alignment's quality in terms of f-measure among EA based [8], MA based [1], PSO based [2] ontology matching approaches and our approach. The statistical comparison is formally carried out by means of a multiple comparison procedure which consists of two steps: in the first one, a statistical technique, i.e. the Friedman's test [5], is used to determine whether the results provided by various approaches present any difference; in the second

Table 2. Friedman's test on ontology alignment's quality. Each value represents the f-measure, and the number in round parentheses is the corresponding computed rank.

ID	EA	MA	PSO	Our approach
101	1.00 (2.5)	1.00 (2.5)	1.00 (2.5)	1.00 (2.5)
103	0.99 (4)	1.00 (2)	1.00 (2)	1.00 (2)
104	0.99 (4)	1.00 (2)	1.00 (2)	1.00 (2)
201	0.50 (3)	0.62 (2)	0.42 (4)	0.72 (1)
202	0.35 (4)	0.41 (3)	0.42 (2)	0.64 (1)
203	0.97 (3)	0.96 (4)	1.00 (1.5)	1.00 (1.5)
204	0.94 (4)	0.97 (3)	0.98 (1.5)	0.98 (1.5)
205	0.83 (1.5)	0.79 (3)	0.73 (4)	0.83 (1.5)
206	0.84 (4)	0.88 (2)	0.85 (3)	0.92 (1)
221	0.99 (3.5)	0.99 (3.5)	1.00 (1.5)	1.00 (1.5)
222	0.99 (2.5)	0.99 (2.5)	0.99 (2.5)	0.99 (2.5)
223	0.99 (2.5)	0.99 (2.5)	0.99 (2.5)	0.99 (2.5)
224	1.00 (2.5)	1.00 (2.5)	1.00 (2.5)	1.00 (2.5)
225	0.99 (4)	1.00 (2)	1.00 (2)	1.00 (2)
228	0.99 (2.5)	0.99 (2.5)	0.99 (2.5)	0.99 (2.5)
230	0.93 (3.5)	0.93 (3.5)	0.98 (2)	1.00 (1)
231	0.99 (3)	0.99 (3)	0.99 (3)	1.00 (1)
232	1.00 (2.5)	1.00 (2.5)	1.00 (2.5)	1.00 (2.5)
233	1.00 (2.5)	1.00 (2.5)	1.00 (2.5)	1.00 (2.5)
234	1.00 (2.5)	1.00 (2.5)	1.00 (2.5)	1.00 (2.5)
248	0.66 (3)	0.69 (2)	0.64 (4)	0.77 (1)
249	0.37 (4)	0.41 (3)	0.46 (2)	0.54 (1)
250	0.51 (2)	0.51 (2)	0.40 (4)	0.51 (2)
301	0.70 (2.5)	0.70 (2.5)	0.64 (4)	0.85 (1)
302	0.61 (3)	0.63 (2)	0.04 (4)	0.72 (1)
303	0.58 (4)	0.72 (2)	0.65 (3)	0.84 (1)
304	0.83 (3)	0.87 (2)	0.72 (4)	0.94 (1)
Average	0.83 (3.07)	0.85 (2.53)	0.81 (2.72)	0.89 (1.66)

one, which method is outperformed is determined by carrying out a post-hoc test, i.e. Holm's test [7], only if in the first step an difference is found.

Friedmans test is a non-parametric statistical procedure which aims at detecting if a significant difference among the behavior of two or more algorithms exists. In particular, under the null-hypothesis, it states that all algorithms are equivalent, hence, a rejection of this hypothesis implies the existence of differences among the performance of all studied algorithms [6]. In order to reject the null hypothesis, the computed value $\mathcal{X}_r^2$ must be equal to or greater than the tabled critical chisquare value at the specified level of significance [12]. In our experimentation, a level of significance $\alpha = 0.05$ is chosen. Since in our case we are comparing four approaches, our analysis has to consider the critical value $\mathcal{X}_{0.05}^2$ for three degrees of freedom that is equal to 7.815. As can be seen from Table 2, in the Friedman's test, the computed $\mathcal{X}_r^2 = 15.87 > 7.815$, the null hypothesis is rejected and it is possible to assess that there is a significant difference between these proposals. According to this result, a post-hoc statistical analysis is needed to conduct pairwise comparisons in order to detect concrete differences among compared algorithms. Holm's procedure is a multiple comparison procedure that works by setting a control algorithm and comparing it with the remaining ones. Normally, the algorithm which obtains the lowest value of ranking in the Friedmans test is chosen as control algorithm. In our experimentation, as shown in Table 3, our proposal is characterized by the lowest value of ranking.

Table 3. Holm's test on ontology alignment's quality.

i	Approach	z value	Unadjusted $p-$value	$\frac{\alpha}{k-i}$, $\alpha = 0.05$
3	MA	2.53	0.0132	0.05
2	PSO	2.72	0.0025	0.02
1	EA	3.07	$5.996 \times e^{-5}$	0.01

Holm's test works on a family of hypotheses where each one is related to a comparison between the control algorithm and one of the remaining algorithms. In details, the test statistic for comparing the ith and jth algorithms named z value is used for finding the corresponding probability from the table of the normal distribution (the so-called p-value), which is then compared with an appropriate level of significance α. In our experimentation $\alpha = 0.05$ and the results of Holm's test are shown in Table 3. As can be seen from Table 3, our approach statistically outperforms other swarm intelligent algorithm based ontology matching approaches on the alignment's quality at 0.05 significance level.

5 Conclusion

One of the main challenges in ontology matching domain is how to select, combine and tune different ontology matchers to obtain the high quality ontology alignment, and swarm intelligent algorithms are appearing as a suitable methodology to face this challenge. To improve the performance of swarm intelligent algorithm based ontology matching technology, in this paper, we propose a HCFO based ontology matching technology to solve the ontology matching problem. The experimental results show that our approach can significantly improve the ontology alignment's quality of existing swarm intelligent algorithm based ontology matching technologies.

Acknowledgment. This work is supported by the National Natural Science Foundation of China (Nos. 61503082 and 61402108), Natural Science Foundation of Fujian Province (No. 2016J05145), Scientific Research Startup Foundation of Fujian University of Technology (No. GY-Z15007), Fujian Province outstanding Young Scientific Researcher Training Project (No. GY-Z160149) and China Scholarship Council.

References

1. Acampora, G., Loia, V., Vitiello, A.: Enhancing ontology alignment through a memetic aggregation of similarity measures. Inf. Sci. **250**, 1–20 (2013)
2. Bock, J., Hettenhausen, J.: Discrete particle swarm optimisation for ontology alignment. Inf. Sci. **192**, 152–173 (2012)
3. Ding, D., Qi, D., Luo, X., Chen, J., Wang, X., Du, P.: Convergence analysis and performance of an extended central force optimization algorithm. Appl. Math. Comput. **219**(4), 2246–2259 (2012)
4. Formato, R.A.: Central force optimization: a new metaheuristic with applications in applied electromagnetics. Prog. Electromagn. Res. **77**, 425–491 (2007)
5. Friedman, M.: The use of ranks to avoid the assumption of normality implicit in the analysis of variance. J. Am. Stat. Assoc. **32**(200), 675–701 (1937)
6. Garcia, S., Molina, D., Lozano, M., Herrera, F.: A study on the use of non-parametric tests for analyzing the evolutionary algorithms behaviour: a case study on the CEC 2005 special session on real parameter optimization. J. Heuristics **15**(6), 617–644 (2009)
7. Holm, S.: A simple sequentially rejective multiple test procedure. Scand. J. Stat. **6**, 65–70 (1979)
8. Martinez-Gil, J., Alba, E., Montes, J.F.A.: Optimizing ontology alignments by using genetic algorithms. In: Proceedings of the First International Conference on Nature Inspired Reasoning for the Semantic, vol. 419, pp. 1–15. CEUR-WS.org (2008)
9. Melnik, S., Garcia-Molina, H., Rahm, E.: Similarity flooding: a versatile graph matching algorithm and its application to schema matching. In: Proceedings of 18th International Conference on Data Engineering, pp. 117–128. IEEE (2002)
10. Miller, G.A.: Wordnet: a lexical database for English. Commun. ACM **38**(11), 39–41 (1995)
11. Nguyen, T.T.A., Conrad, S.: Ontology matching using multiple similarity measures. In: 2015 7th International Joint Conference on Knowledge Discovery, Knowledge Engineering and Knowledge Management (IC3K), vol. 1, pp. 603–611. IEEE (2015)
12. Sheskin, D.J.: Handbook of Parametric and Nonparametric Statistical Procedures. CRC Press, Boca Raton (2003)
13. Shvaiko, P., Euzenat, J.: Ontology matching: state of the art and future challenges. IEEE Trans. Knowl. Data Eng. **25**(1), 158–176 (2013)

14. Stoilos, G., Stamou, G., Kollias, S.: A string metric for ontology alignment. In: International Semantic Web Conference, pp. 624–637. Springer, Heidelberg (2005)
15. Wang, J., Ding, Z., Jiang, C.: Gaom: genetic algorithm based ontology matching. In: IEEE Asia-Pacific Conference on Services Computing, APSCC 2006, pp. 617–620. IEEE (2006)
16. Xue, X., Wang, Y.: Ontology alignment based on instance using NSGA-II. J. Inf. Sci. **41**(1), 58–70 (2015)
17. Xue, X., Wang, Y., Hao, W.: Optimizing ontology alignments by using NSGA-II. Int. Arab J. Inf. Technol. **12**(2), 175–181 (2015)

A Hybrid BPSO-GA Algorithm for 0-1 Knapsack Problems

Jinshui Wang[1,2(✉)], Jianhua Liu[1,2], Jeng-Shyang Pan[1,2], Xingsi Xue[1,2], and Lili Huang[3]

[1] College of Information Science and Engineering, Fujian University of Technology, No. 3 Xueyuan Road, University Town, Fuzhou 350118, Fujian, China
ymkscom@gmail.com

[2] Fujian Provincial Key Laboratory of Big Data Mining and Applications, Fujian University of Technology, No. 3 Xueyuan Road, University Town, Fuzhou 350118, Fujian, China

[3] College of Ecological Environment and Urban Construction, Fujian University of Technology, No. 3 Xueyuan Road, University Town, Fuzhou 350118, Fujian, China

Abstract. 0-1 knapsack problems (KPs) is a typical NP-hard problem in combinatorial optimization problem. For the sake of efficiency, it becomes increasingly popular for researchers to apply heuristic techniques to solve the 0-1 KPs. Due to its simplicity and convergence speed, an increasing number of techniques based on binary particle swarm optimization (BPSO) has been presented. However, BPSO-based techniques suffered from a major shortcoming which is the premature convergence of a swam. To address the problem, this paper proposed a hybrid BPSO-GA algorithm which combines the strengths of BPSO and genetic algorithm (GA). Experimental results show that our proposal is able to find more optimal solutions than BPSO-based algorithm.

Keywords: Knapsack problems · Binary particle swarm optimization (BPSO) · Genetic algorithm · Hybrid algorithm

1 Introduction

Knapsack Problems (KPs) has been extensively studied and widely used in financial and industrial areas, such as investment decision [1], energy minimization [2], resources allocation [3], CPU and GPU implementations [4], and others [5]. As the most common KPs, the objective of 0-1 Knapsack Problem is to choose a subset of the items such that their total profit is maximized, while the total weight does not exceed a given capacity.

Although a large number of algorithms have been presented to solve the 0-1 KPs, there are always some new and more difficult 0-1 KPs emergent appears in

P. Krömer et al. (eds.), *Proceedings of the Fourth Euro-China Conference on Intelligent Data Analysis and Applications*, Advances in Intelligent Systems and Computing 682, DOI 10.1007/978-3-319-68527-4_37

the real world [6]. Therefore, it is necessary to do more in-depth research on 0-1 KPs. Since 0-1 KPs has been proven to be a typical NP-hard problem in combinatorial optimization problem, there are a lot of solving algorithms of 0-1 KPs which can be divided into three categories [6]: exact algorithms, meta-heuristic algorithms, and heuristic algorithms. Although they are able to produce the optimal solutions in solving small-scale problems, exact algorithms perform poorly when comes to actual large-scale problems. Though both meta-heuristics algorithm and heuristics algorithms can solve 0-1 KPs effectively, meta-heuristics algorithms have to undergo the complex iteration process and set different number of populations for each instance, which will undoubtedly increase the difficulty and complexity of solving 0-1 KPs. Therefore, it becomes increasingly popular for researchers to apply heuristic techniques to solve the 0-1 KPs.

Due to its simplicity and convergence speed, an creasing number of techniques based on particle swarm optimization (PSO) has been presented, and a recent study [7] suggests that PSO has become the most prevalent swarm intelligence algorithm. In previous research [8], we have proposed a binary particle swarm optimization (BPSO) algorithm with a new increasing inertia weight scheme to solve 0-1 KPs. However, the proposed technique still suffered from a major shortcoming which is the premature convergence of a swam. The key reason behind this shortcoming is that particles try to converge to a single point, whereas the point probably just located on a line between the global best and the personal best position, and it is not guaranteed for a local optimum [9]. In order to improve the whole performance, the idea of crossover and mutation from genetic algorithm (GA) is introduced, and then a hybrid BPSO-GA algorithm is proposed which combines the strengths of BPSO and GA.

The remainder of the paper is organized as follows. Section 2 introduces the background, including 0-1 Knapsack Problems, genetic algorithm, and binary particle swarm optimization. Section 3 presents the details of the proposed hybrid BPSO-GA algorithm. Section 4 carries out the experiment study on the 0-1 KPs. and Finally, Sect. 5 provides the conclusions.

2 Background

2.1 0-1 Knapsack Problems

0-1 Knapsack Problems is a typical NP-hard problem, it can be defined as follows: Given a knapsack that holds a fixed capacity C, and a set of n items, where each item having a positive profit p_i and a positive weight w_i. The global of 0-1 KPs is to pick a subset of the items such that their total profit is maximized, while the total weight of the item does not exceed C. The problem may be formulated so as to maximize the total profit $f(x)$ as follows:

$$\begin{cases} Maximize & f(x) = \sum_{i=1}^{n} p_i x_i \\ s.t. & \sum_{i=1}^{n} w_i x_i \leq C \end{cases} \tag{1}$$

where $f(x)$ means the objective function value and $x_i \in (0,1)$. If x_i is 1, item i is packed into knapsack; $x_i = 0$, otherwise. Thus, the goal of 0-1 KP is to find a subset of $X = (x_1, x_2, ..., x_n) \in \{0,1\}^n$ which maximizes $f(x)$.

2.2 Genetic Algorithm

Genetic algorithm is inspired by biological evolution process, and its basic view is based on survival of the fittest in natural selection. It was first introduced by Holland [10] and have been widely applied unexpectedly successful in optimization problems, especially in the binary domain [11]. In GA, the problem solutions are expressed as a group of chromosomes. Each chromosome in a binary GA includes several genes with binary values which determine the attributes for each individual, and a fitness function is responsible for evaluating each individual. Then various genetic operations like selection, crossover and mutation are applied on this set of solution. This process continues iteratively until optimized result or maximum number of generations have been reached.

2.3 Binary Particle Swarm Optimization

Swarm intelligence algorithm is a kind of heuristic search algorithm which enlightened by swarm intelligent behaviors in the nature. Swarm intelligence algorithm has been an active research in artificial intelligence (AI), typically in the form of particle swam optimization (PSO) which has become the most prevalent swarm intelligence algorithm [7]. Among a number of discrete PSO variants, the binary PSO (BPSO) is probably the most well known algorithm.

The critical difference between BPSO and PSO is that the position of a particle in BPSO is updated by switching each bit value between 0 and 1 based on the velocity of that bit. Specifically, for the dth bit of the ith particle, the velocity v_{id} is transformed to a probability s(v_{id}) of taking the value of a 1 which is determined by the sigmoid transfer function as following:

$$s(v_{id}) = \frac{1}{1 + e^{-vid}} \tag{2}$$

Based on s(v_{id}), the bit value of x_{id} is updated as

$$x_{id} = \begin{cases} 1, & if \quad rand() \leq s(v_{id}) \\ 0, & otherwise \end{cases} \tag{3}$$

where rand() randomly samples a value from the uniform distribution within the interval of [0,1]. In other words, x_{id} takes 1 with a probability of s(v_{id}).

During the search process of BPSO, v_{id} is updated according to the following rule:

$$v_{id} = wv_{id} + c_1 r_{1d}(p_{id} - x_{id}) + c_2 r_{2d}(g_d - x_{id}) \tag{4}$$

where $0 \leq w \leq 1$ is the inertia weight, p_{id} stands for the dth bit of the personal best position (*pbest*) of the ith particle. g_d denotes the dth bit of the global best position (*gbest*). $c1 \geq 0$ and $c2 \geq 0$ are the acceleration coefficients. r_{1d} and r_{2d} are random variables which follow the uniform distribution between 0 and 1.

3 Methodology

In this section, a hybrid BPSO-GA algorithm is described in Algorithm 1, where $x_i = (x_{i1}, x_{i2}, ..., x_{in})$($n$ is the dimension) and $p_i = (p_{i1}, p_{i2}, ..., p_{in})$ stand for the current position and personal best position of the ith particle, respectively, $g = (g_1, g_2, ..., g_n)$ is the global best position. Specifically, the fitness function used in this paper is defined as

$$f(x) = \sum_{i=1}^{n} p_i x_i + \beta \cdot \sum_{i=1}^{n} \sum_{j=1}^{m} min(C_j - r_{ij} x_i, 0) \tag{5}$$

where β is the penalty coefficient to control the trade-off between the objective value and the violation to the constraints. In our experimental study, β is simply set to 10^{100}, which is a sufficiently large number to eliminate the infeasible solutions.

In hybrid BPSO-GA algorithm, multiple solutions are generated randomly as an initial population, a set of velocities are generated randomly used to update population, and then objective function values are evaluated for each solution. After the evaluation, the population is evolved by a two-steps operation, i.e., PSO operation and GA operation.

First, each particle is represented as a bit string. Each bit is updated by (2)–(4). To simplify analysis, we adopt the stagnation assumption as we had done in our previous work [6], which is commonly used for analyzing PSO. In practice, problem-dependent velocity clamping techniques are often used to prevent too large velocities. In our algorithm, after been updated by (4), the velocity v_{id} is bounded by a threshold $\hat{v}$ as

$$v_{id} = \begin{cases} \hat{v}, & if \quad v_{id} \geq \hat{v} \\ -\hat{v}, & if \quad v_{id} \leq -\hat{v} \end{cases} \tag{6}$$

Furthermore, our previous research findings [6] suggested that in the binary case, a smaller inertia weight enhances the exploration capability while a larger inertia weight encourages exploitation. Therefore, our proposed BPSO-GA hybrid algorithm adopt an adaptive inertia weighted scheme for BPSO. This scheme allows the search process to start first with exploration and gradually move towards exploitation by linearly increasing the inertia weight is proposed as

$$w = \begin{cases} \bar{w} - \dfrac{\pi \cdot (\bar{w} - \underline{w})}{p \cdot \bar{\pi}}, & if \quad \pi \leq p \cdot \bar{\pi} \\ \underline{w}, & if \quad p \cdot \bar{\pi} \leq \pi \leq \bar{\pi} \end{cases} \tag{7}$$

where π and $\bar{\pi}$ stand for the number of iterations elapsed so far and the maximal number of iterations, $\bar{w}$ and $\underline{w}$ are the predefined upper and lower bounds of the inertia weight, and p is the fraction of iterations for changing w, which is set to 0.9 in our experimental study.

After the position and velocity of each particle were updated, our algorithm take all of the particles as chromosomes and evaluate theirs fitness values again for subsequent GA operations. In GA of the presented hybrid algorithm, each operation is selected to enhance the diversity of solutions. In order to identify, inherit and protect good common genes shared by chromosomes, a crossover operator is applied to help our algorithm to converge to optima. We check if the crossover could be applied according to the crossover probability *pc*, and if it is, two new children are then generated from their parents. In order to reduce the bias associated with the length of the binary representation, an uniform crossover operator is applied. Firstly, a binary mask is constructed randomly, and then the children inherits the allele from their parents according to the value of the mask. For example, the first child inherits the allele from the first parent when the value of locus in mark equal to 1, and from the second parent when the value of locus in mark equal to 0.

After that, a mutator operator is carried out to prevents irreversible loss of certain patterns by introducing small random changes into chromosomes. Mutation operator assures diversity in the population and prevents premature convergence. In our experiment study, we check each bit in the individual if the mutation could be applied according to the mutation probability and if it is, the value of that bit is then flipped. Finally, the *pbest* and *gbest* were updated according to the offspring's fitness value.

4 Experimental Study

In this section, the proposed hybrid BPSO-GA algorithm is compared with standard BPSO algorithm on the 0-1 knapsack problems. For the sake of convenience, the proposed BPSO-GA algorithm and standard BPSO algorithm are denoted as "BPSO-GA" and "BPSO", respectively. In our experiments, the data set is obtained from an academic webpage of Michigan Technological University[1], which contains 25 randomly generated instances.

The parameter settings in the experiments in detail is described in Table 1. It should be noted that, with respect to the values of the above parameters, such as number of particles and max iteration, they should be higher if the scale of the data set is large, and the data scale in our experimental study is medium.

In our experimental study, we compared the results which produced by the BPSO-GA and BPSO strategies. For each test instance and each compared algorithm, 20 independent runs were conducted, and the average and standard deviation of the 20 corresponding results were calculated. The column BK gives the best known profit for each instance, which is the maximal profit obtained by all the tested algorithms. If an algorithm is able to **consistently** achieved the optimal value, the corresponding entry in Table 2 is marked with *. In addition, if

[1] http://www.math.mtu.edu/kreher/cages/Data.html.

Algorithm 1. Framework of hybrid BPSO-GA algorithm

Input: set of n items, where item i owns profit p_i and weight w_i and a knapsack that holds a fixed capacity C
Output: subset of the set of n items such that their total profit is maximized, while the total weight of the item is not larger than C
1: Randomly generate an initial population;
2: Randomly generate the initial velocities within the velocity bound;
3: **repeat**
4: **for** $i = 1$ to Population Size **do**
5: **if** $f(x_i) \leq f(p_i)$ **then** $p_i = x_i$
6: **end if**
7: **if** $f(p_i) \leq f(g)$ **then** $g = p_i$
8: **end if**
9: **end for**
10: **for** $i = 1$ to Population Size **do**
11: **for** $d = 1$ to Dimension Size **do**
12: Calculate w using Eq. (7);
13: Update velocity with Eq. (4);
14: Update position using Eq. (2) and Eq. (3);
15: **end for**
16: **end for**
17: Take every particle as chromosome, and evaluate the fitness value of each chromosome;
18: Select two parents and carry out (GA) crossover operation;
19: Apply (GA) mutation operator;
20: update *pbest* and *gbest*;
21: **until** termination criterion is met;

Table 1. Parameter setting of the experiments on the 0-1 knapsack problems

Parameter	Description	Value
$\bar{w}$	Upper bound of w	1
$\underline{w}$	Lower bound of w	0.4
$\bar{p}$	Fraction of iterations for changing w in Eq.	0.9
β	Penalty coefficient in Eq.	10^{100}
N	Number of particles	20
$\bar{pi}$	Maximal number of iterations	1000
pc	Crossover probability	0.6
pm	Mutation probability	0.05
MaxIter	Max iteration	500
Repetition	Repetition run times	20

an algorithm achieved better results than the other one, then the corresponding entry in Table 2 is marked in bold.

Table 2 shows the average and standard deviation of the results of the compared algorithms over 20 independent runs on the data set. In terms of the average profit, the BPSO-GA performed better than BPSO on ten out of the total 25 instances, while the BPSO performed better on six instances, which means BPSO-GA achieved four better results than BPSO. In terms of success rate, BPSO-GA also obtained better success rate, e.g., the instance ky_12a for which BPSO failed to find the global optimum. It can be seen clearly from the table that the BPSO-GA outperforms than BPSO.

Table 2. Results of BPSO and BPSO-GA over 20 independent runs

FileName	BK	BPSO	BPSO-GA
ks_8a	3.9244E+06	3.9244E+06(0.0000E+06) *	3.9244E+06(0.0000E+06) *
ks_8b	3.8137E+06	3.8137E+06(0.0000E+06) *	3.8137E+06(0.0000E+06) *
ks_8c	3.3475E+06	3.3475E+06(0.0000E+06) *	3.3475E+06(0.0000E+06) *
ks_8d	4.1877E+06	4.1877E+06(0.0000E+06) *	4.1877E+06(0.0000E+06) *
ks_8e	4.9556E+06	4.9556E+06(0.0000E+06) *	4.9556E+06(0.0000E+06) *
ks_12a	5.6889E+06	5.6889E+06(0.0000E+06) *	5.6889E+06(0.0000E+06) *
ks_12b	6.4986E+06	6.4960E+06(7.6740E+03)	**6.4986E+06** (0.0000E+06) *
ks_12c	5.1706E+06	5.1706E+06(0.0000E+06) *	5.1706E+06(0.0000E+06) *
ks_12d	6.9924E+06	6.9924E+06(0.0000E+06) *	6.9924E+06(0.0000E+06) *
ks_12e	5.3375E+06	5.3375E+06(0.0000E+06) *	5.3375E+06(0.0000E+06) *
ks_16a	7.8510E+06	7.8397E+06(1.1771E+04)	**7.8472E+06**(9.2640E+03)
ks_16b	9.3530E+06	9.3491E+06(1.3402E+04)	**9.3525E+06**(1.5780E+03)
ks_16c	9.1511EE+06	9.1294E+06(2.6283E+04)	**9.1389E+06**(1.8243E+04)
ks_16d	9.3489E+06	**9.3405E+06**(1.4255E+04)	9.3402E+06(1.4999E+04)
ks_16e	7.7691E+06	**7.7667E+06**(5.7915E+03)	7.7632E+06(8.1868E+03)
ks_20a	1.0727E+07	**1.0708E+07**(1.9945E+04)	1.0705E+07(2.1847E+04)
ks_20b	9.8183E+06	**9.7959E+06**(2.7020E+04)	9.7844E+06(2.2322E+04)
ks_20c	1.0714E+07	**1.0701E+07**(2.5557E+04)	1.0696E+07(3.2095E+04)
ks_20d	8.9292E+06	8.9067E+06(3.1765E+04)	**8.9134E+06**(1.8071E+04)
ks_20e	9.3580E+06	9.3437E+06(1.5220E+04)	**9.3456E+06**(1.5231E+04)
ks_24a	1.3549E+07	1.3489E+07(3.6370E+04)	**1.3494E+07**(3.6093E+04)
ks_24b	1.2234E+07	1.2175E+07(2.6116E+04)	**1.2176E+07**(3.5804E+04)
ks_24c	1.2449E+07	1.2397E+07(3.1116E+04)	**1.2402E+07**(2.3879E+04)
ks_24d	1.1815E+07	**1.1773E+07**(2.1078E+04)	1.1770E+07(2.6415E+04)
ks_24e	1.3940E+07	1.3882E+07(3.6330E+04)	**1.3900E+07**(2.5374E+04)

5 Conclusions

KPs plays an very important role in industry, financial management, and various techniques are presented to analyze and solve this problem. In this paper, 0-1 KPs is regarded as a combinatorial optimization problem and a hybrid BPSO-GA algorithm are proposed to solve it. The hybrid algorithm were developed by combining BPSO and GA in order to combines the strengths of them. To evaluate whether our algorithm can improve performance and achieving better results, we conducted an experimental study on a data set which contains 25 randomly generated instances. The results show that our proposal is able to produces better result as compared to BPSO.

Acknowledgment. This work is supported by National Natural Science Foundation of China (61402108, 61503082), Foundation for Scientific Research of Fujian Education Committee (GY-Z15121, JA13211), Foundation of Fujian University of Technology (GY-Z15101, GY-Z14068, GY-Z13113), Key Project of Fujian Province Department of Science & Technology (2013H0002).

References

1. Rooderkerk, R.P., Heerde, H.J.V.: Robust optimization of the 0-1 knapsack problem: balancing risk and return in assortment optimization. Eur. J. Oper. Res. **250**(3), 842–854 (2016)
2. Muller, S., Al-Shatri, H., Wichtlhuber, M., Hausheer, D.: Computation offloading in wireless multi-hop networks: energy minimization via multi-dimensional knapsack problem. In: IEEE International Symposium on Personal, Indoor, and Mobile Radio Communications, vol. 51, no. 43, pp. 1717–1722 (2015)
3. Jacko, P.: Resource capacity allocation to stochastic dynamic competitors: knapsack problem for perishable items and index-knapsack heuristic. Ann. Oper. Res. **241**(1), 83–107 (2016)
4. Li, K., Liu, J., Wan, L., Yin, S., Li, K.: A cost-optimal parallel algorithm for the 0–1 knapsack problem and its performance on multicore CPU and GPU implementations. Parallel Comput. **43**(C), 27–42 (2015)
5. Kuchta, D., Ryca, R., Skorupka, D., Duchaczek, A.: The use of the generalised knapsack problem in computer aided strategic management. In: International Conference on Enterprise Information Systems, Wroclaw, Poland, pp. 39–46 (2016)
6. Lv, J., Wang, X., Huang, M., Cheng, H., Li, F.: Solving 0–1 knapsack problem by greedy degree and expectation efficiency. Appl. Soft Comput. **41**(C), 94–103 (2016)
7. Zhang, Y., Wang, S., Ji, G.: A comprehensive survey on particle swarm optimization algorithm and its applications. Math. Probl. Eng. **2015**(1), 1–38 (2015)
8. Liu, J., Mei, Y., Li, X.: An analysis of the inertia weight parameter for binary particle swarm optimization. IEEE Trans. Evol. Comput. **39**, 43–47 (2015)
9. Frans, V.D.B., Engelbrecht, A.P.: A cooperative approach to particle swarm optimization. IEEE Trans. Evol. Comput. **8**(3), 225–239 (2004)
10. Holland, J.H.: Adaptation in Natural and Artificial System. MIT Press, Cambridge (1975). vol. 6, no. 2, pp. 126–137
11. Settles, M., Soule, T.: Breeding swarms: a ga/pso hybrid. In: Conference on Genetic and Evolutionary Computation, Washington DC, USA, pp. 161–168 (2005)

Nearest Neighbor Search Techniques Applied in the Nearest Feature Line Classifier

Fang Guo[1,2(✉)] and Jeng-Shyang Pan[1,2]

[1] Department of Information Science and Engineering, Fujian University of Technology, Fuzhou 350118, Fujian, China
Davidace@fjut.edu.cn

[2] Fujian Provincial Key Laboratory of Big Data Mining and Applications, Fujian University of Technology, Fuzhou 350118, Fujian, China

Abstract. Pointing to the computational complexity to find the minimum distance in the nearest feature line (NFL) classification algorithm, the nearest neighbor search methods with Full Search (FS), Partial Distortion Search (PDS), Absolute Error Inequality (AEI) and Equal-average Nearest Neighbor Search (ENNS) is used to evaluate the calculated performance on NFL. The experimental results demonstrate that the computational complexity on NFL using these search techniques is different and some of the nearest neighbor search methods could improve the calculated performance on finding out the minimum distance applied in the NFL classification.

Keywords: NFL · Minimum distance · Nearest neighbor search · FS · PDS · AEI · ENNS

1 Introduction

Object classification was widely applied in the fields of face recognition and speaker identification [1–5]. Among the algorithms of face recognition, nearest feature line (NFL), as it is simple for training and is high reliability to classify object, is being concerned by many researchers [6, 7]. The NFL was first proposed by Li *et al*. For the NFL, the samples in the same class are seen as a group which are correlation with each other. Any two feature points expressing the samples could be connected with a line called feature line. All points on the line indicate the gradual change from one feature point to the other feature point. Therefore, the NFLS could include all points within the feature space with limited samples. The NFL classifies the images into different classes by finding the smallest distance between the query sample and every feature line within each of classes. Then the query sample would be divided into the class which the feature line of the smallest distance belongs to.

However, in the NFL, extensive computer operation is required to find the smallest square Euclidean distance. For query samples R, classes M, sample points N within every class and the dimension K of every vector point, it needs to perform R * K * M * N(N−1)/2 multiplications, (2 * K−1) * R * M * N(N−1)/2 additions, and R * M * (N(N−1)/2−1) comparisons to classify the R query samples. Thus, it is a time-consuming process when K, N, M, R is large.

P. Krömer et al. (eds.), *Proceedings of the Fourth Euro-China Conference on Intelligent Data Analysis and Applications*, Advances in Intelligent Systems and Computing 682, DOI 10.1007/978-3-319-68527-4_38

The problem is similarly faced with in the vector quantization (VQ). It is described as Nearest Neighbor Search (NNS). In VQ, it need find the minimum distortion among the distortions of the input vector and all codewords in codebook. Fortunately, a lot of algorithms for NNS have been proposed by many researchers such as FS, PDS, AEI, ENNS, *et al.* Maybe, the algorithms could be applied in the NFL and improve the calculate performance finding the minimum distances. It has important value in the real-time system. Next, the NFL theory and the NNS algorithms are elaborated.

2 Nearest Feature Line Classifier

For the NFL, the features is first extracted from each of the prototype images. Then the vector (feature point) representing the image is created. All of the feature points belongs to the same class constitute the feature space. The line passing through every two feature points belonging to the same feature space includes the continuous change of the prototype images [8]. For easy of calculation, the two points, such as y_i^c and y_j^c, are connected with one straight line called feature line L. This is shown in Fig. 1.

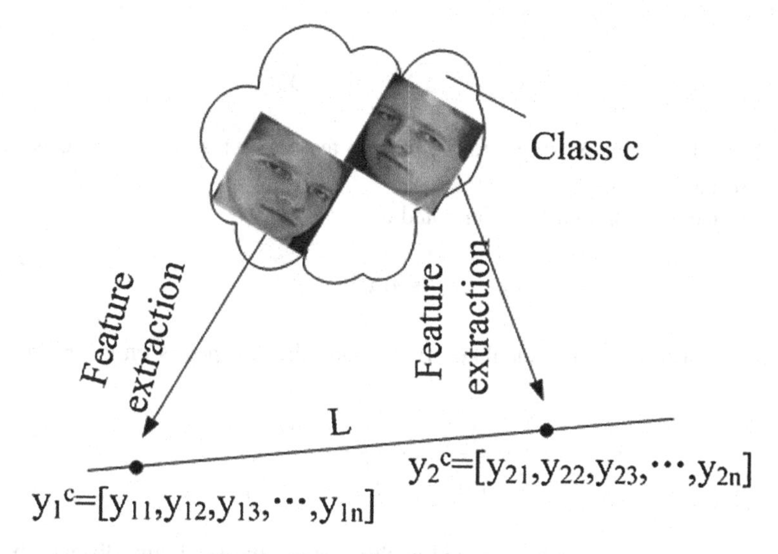

Fig. 1. Feature point and feature line

There would be N(N − 1)/2 feature lines for N vector points in one class. Compared with NN classifier, the minimum vertical distance from the query sample y to projection point of the query sample is used as the mark to classify the query sample, such as the Fig. 2.

The distance between the query sample y and its projection point y_p is given as

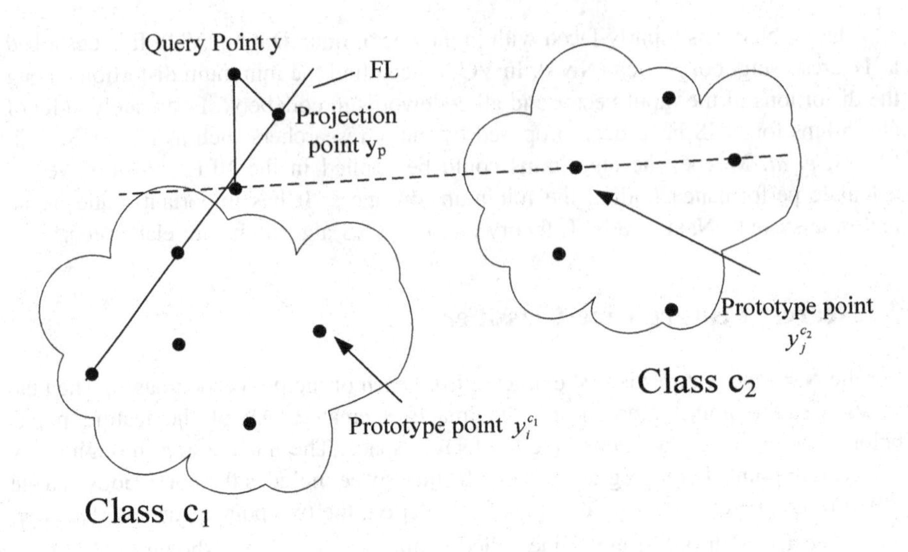

Fig. 2. Classification with NFL

$$d\left(y, \overline{y_i^{c_k}, y_j^{c_k}}\right) = \|y - y_p\| \tag{1}$$

where $y_i^{c_k}, y_j^{c_k}$ is the line passing through the two feature points in the same class, and y_p is the projection point.

The projection point can be computed as

$$p = y_i^{c_k} + u\left(y_j^{c_k} - y_i^{c_k}\right)$$

where $u \in R$. The u can be calculated from y and the two points on the FL as

$$u = \frac{\left(y - y_i^{c_k}\right)^T \left(y_j^{c_k} - y_i^{c_k}\right)}{\left(y_j^{c_k} - y_i^{c_k}\right)^T \left(y_j^{c_k} - y_i^{c_k}\right)} \tag{2}$$

The query point would be grouped as the class of the minimum distance between the query feature point and the FL's by

$$d\left(y, \overline{y_{i*}^{c*}, y_{j*}^{c*}}\right) = \min_{1 \le c \le m} \min_{1 \le i < j \le n_c} d\left(y, \overline{y_i^c, y_j^c}\right) \tag{3}$$

For query samples 100, classes 100, sample points 100 within every class and the dimension 100 of every vector point, it needs to perform 4.95×10^9 multiplications, 9.85×10^9 additions, and 4.95×10^5 comparisons to classify the 100 query samples. So, it would consume mass time to find out the minimum distance.

3 Nearest Neighbor Search

The Nearest Neighbor Search (NNS) algorithms are applied in the vector quantization (VQ) to search for the closest codeword through the codebook.

Let codebook $C = \{y_1, y_2, \cdots, y_n\}$, where y_i is the codeword and n is the codebook size. For each input vector $x = (x_1, x_2, \ldots, x_k)$, The nearest codeword $y_j = (y_{j1}, y_{j2}, \ldots, y_{jk})$ is the smallest among all codewords in the codebook. The squared Euclidean distance is measured between x and y_i as follow:

$$d(x, y_i) = \sum_{l=1}^{k} (x_l - y_{il})^2 \tag{4}$$

where k is the dimension number.

For the sake of large codebook size and high dimension in the VQ system, a full search algorithm is a time-consuming process. To reduce the computational burden, many fast algorithms have been proposed [9–15].

3.1 Partial Distortion Search

The partial distortion search (PDS) algorithm [16] can reduce the number of multiplications and additions by comparing the partial accumulated distance with the current minimum distance as:

if

$$\sum_{l=1}^{s} (x_l - y_{il})^2 \geq d_{\min} (1 \leq s \leq k) \tag{5}$$

then

$$d(x, y_i) \geq d(x, y_p) \tag{6}$$

where $d(x, y_i)$ is the distance between the ith codeword and input vector x with $1 \leq i \leq k$ and $d_{\min} = d(x, y_p)$ is the current minimum distance.

3.2 Absolute Error Inequality Search

The nearest neighbor codeword search algorithm based on absolute error inequality (AEI) [17] can also reduce the number of multiplications and additions, but the increase in the number of comparisons is moderate. For the AEI, the absolute error associated with each component of the code vector y_i, $|x_l - y_{il}|$ is compared to the square root of the minimum distortion found, where x_l and y_{il} denote respectively the lth component of the input vector x and the code vector y_i. This is described as follows:

Let $d_{\min} = d(x, y_p), 1 \leq s \leq k$

if

$$\sum_{l=1}^{s} e_{il} = \sum_{l=1}^{s} |x_l - y_{il}| \geq \sqrt{k.d_{\min}} \tag{7}$$

then

$$d(x, y_i) \geq d(x, y_p) \tag{8}$$

3.3 Equal-Average Nearest Neighbor Search

The equal-average nearest neighbor search (ENNS) use the concept of the central line in an Euclidean space R^k. The mean values on the hyperplane orthogonal to l are the same [18]. Assume the current smallest distortion is $d_{min=}$ $d(x, y_p)$, and m_i is the mean value of codeword. Then the search range can be bounded by two equal-average hyperplanes with mean values

$$m_{\min} = m_i + \sqrt{d_{\min}/k} \text{ and } m_{\min} = m_i - \sqrt{d_{\min}/k}.$$

The codewords can be rejected as:

If

$$|\mathrm{m_i} - \mathrm{m_x}| \geq \sqrt{\mathrm{d_{min}}/\mathrm{k}} \tag{9}$$

Then

$$d(x, y_i) \geq d(x, y_p) \tag{10}$$

Because the codeword search is to find the nearest neighbor codeword of input vector x from codebook, this question is similar to the problem faced with by the NFL how to find the minimum vertical distance.

4 Experimental Results

To evaluate the computing performance on the NFL with the NNS algorithms, the experiments are performed on a Intel Core i5 PC with the 2.50 GHZ of CPU using the face database of ORL, Yale, AR and GTdb. Matlab 2008 is also used as the simulating software.

The ORL face database is comprised of 400 images of 40 people, 112 * 92 of the pixels for every image. There are 10 images per subject. Five images are used as training samples and the others are used as testing set. The Yale face database is comprised of 165 images of 15 people, 100 * 100 of the pixels for every image. There are 11 images per subject. Six images are used as training samples and the others are used as testing set. The AR face database is comprised of 1680 images of 120 people, 50 * 40 of the pixels for every image. There are 14 images per subject. Seven images are used as training samples and the others are used as testing set. The GTdb face database is comprised of 750 images of 50 people, 80 * 64 of the pixels for every image. There are 15 images per subject. Eight images are used as training samples and the others are used as testing set.

The NFL is used to classify the testing set into the sampling class with formulation (3), and the algorithms of FS, PDS, AEI and ENNS are adopted respectively with formulation (5), (6), (7), (8), (9), (10) to find the shortest distance between every test sample and each of the feature line. The results are shown in Table (1).

Table 1. Performances of the nearest neighbor search algorithm's runtime (sec) using FS, PDS, AEI, ENNS to classify the faces respectively in the face databases of Yale, AR, ORL, GTdb using NFL algorithm with the same results

Method	FS	PDS	AEI	ENNS
Yale	0.023	0.022	0.018	0.026
AR	11.76	9.24	6.05	12.6
ORL	0.22	0.19	0.16	0.21
GTdb	2.84	2.36	2.20	2.88

The results show that the nearest neighbor search algorithms can be applied in the images classification of NFL, but not all the nearest neighbor search algorithms can reduce the runtime of NFL. The runtime using PDS and AEI can save more runtime than ENNS and FS algorithm, but the ENNS algorithm consume even more runtime than FS algorithm. The reason is that the consuming time using ENNS almost is the same with the algorithm using FS when $\sqrt{d/k}$ is much larger than $|m_i - m_x|$. However, the PDS and AEI can save a large number of calculations by comparing the partial sum of vector distance with current d_{min}.

5 Conclusions

The nearest neighbor search algorithms applied in the NFL classifier were analyzed. By the simulation experiment with the face databases of Yale, AR, ORL, GTdb, the results show that the PDS and AEI algorithm can consume less runtime than the FS and ENNS to find the minimum distance in the NFL classifier. However, there are still others search algorithms in the fields of VQ [19–21], and maybe the PDS and AEI is not the best algorithm applied in the NFL, the work analyzing how the others algorithms influence the runtime of NFL will be researched in the further.

Acknowledgment. The author wishes to thank Science Foundation (2017J01732) of the Fujian Province, China

References

1. Saeed, K., Nammous, M.K.: A speech-and-speaker identification system: feature extraction, description, and classification of speech-signal image. IEEE Trans. Ind. Electron. **54**(2), 887–897 (2007)

2. Wang, Y., Chen, L.: Image-to-image face recognition using dual linear regression based classification and electoral college voting. In: 2016 IEEE 15th International Conference on Cognitive Informatics & Cognitive Computing (ICCI*CC), pp. 22–23, February 2017
3. Ma, J., Zheng, L., Yaguchi, Y.C., Dong, M.X., Oka, R.C.: Image classification based on segmentation-free object recognition. In: 2010 17th IEEE International Conference on Image Processing (ICIP), pp. 26–29, December. 2010
4. Li, S.Z.: Content-based audio classification and retrieval using the nearest feature line method. IEEE Trans. Speech Audio Process. **8**(5), 619–625 (2000)
5. Xu, Y., Zhang, D., Yang, J., Yang, J.Y.: A two-phase test sample sparse representation method for use with face recognition. IEEE Trans. Circ. Syst. Video Technol. **21**(9), 1255–1262 (2011)
6. Chien, J.T., Wu, C.C.: Discriminant waveletfaces and nearest feature classifiers for face recognition. IEEE Trans. Pattern Anal. Mach. Intell. **24**(12), 1644–1649 (2002)
7. Chen, K., Wu, T.Y., Zhang, H.J.: On the use of nearest feature line for speaker identification. Pattern Recogn. Lett. **23**(14), 1735–1746 (2002)
8. Li, S.Z., Lu, J.W.: Face recognition using the nearest feature line method. IEEE Trans. Neural Netw. **10**(2), 439–443 (1999)
9. Friedman, J.H., Baskett, F., Shustek, L.J.: An algorithm for finding nearest neighbors. IEEE Trans. Comput. **24**(10), 1000–1006 (1975)
10. Friedman, J.H., Bentley, J.L., Finkel, R.A.: An algorithm for finding best matches in logarithmic expected time. ACM **3**(3), 209–226 (1977)
11. Cheng, D.Y., Gersho, A.: A fast codebook search algorithm for nearest-neighbor pattern matching. In: IEEE International Conference on Acoustics, Speech, and Signal Processing (ICASSP 1886), vol. 11, pp. 265–268, April 1986
12. Lo, K.T., Cham, W.K.: Subcodebook searching algorithm for efficient VQ encoding of images. IEE Proc. I – Commun. Speech Vis. **140**(5), 327–330 (1993)
13. Ra, S.W., Kim, J.K.: A fast mean-distance- ordered partial codebook search algorithm for image vector quantization. IEEE Trans. Circ. Syst. II: Analog Digit. Signal Process. **40**(9), 576–579 (1993)
14. Lee, C.H., Chen, L.H.: Fast closest codeword search algorithms for vector quantisation. IEE Proc. – Vis. Image Signal Process. **141**(3), 143–148 (1994)
15. Chen, S.X., Li, F.W., Zhu, W.L.: Fast searching algorithm for vector quantisation based on features of vector and subvector. Let Image Process. **2**(6), 275–285 (2008)
16. Bei, C.D., Gray, R.: An improvement of the minimum distortion encoding algorithm for vector quantization. IEEE Trans. Commun. **33**(10), 1132–1133 (1985)
17. Soleymani, M.R., Morgera, S.D.: An efficient nearest neighbor search method. IEEE Trans. Commun. **35**(6), 677–679 (1987)
18. Guan, L., Kamel, M.: Equal-average hyperplane partitioning method for vector quantization of image data. Pattern Recogn. Lett. **13**(10), 693–699 (1992)
19. Lu, Z.M.: Equal-average equal-variance equal-norm nearest neighbor search algorithm for vector quantification. Ieice Trans. Inf. Syst. **86**(3), 660–663 (2003)
20. Chu, S.C., Lu, Z.M., Pan, J.S., Huang, K.C.: Hadamard transform based fast codeword search algorithm for high-dimensi- onal VQ encoding. Inf. Sci. **177**(3), 734–746 (2007)
21. Xie, Y.F., Liu, J.H., Zhang, C.F., Kong, L.S., Yi, J.L.: Codewords Distribution-Based Optimal Combination of Equal-Average Equal-Variance Equal-Norm Nearest Neighbor Fast Search Algorithm for Vector Quantization Encoding. IEEE Trans. Image Process. **25**(12), 5806–5813 (2016)

The Redundancy Problem in Composition of Parallel Finite Automata and Its Optimization Method

Zheng-Yi Tang[1,2(✉)], Jin-Shui Wang[1,2], Yi Chen[1,2], and Xing-Si Xue[1,2]

[1] School of Information Science and Engineering, Fujian University of Technology, Fuzhou 350118, Fujian, China
tangzy84@126.com

[2] Fujian Provincial Key Laboratory of Big Data Mining and Applications, Fuzhou 350118, Fujian, China

Abstract. Parallel finite automata is a formal model which can describe the complex control structure, but its composition model may exist redundant elements. This problem will increase the complexity of model. This paper researched the redundancy problem in the *Split-Join* composition of parallel finite automata and gave the condition which lead to produce redundant location and transfer. Furthermore, the algorithms which are used to eliminate the redundant location and transfer also was given. Finally, it proved the equivalence of parallel finite automata and finite automata. This conclusion ensures that he verification technology of system which base on the finite automata can be used to parallel finite automata by extension.

Keywords: Finite automata · Parallel · Split-join composition · Redundant transfer/location · Equivalence

1 Introduction

Finite automata (FA) is an widely-used computational model. It is very useful in many domains such as text processing, program compilation and hardware design. In the system modeling, FA is suitable for describing discrete system. The state of FA represents the status of system and the transfer of FA represents the action of system, thereby the trajectory of system can be described by them [1]. Based on the model of finite automat, many formal method such as model checking and deductive inference can be used to verify the system property [2–4]. At present, this method has been used to ensure the reliability of software system widely [5–7].

For describing the features of deferent systems, there appears many variants of finite automata. For example, the timed automata which is suitable for describing real-time system [8], the probabilistic automata which is suitable for describing probabilistic system [9]. These expansions strengthen the ability of finite automata to describe the system feature effectively. However, these expansions are all for the external features of system and not the inner execution flow. Because of this, these

P. Krömer et al. (eds.), *Proceedings of the Fourth Euro-China Conference on Intelligent Data Analysis and Applications*, Advances in Intelligent Systems and Computing 682, DOI 10.1007/978-3-319-68527-4_39

expansions can only describe simple workflow, such as sequence, selection, cycle, and the ability of modeling system is limited.

For the above problem, we proposed the parallel finite automata (PFA) which has more types of trawnsfer [10]. The PFA can describe the complex workflow and support all nine control structures by basic unit and composition rules. This method can construct the formal model of any complex control structure automatically.

But in some cases, the workflow model which is construct by composition rules may have redundant transfers and locations, and the complexity of model thereby is increased. For this problem, this paper gives the production condition of redundant transfers and locations, and the algorithms which are used to optimize workflow model also are given. Finally, we prove the equivalence of parallel finite automata and finite automata. By the proof, the verification technology of system which base on the finite automata can be used to parallel finite automata.

2 Parallel Finite Automata

2.1 The Syntax and Semantics of Parallel Finite Automata

The parallel finite automata is the expansion of finite automata, it has two new transfers which are used to describe parallel flow. Its syntax and semantics are defined bellow:

Definition 1 (The Syntax of Parallel Finite Automata). A parallel finite automata $PFA = (L, IL, FL, \Sigma, E)$ where L is the set of locations, $IL \in L$ is the set of initial locations, $FL \subseteq L$ is the set of acceptable locations, Σ is the set of transfer labels, $E = \{e_{sin}\} \cup \{e_{con}\} \cup \{e_{syn}\}$ is the set of transfers. The edges are divided into three types:

- **Step edge** $e_{sin} : L \times \Sigma \to L$
- **Parallel edge** $e_{con} : L \times \Sigma \to 2^L$
- **Synchronous edge** $e_{syn} : 2^L \times \Sigma \to L$

Parallel edge is an edge from one location to multiple locations, and the synchronous edge is an edge from multiple locations to one location. So the transfer in PFA is the move between location set which is named state. For convenience, the set of transfer labels can include empty label ε except there is special instruction.

In the definition of semantics, a new operator $S[y/x]$ is need. It means the new element y replace the x which is in set S.

The semantics of PFA is following:

Definition 2 (The Semantics of Parallel Finite Automata). $PFA = (L, IL, FL, \Sigma, E)$ is a parallel finite automata. Its semantics is a labeled transfer system (S, s_0, FS, Δ) where $S \subseteq 2^L$ is the set of states, $s_0 = IL$ is the initial sate, $FS \subseteq 2^{FL}$ is the set of acceptable states, the transfer function $\Delta : S \times \Sigma \to S$ is defined following:

- **Step transfer** $\Delta_{sin}(s_k, a) = s_k[l_j/l_i]$, if there is $e_{sin}(l_i, a) = l_j \in E$ and $l_i \in s_k$.
- **Parallel transfer** $\Delta_{con}(s_k, a) = (s_k - \{l_i\}) \cup \{l_1, l_2, \ldots, l_n\}$, if there is $e_{con}(l_i, a) = \{l_1, l_2, \ldots, l_n\} \in E$ and $l_i \in s_k$.

- **Synchronous transfer** $\Delta_{con}(s_k, a) = (s_k - \{l_1, l_2, \ldots, l_n\}) \cup \{l_i\}$, if there is $e_{syn}(\{l_1, l_2, \ldots, l_n\}, a) = l_i \in E$ and $\{l_1, l_2, \ldots, l_n\} \subseteq s_k$.

Figure 1 is a Parallel finite automata, the initial location is identified by an arrow without origin, the acceptable location is identified by double circle.

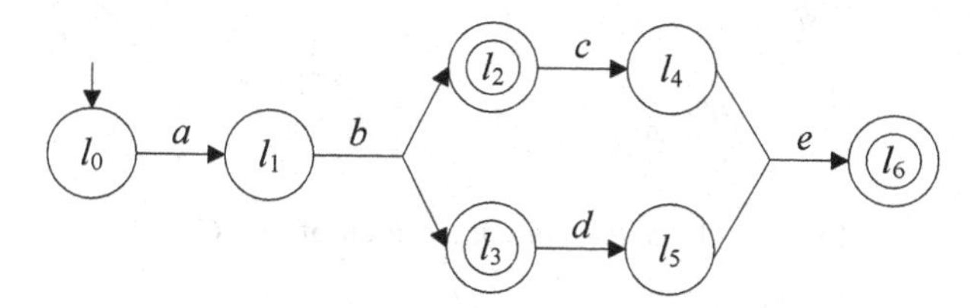

Fig. 1. A parallel finite automata.

For convenience, the bifurcate arrow is used to identify the parallel transfer (the edge with label *b*), the arrow with bifurcate tail is used to identify the synchronous transfer (the edge with label *e*).

2.2 The *Split-Join* Composition Rule of PFA

In the composition for CFA, the basic unit is a *CFA* with an only initial location and an only terminal location, and there are only step transfer in basic unit. The basic unit is called atomic process automata (APA). There are two *APA* in Fig. 2, the initial location is identified by an arrow without origin, and the terminal location is identified by double circle.

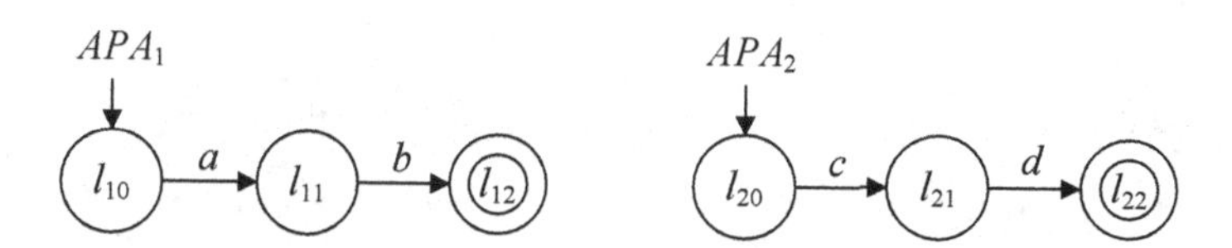

Fig. 2. Two atomic process automata.

For describing the operation in control structure better, the transfer label is extended to a quadruple $\sigma = (In, Out, Pre, Eft)$ where *In* and *Out* are the set of inputs and outputs, the *Pre* and *Eft* are preconditions and effects.

The composition of *APA* is named composition process automata (CPA). The composition rules ensure that a *CPA* has the only initial location and the only terminal location. Thus the CPA is same as APA from outside, a *APA* is also a *CPA*. The *Split-Join* composition rule is following, other rules are referred in literature [10].

Split-Join workflow activate several *CPA*s, they will run parallel and join together when all *CPA*s terminate. The *Split-Join* composition for several *CPA*s can be realized by adding a parallel transfer and a synchronous transfer. The set of target locations of parallel transfer includes initial locations of all *CPA*s, and the set of origin locations includes terminal locations of all *CPA*s. Figure 3 is the *Split-Join* composition of two *CPA*s in Fig. 2.

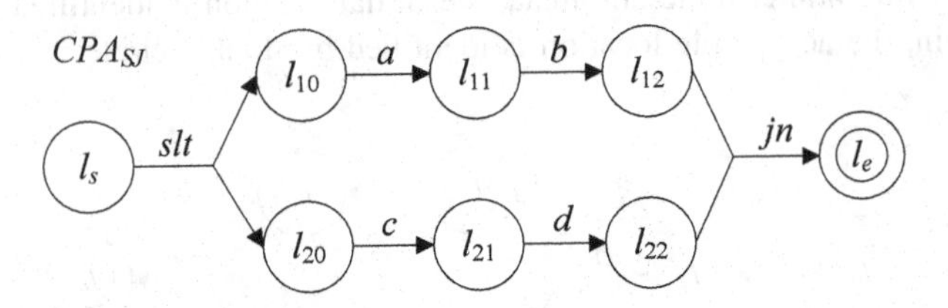

Fig. 3. The *Split-Join* composition of two *CPA*s.

In Fig. 3, $slt = (\varnothing, \varnothing, \varphi_s, \varnothing)$ where φ_s is the condition of split, $jn = (\varnothing, \varnothing, \varphi_j, \varnothing)$ where φ_j is the condition of join. These two conditions can be default.

3 Redundant Transfer and Optimization Algorithm

3.1 Redundant Parallel/Synchronous Transfer

There are two *CPA*s with *Split-Join* structure, and their parallel conditions are same, and their synchronous conditions also are same. If they are composited by *Split-Join* rule, and the split condition and then join condition are same as the corresponding conditions of *CPA*s, then the composition will produce redundant transfers and locations. The process is shown in Fig. 4. The location l_{10}, l_{20}, l_{17}, l_{27} and the corresponding transfers are all redundant. The location l_{11}, l_{12}, l_{21}, l_{22} can connect to l_s directly, and the location l_{15}, l_{16}, l_{25}, l_{26} also can connect to l_e directly.

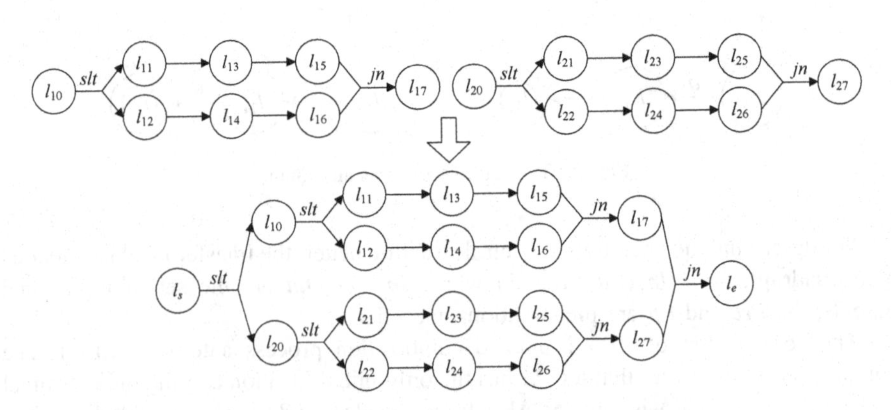

Fig. 4. The production of redundant transfers and locations.

For a parallel transfer $\Delta_{con}(l, slt) = tL = \{l_1, \ldots, l_n\}$, if the following conditions are all held:

(1) The origin location l is in the set of target locations of another parallel transfer, and the split condition of this transfer is same as slt.
(2) There is not step transfer whose origin location or target location is l.

Then the location l is redundant, and the transfer $\Delta_{con}(l, slt) = tL$ is redundant too. The Algorithm of eliminating redundant parallel transfer is following:

Algorithm 1 ERPT (**E**liminate **R**edundant **P**arallel **T**ransfer)

Input: A Composite Process Automata $CPA = (L, IL, FL, sLoc, eLoc, \Sigma, E)$.
Output: The Modified CPA.

```
Set modi as a bool variable;
do {
  modi := false;
  for each e_con(l, slt) = tL ∈ E {
    for each l_i ∈ tL {
      if(∃e_con(l_i, slt) = tL' ∈ E) then {
        if(e_sin(l_i, a) = l' ∉ E ∧ e_sin(l', a) = l_i ∉ E) then {
          L := L − {l_i};
          E := E − {e_con(l, slt) = tL, e_con(l_i, slt) = tL'} ∪
                   {e_con(l, slt) = tL − {l_i} ∪ tL'};
          modi := true;
        }
      }
    }
    if(modi = true) then break;
  }
} while(modi = true);
return CPA;
```

In the Algorithm 1, there is no need to judge a location l whether is origin or target state of a transfer when judging l whether is redundant location. This conclusion can be proved by the following lemmas and theorems.

Lemma 1. For any $CPA = (L, IL, FL, sLoc, eLoc, \Sigma, E)$, if it is composited by any composition rules and the result is $CPA' = (L', IL', FL', sLoc', eLoc', \Sigma', E')$, then $sLoc' \neq eLoc$ and $eLoc' \neq sLoc$.

PROOF: The lemma can be proved by compositon rules directly [10]. In all the rules, the initial location of a CPA will never be a terminal location of another CPA, and the terminal location of a CPA will never be a initial location of another CPA too. □

Theorem 1. For any operand $CPA_i = (L_i, IL_i, FL_i, sLoc_i, eLoc_i, \Sigma_i, E_i)$ in any compositon rules, the compsotion result don't have the synchronous transfer whose origin location or target location is $sLoc_i$.

PROOF: The syschronous transfer is produced only with *Split-Join* rule. This rule uses all terminal locations of operand as the initial state of syschrounous transfer, and the terminal state of syschrounous transfer is newly added. So the operand's intital location $sLoc_i$ will never be part of initial state of a synchronous transfer.

By Lemma 1, any *CPA*'s initial location $sLoc_i$ will never be the terminal location of another *CPA*. □

Lemma 2. For any Parallel transfer $\Delta_{con}(l, slt) = \{l_1, \ldots, l_n\}$, its origin state l used to be the initial location of a *CPA*.

PROOF: The parellel transfer is produced only with *Split-Join* rule. This rule uses a new location as the origin state of the parellel transfer, and the new location also is also the initial location of the new *CPA*. □

Through the Algorithm 1, if location l is rudundatn, it is a origin state of a parallel transfer. By Lemma 2 and Theorem 1, it is obvious that there is no synchronous transfer whose origin state or target state includes location l.

For a synchronous transfer $\Delta_{syn}(tL = \{l_1, \ldots, l_n\}, jn) = l$, if the target state l is part of the origin state of another synchronous transfer, and the synchronous conditions of these two tranfers are same, then location l and transfer Δ_{syn} are redundant.

The Algorithm of eliminating redundant synchronous transfer is following:

Algorithm 2 ERST(**E**liminate **R**edundant **S**ynchronous **T**ransfer)

Input: A Composite Process Automata $CPA = (L, IL, FL, sLoc, eLoc, \Sigma, E)$.
Output: The Modified *CPA*.

Set *modi* as a bool variable;
do {
  modi := *false*;
  for each $e_{syn}(tL, jn) = l \in E$ {
    for each $l_i \in tL$ {
      if($\exists e_{syn}(tL', jn) = l_i \in E$) **then** {
        if($e_{sin}(l_i, a) = l' \notin E \wedge e_{sin}(l', a) = l_i \notin E$) **then** {
          $L := L - \{l_i\}$;
          $E := E - \{e_{syn}(tL, jn) = l, e_{syn}(tL', jn) = l_i\} \cup$
          $\{e_{syn}(tL - \{l_i\} \cup tL', jn) = l\}$;
          modi := *true*;
        }
      }
    }
    if(*modi* = *true*) **then break**;
  }
}**while**(*modi* = *true*);
return *CPA*;

In Algorithm 2, there is also no need to judge a location l whether is origin or target state of a parallel transfer. The proof procedure is similar.

Theorem 2. For any operand $CPA_i = (L_i, IL_i, FL_i, sLoc_i, eLoc_i, \Sigma_i, E_i)$ in any compositon rules, the composition result don't have the parallel tansfer whose origin or target state includes $eLoc_i$.

PROOF: By the *Split-Join* rule, the target location set of parallel transfer consists of initail locations of all operands, and the origin location is newly added.

By the Lemma 1, the terminal location of CPA_i will never be the initail location of another *CAP*. Thus there will never be a synchronous transfer whose initail or target location is $sLoc_i$. □

Lemma 3. For any synchronous transfer $\Delta_{syn}(\{l_1, \ldots, l_n\}, jn) = l$, its target location used to be the terminal location of a *CPA*.

PROOF: Its proof is similar to Lemma 1. □

Through the Algorithm 2, if location l is redundant, it is the target location of a synchronous transfer. By the Theorem 2 and Lemma 3, it is obvious that there is no parallel transfer whose origin or target location is l.

Using Algorithm 1 and Algorithm 2 to optimize the *CPA* in Fig. 4, the result is showed in Fig. 5.

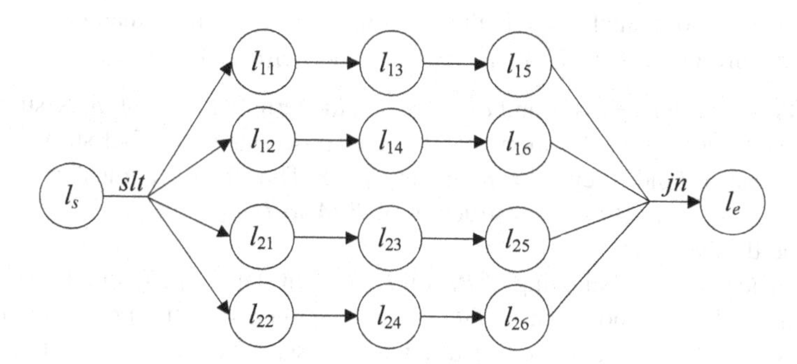

Fig. 5. The optimization result.

4 The Equivalence of Parallel Finite Automata and Finite Automata

The following is the proof of the equivalence of parallel finite automata and finite automata. The conclusion means that the verification technology of system which base on the finite automata can be used to parallel finite automata by extension.

The idea of the proof based on the technology of subset construction [11]. The method is that The $PFA = (L, IL, FL, \Sigma_C, \Delta_C)$ converts to a $DFA = (Q, q_0, F, \Sigma_D, \Delta_D)$ which accepts the same language:

- $Q = 2^L, q_0 = IL, F = \{q | q \in Q \wedge q \subseteq FL\}$
- $\Sigma_D = \Sigma_C$
- $\Delta_C = \{\delta(q, \sigma) = q' | \exists \delta(s, \sigma) = s' \in \Delta_C \wedge s = q \wedge s' = q'\}$

This method can get an equivalent *DFA*, but the *DFA* may have too many states. If the *PFA* has n locations, the corresponding *DFA* will have 2^n states. But in fact many states are unreachable, the states of *DFA* are much less than 2^n. According to the following computational method, the reachable states can be got.

BASIC: The set of initial locations *IL* is reachable state.

INDUCTION: If the state s is reachable, then $s' = \delta(s, \sigma)$ is reachable where $\sigma \in \Sigma_C$.

By the construction process of *PFA* to *DFA*, a state of *DFA* correspond to a set of locations of *PFA*.

The following theorem shows that the *PFA* which is constructed by *DFA* can accept the same language with *DFA*.

Theorem 3. Using a *PFA* to construct a *DFA* by the above method, they have the same acceptable language, namely $L(C) = L(D)$.

PROOF: Firstly, it is need to prove a thesis:

If a *PFA* and a *DFA* get a same label string ω, they will move to the same state after processing the ω.

The proof is following:

BASIC: If $|\omega| = 0$, namely $\omega = \varepsilon$, then $\Delta_C(s_0, \varepsilon) = s$. By the semantics of *PFA* and the constructure process of *DFA*, $s_0 = q_0 = IL$ and $\Delta_D(q_0, \sigma) = q'(q' = s)$.

INDUCTION: If $|\omega| = n+1$ and $\omega = xa$, a is the terminal label of ω. Assuming that the *DFA* and *PFA* are on location q and s after processing the label string x. By the basic, the thesis is hold when $|\omega| = n$, namely $q = s$. Thus there are transfers $\Delta_C(s, a) = s'$ and $\Delta_D(q, \varepsilon a) = q'$. By the construction of *PFA* to *DFA*, $s' = q'$.

So, the thesis is true.

ω is an acceptable label string of *PFA* if PFA is on state $s \subseteq FL$ after processing the label string ω. By the above thesis, *DFA* is on location $q = s$ after processing the label string ω, so $q \subseteq FL$. Through the construction process, q is the acceptable state of *DFA*, so ω is acceptable label string of *DFA*. Thus the set of label string which can be accepted by *PFA* is also accepted by *DFA*, namely $L(C) = L(D)$. □

The Theorem 1 proves that any *CFA* can be converted to a *DFA* which can accept the same language.

Conversely, a *DFA* also can convert to a *PFA* easily. DFA can be deemed a special PFA which only has step transfer. Thus the state of DFA always have only one location. A *DFA* can be converted to a *PFA* by the following method: Using $\{q_0\}$ as the initial state of *PFA* where q_0 is the initial state of *DFA*, and the acceptable state of *PFA* is the set of acceptable states of *DFA*, namely $FS = \{\{q\}|q \in F\}$. It can be proved that their acceptable language coincident is by induction method.

In summary, a language L can be accepted by a *DFA* if and only if L can be accepted by a *CFA*. It means that PFA and DFA are equivalent.

5 Summary

This paper introduces the syntax and semantics of Parallel automata, and the construction method of complex control structure by parallel finite automata. Then it researches the *Split-Join* rule of construction method. It points out that this rule may produces redundant location and transfer. This problem will increase complexity of composition model. This paper gives one condition which may lead to produce redundant location and transfer by *Split-Join* rule and the algorithms which are used to eliminate the redundant location and transfer. Finally, it prove the equivalence of parallel finite automata and finite automata by subset construction method. This conclusion ensures that the verification technology of system which base on the finite automata can be used to parallel finite automata by extension.

Acknowledgment. This work is supported by Natural Science Foundation of Fujian Province (Nos. 2016J05146 and 2016J05145), National Natural Science Foundation of China (Nos. 61402108 and 61503082), Educational Research Project for Young and Middle-Aged Teachers of Education Department of Fujian Province (No. JA15336), and Scientific Research Development Foundation of Fujian University of Technology (No. GY-Z15087).

References

1. Maneva, S., Manev, K.: Finite automata models in agro-ecosystem and plant protection. Int. J. Comput. Appl. **32**(11), 1–6 (2015)
2. Neider, D., Jansen, N.: Regular model checking using solver technologies and automata learning. In: 6th International Proceedings on NASA Formal Method, CA, USA, pp.16–31. Springer, Heidelberg (2013)
3. Armbrust, C., Kiekbusch, L., Ropertz, T., Berns, K.: Verification of behaviour networks using finite-state automata. In: German Conference on Advances in Artificial Intelligence, Koblenz, German, pp. 1–12. Springer, Heidelberg (2012)
4. Li, Y., Li, L.: Model checking of linear-time properties based on possibility measure. IEEE Trans. Fuzzy Syst. **21**(5), 842–854 (2013)
5. Cambronero, M.E., Díaz, G., Valero, V., et al.: Validation and verification of web services choreographies by using timed automata. J. Logic Algebraic Program. **80**(1), 25–49 (2011)
6. Damm, W., Ihlemann, C., Sofronie-Stokkermans, V.: Decidability and complexity for the verification of safety properties of reasonable linear hybrid automata. In: International Conference on Hybrid Systems: Computation and Control, Chicago, USA, pp. 73–82. ACM (2011)
7. Ruiz, V., Diaz, G., Cambronero, M.E.: Timed automata modeling and verification for publish-subscribe structures using distributed resources. IEEE Trans. Softw. Eng. **43**(1), 76–99 (2017)
8. Waez, M.T.B., Dingel, J., Rudie, K.: A survey of timed automata for the development of real-time systems. Comput. Sci. Rev. **12**(9), 1–26 (2013)
9. Delahaye, B., Katoen, J.P., Larsen, K.G., et al.: Abstract probabilistic automata. Inf. Comput. **232**(6), 66–116 (2013)
10. Tang, Z.-Y., Wang, J.-S., Wei, L., et al.: Formal description and compatibility analysis of OWL-S process model **38**(3), 478–485 (2016)
11. Noord, G.V.: Treatment of epsilon moves in subset construction. Comput. Linguist. **26**(1), 61–76 (2000)

An Intelligent Data Analysis of the Structure of NP Problems for Efficient Solution: The Vehicle Routing Case

Esteban Perez-Wohlfeil(✉), Francisco Chicano, and Enrique Alba

University of Malaga, Boulevard Louis Pasteur 35, Malaga, Spain
estebanpw@uma.es, {chicano,eat}@lcc.uma.es

Abstract. The Vehicle Routing Problem is a combinatorial problem with considerable industrial applications such as in traditional logistics and transportation, or in modern carpooling. The importance of even small contributions to this problem is strongly reflected in a significant cost savings, pollution, waste, etc., given the high impact of the sector in almost any economic transaction. The VRP is often treated as an optimization problem, however, the fitness function converges quickly and the algorithms become stagnant in late steps of the executions, which is a recurrent problem. In this work, we perform an analysis of the structure of solutions to identify potential use of existing ideas from other domains to achieve higher efficiency. In this sense, the feasibility of applying the Partition Crossover –an operator initially designed to tunnel through local optima for the Travelling Salesman Problem– to the Capacitated Vehicle Routing Problem is studied in order to escape local optima. Moreover, an implementation is provided along with an analysis applied to real use-cases, which show a promising rate of local optima tunneling.

Keywords: Optimization · Data analysis · Crossover operator · Genetic algorithms · Local optima · Graph theory · Smart cities

1 Introduction

In this work, we research the similitude of the Vehicle Routing Problem (VRP) to the Travelling Salesman Problem (TSP), and try to make an exercise of cross-fertilization by analyzing the underlying structure of their solutions. We try to find a way of using this analysis to port recent advances in TSP, the *partition crossover*, to VRP. In particular, we will try to apply such methodology to the Capacitated Vehicle Routing Problem (CVRP). The CVRP is an extension of the VRP, an NP-hard problem –that is, the best known algorithm to solve it requires polynomial time in a non-deterministic Turing Machine– which can be seen as a n-dimensional generalization of the TSP. The CVRP is formalized as

P. Krömer et al. (eds.), *Proceedings of the Fourth Euro-China Conference on Intelligent Data Analysis and Applications*, Advances in Intelligent Systems and Computing 682, DOI 10.1007/978-3-319-68527-4_40

an undirected graph $G = (\mathbf{V}, \mathbf{E})$ where $\mathbf{V}$ is a set of vertices $\{v_0, v_1, ..., v_n\}$ and $\mathbf{E}$ is a set of edges (v_i, v_j). Each edge (v_i, v_j) has an associated travelling cost, and each vertex v_i has an associated customer demand. A fleet of m vehicles with uniform capacity must determine the set of m non-overlapping routes except for the initial vertex (depot) of minimum cost and while holding the capacity constraint, incremented by the customer demand as each vertex is travelled.

Whitley et al. [1] showed that jumping between local optimal solutions in genetic algorithms is possible by applying a recombination method called partition crossover that respects and maintains certain properties of the parents across offspring (i.e. a respectful operator [2]). Additionally, it was shown that local optima is often found between the offspring, thus making the partition crossover (PX from now on) a powerful operator to guide the search in latter stages, where traditional evolutionary algorithms get stuck and do not generate new solutions [3]. PX and its generalizations (e.g. [4]) are based on searching in the space of a partition of the solution components (i.e. vertices). If the cardinality of the partition is q, PX returns the best solution among 2^q solutions in linear time. Therefore, the operator is able to jump through an exponential number of possible solutions to find new local optima when certain conditions in the solutions are met.

Such landscape jumps can be performed to quickly produce better solutions and thus improve graph-like problems, which can be found commonly within the design of Smart Cities: from vehicle parking applications or smart carpooling to modeling urban water-supplies networks. Moreover, in this manuscript, we first propose an analysis and then an operational method to port PX from TSP to CVRP while hopefully maintaining the same properties, which finally leads to unseen performances. Furthermore, we perform an in-depth analysis of the relation between the geometry of the solutions that are being recombined and the success rate of yielding new local optima. At last, we apply the algorithm to state-of-the-art instances that could depict recurrent scenarios to deal with in Smart Cities.

2 Analysis and Operational Proposal

In this section, we start our analysis of the problem solutions within TSP and CVRP, so that the different methods that compose the Partition Crossover (PX) operator are highlighted and discussed. In particular, this section will address: (1) Original PX operator applied to the TSP, (2) Modifications performed to the PX operator in order to apply it to CVRP, (3) Encoding of chromosomes in the CVRP, (4) Generation of initial solutions and local search methods and (5) Application of the modified PX operator to the CVRP.

2.1 Original Partition Crossover for the Travelling Salesman Problem

In the original paper by Whitley et al., a first version of the partition crossover (PX) is presented. In short, a series of random tours in the Symmetric Travelling

Salesman Problem (STSP) are converted into local optima under the heuristic operator 2-OPT [5]. PX is applied between all possible combinations of the generated local optima (which accounts for $\frac{m(m-1)}{2}$, being m the number of solutions used in the TSP) by finding a partition of cost 2 of the original graph G. Such partition can be found in polynomial time as described below:

1. First, the reduced graph G^* is constructed by removing common edges (common tours between parent solutions P_1 and P_2). If more than two partitions exists, the crossover operator fails.
2. Secondly, the two tours contained in the partitions are swapped with each other. Thus two offspring are produced, O_1 and O_2, where O_1 is equal to P_1 except for a tour in one of the partitions. Likewise, O_2 is equal to P_2 except for the contrary tour to that of O_1.

Nonetheless, the original PX could only be applied when just two partitions along with the residual partition (complement of the two partitions) coexisted. This is, a recombination graph with more than two partitions could not be recombined, and thus $2^q - 2$ possible recombinations would be missed (being q the number of partitions). On a second paper [6], Whitley et al. described an improvement upon the PX operator such that reduced graphs G^* with more than two partitions could be recombined, and additionally, the procedure was developed not only for the Symmetric TSP but for the Asymmetric one. In particular, this was accomplished by splitting nodes with degree 4 into two nodes of degree 3 with a common edge of cost zero between them. Such procedure enables to partition the space of nodes, and more important, allows to detect entries and exits within partitions. Entries and exits depend on the direction of the edges (i.e. the order within a partition matters), which has to be accounted for in the case of the asymmetric TSP. The detection of entries and exits is crucial for the operator since it allows to recombine partitions on their own, without the limitation imposed on the previous algorithm. Partitions will be feasible if their entries and exits are equal, since a partition is composed of the same nodes in both parent solutions, therefore if both solutions enter and exit the partition in the same fashion, they have travelled all nodes but using a distinct route. In short, the described PX operator produces more partitions since:

1. Nodes with degree 2 are common, and either belong to a partition, or separate two partitions.
2. Nodes with degree 3 belong to a partition, since these are connected to common edges in one side, and are the entry or exit to a partition on the other.
3. Nodes with degree 4 are converted to degree 3 by adding a ghost vertex. Thus, these nodes also belong to a partition given the above argument.

Therefore, all nodes either belong to a partition or separate two partitions because they are common edges shared by the parent solutions.

2.2 Modifications in the PX Operator to Fit CVRP Constraints

The Vehicle Routing Problem includes unique characteristics that are not present in the Travelling Salesman Problem; although several ones are shared, the porting of the PX from TSP to VRP is not straightforward. In this section we will review which changes have to be addressed in order to transit from one operator to the other.

Starting Route Node. As reflected in the encoding of the chromosome, all routes are required to start from the depot and to return back once their capacity is reached. This limitation forces that a subset of edges between the ending nodes (i.e. last customers visited before return to depot) and the depot can not be used for the generation of partitions. Briefly, the depot node could be considered as being part of *any* partition, thus making it a common edge available to connect petals. However, a recombination performs a rearrangement (permutation) on the nodes that belong to the partition, and thus the order in which the depot is visited would be affected and the capacity constraints would not hold for the resulting route. Hence, the depot node must be considered as a partition breakpoint, and recombination must be performed between overlapping petals without including the depot.

Capacity. The biggest drawback in the generation of recombinable partitions is the capacity limitation. When the capacity constraint is reached, the vehicle is forced to return to the depot. From the perspective of optimal routing (as in TSP), this limitation abruptly breaks (and starts) a route that follows a greedy (or even near-optimal) path, which has a higher probability of being included in the optimal solution. Therefore, the capacity constraint produces two main effects:

1. Common edges that could connect partitions are removed, since these are forced to return to the depot.
2. The number of partitions is larger than that of the same instance treated as a TSP. However, these partitions are of smaller size, and since there is a reduction of common edges that tend to form part of the optimal solutions, entries and exits into these make a considerable amount of partitions become unfeasible.

Moreover, to improve the feasibility of the smaller partitions, subroutes can also be evaluated inside unfeasible partitions (partitions whose entries and exits do not match), since an unfeasible partition can contain feasible subroutes, i.e. a route in each solution and partition shares the same entry and exit. Such route can be used to recombine the solutions, although the partition is not fully recombinable.

2.3 Application of the Modified PX Operator to the CVRP

This section describes how the PX is applied to CVRP while addressing the constraints depicted in the previous section.

Encoding of CVRP Solutions. In order to enable manipulation of the CVRP instance, the solutions have to firstly be encoded in chromosomes. This is done by numbering nodes (customers and the depot) from 0 to n, being n the number of customers, and 0 the depot, from which all vehicles start their routes. Such encoding is accomplished by using a string of integer numbers. Therefore, a single chromosome for a CVRP with k vehicles and n nodes is a string containing all of the n nodes, separated with the depot marker 0. Hence, the length of a chromosome with n nodes and k vehicles is $n + k - 1$. For instance, a representation of a CVRP instance with $n = 8$ and $k = 3$ could be depicted as shown in the chromosome in Fig. 1:

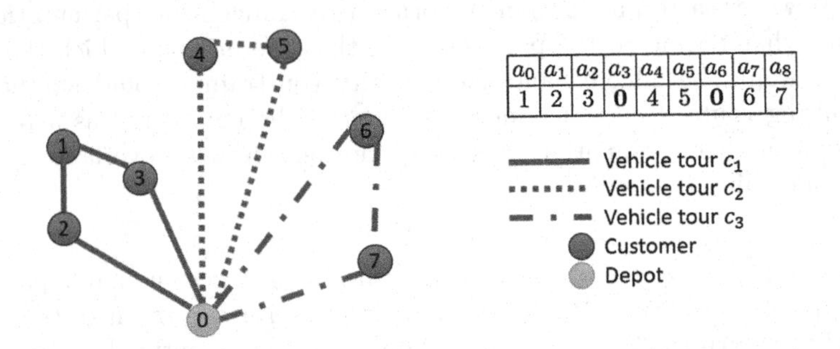

Fig. 1. Chromosome encoding (top right) of a trivial CVRP solution (left) with seven customers and one depot. The depot is labeled **0** and is used as separator of each individual vehicle route. Each route followed by a different car is depicted with either a solid, dotted or dashed line.

2.4 Generation of Initial Solutions

Initial solutions are generated using the Petal algorithm [7]. In particular, the Petal algorithm has proved to be a good starting point in the resolution of the VRP [8], since optimal solutions often show a petal-like structure. In order to produce the petal structures, all nodes are firstly labeled radially in polar coordinates r ρ with respect to the cartesian coordinates of the depot. The ray r is discarded and the nodes are sorted by their angle ρ. This procedure enables the construction of a sequence of sorted nodes to generate petal-like structured routes in the CVRP. In later sections, we will address the use of this sequence to improve the number of recombinations in the PX operator.

A 2-OPT local search is applied to each generated solution. Notice that under the 2-OPT operator, these improved solutions will be locally optima. A local optima under 2-OPT occurs when there exists no path of edges,

$$(v_i, v_{i+1}), (v_{i+1}, v_{i+2}), ..., (v_{j-2}, v_{j-1}), (v_{j-1}, v_j)$$

such that the reversion of the succession in between starting and ending node,

$$(v_i, v_{j-1}), (v_{j-1}, v_{j-2}), ..., (v_{i+2}, v_{i+1}), (v_{i+1}, v_j)$$

does not produce an improved solution. Additionally, a Don't Look Bits, Neighborhood Lists and delta-evaluations schemes are applied to reduce the computational complexity of the 2-OPT operator from cubic to nearly linear.

2.5 Improving the Recombination Rate Using Node Shifting

As argued in Sect. 2.2, the capacity constraint along with the fixed starting and ending node produce a lack in common edges that promote the generation of larger and feasible partitions. To overcome this limitation, we propose using an incremental generation of initial petal solutions by shifting the starting angle ρ in polar coordinates from which solutions are created. This procedure aims to force the creation of overlapping petals that take advantage of the radial order of nodes to find common edges that will generate partitions in the extremes of such common edges. However, creating initial solutions shifted by ρ degrees will produce long common edges, since (assuming uniform demands in customers) a petal solution A visiting first customer c_1 and ending in c_n will share $n-1$ nodes with a solution B that visits first customer c_2 and ends in c_n+1. The partitions generated by this method rely on customer demands not following a uniform distribution and hence sharing less nodes between overlapping solutions. To overcome this limitation, a recurrence equation to visit nodes is introduced as a modification of the petal algorithm. This modification is based on two principles:

1. The ideal partition has only one entry and one exit, and the rest of nodes is within these two common edges.
2. The nodes within the common edges (inside the partition) should have the minimum number of common edges in order to favor recombination.

Therefore the proposed recurrence to visit nodes attempts to only share an initial and ending edge between the first and last customer and the depot, whereas the rest of the route between these nodes crosses nodes alternatively. To do so, the following recurrence equation is proposed:

$$T_k = \begin{cases} 0, & \text{if } k = 0 \\ (T_{k-1} - 1) \bmod n, & \text{if } k \equiv 2 \bmod 3 \\ (T_{k-1} + 2) \bmod n, & \text{otherwise} \end{cases} \tag{1}$$

where 0 is the initial customer corresponding given the reordering of the shift angle ρ and n is the number of nodes in the instance. This recurrence generates the succession $s_i, s_{i+2}, s_{i+1}, s_{i+3}, s_{i+5}, s_{i+4}, s_{i+6}, \ldots$ which is represented graphically in Fig. 2. Notice that no initial shift is used in the generation of the modified solution in order to enable a visual appreciation of the recurrence. However, a shift in the starting node by ρ would discard nodes from c_0 to c_i and end in nodes from c_j to c_{j+k}, thus modifying the overlapping of the structure.

Additionally, notice that the routes inside the common edges ($\{d, c_0\}$ and $\{c_{10}, d\}$) share no common edges and thus pose the most adequate scenario for the PX operator.

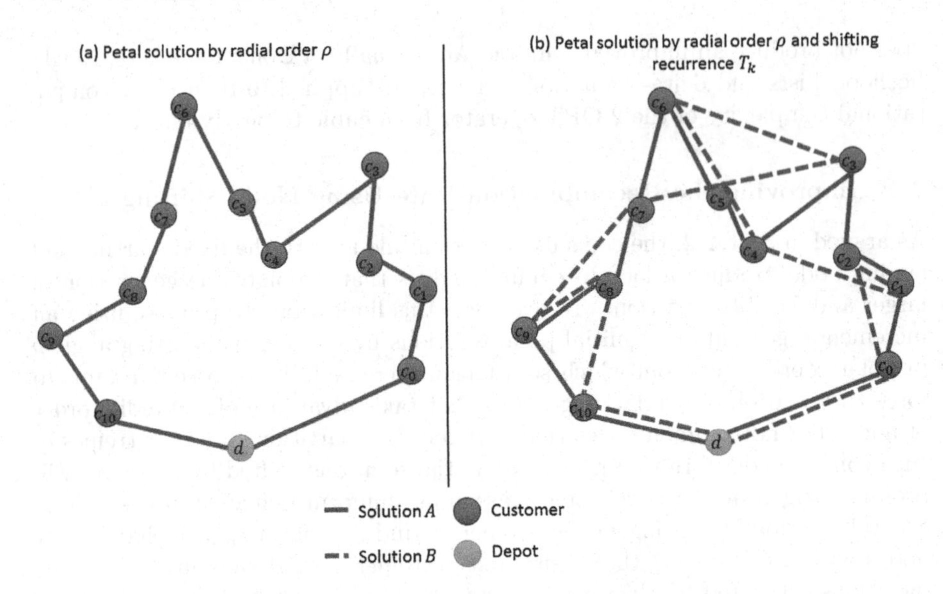

Fig. 2. Petal shifting with alternating partition routes. (a) On the left, initial solution generated by the Petal algorithm, in blue solid line. (b) On the right, the modified petal structure using a shift of zero (i.e. no shift), in dashed red. (Color figure online)

2.6 Application of the Modified PX Operator to the CVRP

In order to tunnel through local optima in the CVRP, a number of initial solutions are generated following the algorithm described in Sect. 2.5, which are then recombined using the modified PX operator. That is, each solution is recombined with all other solutions, hence generating $\frac{m(m-1)}{2}$, being m the number of initial solutions. For each offspring O_{ij} from parents P_i and P_j, the fitness value is stored along with the number of recombinations obtained. Algorithm 1 shows the pseudo code for the incremental recombination of shifted solutions.

Algorithm 1. Incremental recombination using shifted petals

```
1: procedure VRP_PX_PETALS(Solutions S, Integer shift)
2:     for Solution s_i in S do
3:         petal_shifted[i] ← generate_petal(s_i, shift)
4:     end for
5:     for Solution s_i in petal_shifted do
6:         for Solution s_j in petal_shifted[i : end] do
7:             offspring ← apply_PX_VRP(s_i, s_j)
8:         end for
9:     end for
10: end procedure
```

3 Results and Discussion

In this section, we perform an analysis over the proposed PX operator for the CVRP. Briefly, we will:

1. Compare the petal structure of the shifting strategy with and without local search.
2. Compare the number of recombinations and tunnels between local optima as a function of the shift across several CVRP instances.

As discussed throughout this manuscript, the main contribution of this work is a first approach to the PX operator in the CVRP and to analyze the underlying structures that appear near local-optima. Additionally, we propose a procedure to port the PX operator from TSP to VRP, and a procedure to obtain a higher ratio of recombination between solutions along with the tunneling of local optima. In this sense, it is important to remark that this work, in its current state, does not aim to compete with state of the art methods in terms of fitness and performance. At this point, we present a framework to perform tunneling between local optima in the CVRP. Moreover, relatively simple changes such as the introduction of Iterated Local Search in the post-generation of initial solutions could yield considerably better results.

3.1 Petal Structure Comparison

Using a small sized instance, namely A-n80-k10 (extracted from CVRPLIB[1]), we show the petal structure generated from a node shift of 1. Figure 3 illustrates such structure showing two random solutions with and without 2-OPT local search optimization. Notice for the top plot on Fig. 3 that there is a considerably small amount of common edges, except for those connecting with the depot. This enables larger partitions with less entries and exits. Moreover, on the bottom plot of Fig. 3, it can be seen that even after 2-OPT several edges are now common for both solutions, since these are now optima under the operator. However, still a significant amount of uncommon edges can be detected inside vehicle routes, such as those in nodes $15, 33, 55, 56, 64$. On the other hand, the edge between $4, 71$ is clearly not a good choice for this instance, and should be addressed in further work.

3.2 Recombinations and Tunnels as a Function of the Shift ρ

In this section, we compare the number of recombinations generated by the proposed operators (Sect. 2.5) as a function of the node shift using the recurrence described in Sect. 2.4. Figure 4 shows such function for three selected instances, namely Tai-n150a-k15, X-n459-k26 and M-n200-k17 (in advance, Tai-150, X-459 and M-200, respectively). Firstly, notice that the number of recombinations appears to not be related to the size of the problem, since in Tai-150 there is

[1] http://vrp.atd-lab.inf.puc-rio.br/index.php/en/.

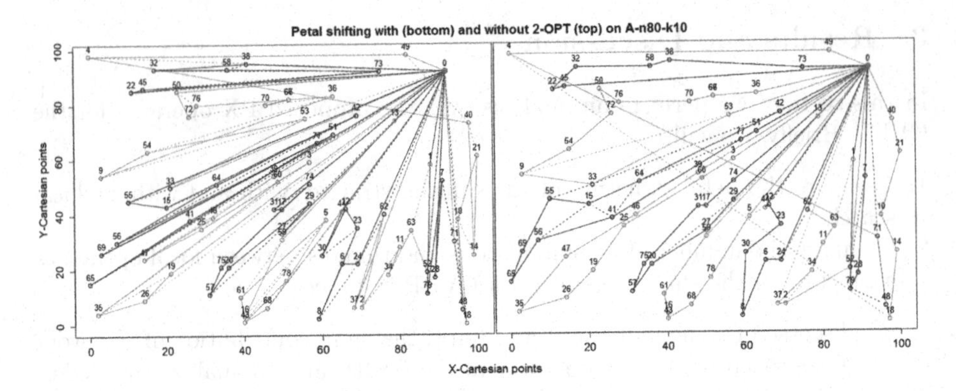

Fig. 3. Comparison of the petal structure of two solutions generated by the proposed method. Blue and red colors are used to alternate between vehicle routes, whereas dashed or solid lines differentiate the two solutions. On top, the generated structure without 2-OPT. Bottom, the same structure with 2-OPT. (Color figure online)

a somewhat continuous rate of 40 recombinations in most of the node shifts, except for the interesting peaks which appear to happen at the node shifts that are multiples of 3 (i.e. $\{3, 6, 9, 12, \ldots, 24, 27, 30\}$). This result suggests that the optimal recombination rate is intrinsically related to the node shift and other parameters, such as capacity, or the geometry of the instance itself. However, this evidence seems to be specific to each instance (as can be observed on the other instances, where the peaks differ).

On the other hand, in X-459 a clear tendency of alternating recombination rates is observed, except for the case of the shift $\rho = 15$ that produces the highest peak at 78. However, the repeated peaks alternating between higher and lower recombination rates, suggest that a single node shift can displace the solutions enough to break common edges between partitions. Furthermore, on M-200, the pattern alternates less, and a surprising peak shows with a much higher recombination rate than the rest at $\rho = 12$. The geometrical analysis of such finding is left for further studies.

An average recombination rate of 65, 59 and 55 is found respectively for each instance, which accounts for a recombination rate of approximately 30%. On the best node shifts, a promising rate of 66%, 41% and 51% is found for each instance. Moreover, when applying 2-OPT to the offspring, 22%, 47% and 10% (respectively) of the recombined offspring were found to be local optima with improved fitness. Although the results shown herein are for the shift interval of $\rho = [1, 30] \in \mathbb{N}$, we have verified that the patterns hold for larger ranges.

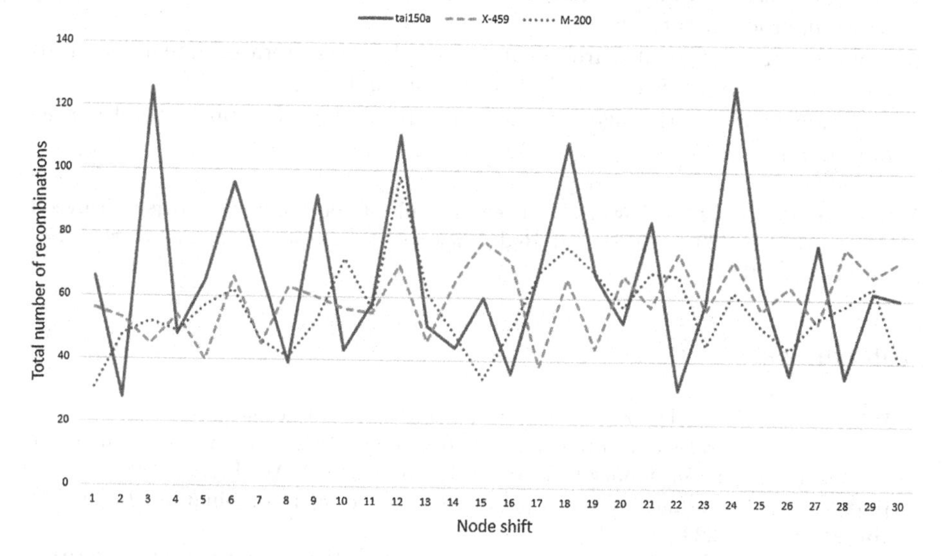

Fig. 4. Number of recombinations found as a function of the node shift ρ used to generate the solutions. On the *x-axis*, the ρ shift. On the *y-axis*, the number of new local optima found out of 190 recombination. The selected instances are plotted in different colors and lines: blue solid line for Tai-150a, red dashed line for X-459 and purple dotted line for M-200. (Color figure online)

4 Conclusions

In this manuscript, we have presented an initial approach to tackle the Partition Crossover operator in the Capacitated Vehicle Routing Problem. At a first step, we have discussed the initial operator through its different improvements, and at a second step, we have argued about the difficulties of adapting such operator to the CVRP. The main contribution of this work is our analysis of underlying structures that appear in local optima, which shows that a natural pattern exists. This experimental analysis suggests a strong dependence between the vehicle capacity and the number of shifts and that this information can be used for an efficient recombination and improvement of solutions. Additionally, we have proposed a combined PX operator which obtains a recombination rate of up to 66% depending on the node shift ρ. These recombination rates are promising and could be applied in evolutionary programming to escape local optima in latter stages of execution, thus avoiding fitness stagnancy. However, more work needs to be carried out in this area, which we plan to continue, such as:

1. Study the peaks and patterns discovered in the recombination rate as a function of the node shift.
2. Further improve the PX operator to enable the swapping of nodes between different vehicle routes.

3. Study the usage of meta-algorithms to automatically discover features on the above mentioned patterns.
4. Inclusion of other local search heuristics such as the Iterated Local Search to improve the overall fitness and goodness of solutions.
5. Addition of a genetic algorithm scheme using the recombined solutions as population.

Acknowledgements. This work has been partially supported by the projects Moveon TIN2014-57341-R (2015–2018) and Red Nacional de Investigación en Smart Cities TIN2016-81766-REDT.

References

1. Whitley, D., Hains, D., Howe, A.: Tunneling between optima: partition crossover for the traveling salesman problem. In: Proceedings of the 11th Annual Conference on Genetic and Evolutionary Computation, pp. 915–922. ACM, July 2009
2. Radcliffe, N.J.: Forma analysis and random respectful recombination. In: ICGA, vol. 91, pp. 222–229 , July 1991
3. Andre, J., Siarry, P., Dognon, T.: An improvement of the standard genetic algorithm fighting premature convergence in continuous optimization. Adv. Eng. Softw. **32**(1), 49–60 (2001)
4. Whitley, D., Hains, D., Howe, A.: A hybrid genetic algorithm for the traveling salesman problem using generalized partition crossover. In: Parallel Problem Solving from Nature, PPSN XI, pp. 566–575 (2010)
5. Croes, G.A.: A method for solving traveling-salesman problems. Oper. Res. **6**(6), 791–812 (1958)
6. Tinós, R., Whitley, D., Ochoa, G.: Generalized asymmetric partition crossover (GAPX) for the asymmetric TSP. In: Proceedings of the 2014 Annual Conference on Genetic and Evolutionary Computation, pp. 501–508. ACM, July 2014
7. Ryan, D.M., Hjorring, C., Glover, F.: Extensions of the petal method for vehicle routeing. J. Oper. Res. Soc. **44**, 289–296 (1993)
8. Renaud, J., Boctor, F.F., Laporte, G.: An improved petal heuristic for the vehicle routeing problem. J. Oper. Res. Soc. **47**(2), 329–336 (1996)

Evaluation of Traveling Salesman Problem Instance Hardness by Clustering

Pavel Krömer(✉) and Jan Platoš

Department of Computer Science, VŠB Technical University of Ostrava,
Ostrava, Czech Republic
{pavel.kromer,jan.platos}@vsb.cz

Abstract. Traveling salesman problem (TSP) is a well-known NP-hard combinatorial optimization problem. It has been solved by a number of exact and approximate algorithms and serves as a testbed for new heuristic and metaheuristic optimization algorithms. However, it is often not easy to evaluate the hardness (complexity) of a TSP instance. Simple measures such as the number of cities or the minimum (maximum) route length do not capture the internal structure of a TSP instance sufficiently. In this work, we propose a new method for the assessment of TSP instance complexity based on clustering. The new approach is evaluated on a set of randomized TSP instances with different structure and its relation to the performance of a selected metaheuristic TSP solver is studied.

1 Introduction

The traveling salesman problem is an iconic hard combinatorial optimization problem with a long history, a number of different variants, and countless real-world applications [1,13]. In the past, a number of exact [1,7] and approximate [2] methods has been proposed to address the TSP. The interest in approximate TSP solvers is motivated by its complexity. It is known to be NP-hard and, therefore, no polynomial time algorithm for exact TSP computation is known and none is expected to exist. That limits the use of exact methods to TSP instances with small or moderate number of cities only [1]. This is a severe restriction that prevents their use in many real-world applications that can be cast as a TSP.

Nature-inspired methods, on the other hand, are able to find approximate (optimal or sub-optimal) TSP solutions in a reasonable time. For solving TSP, global metaheuristic search strategies are often coupled with certain local search algorithms in order to improve the performance of the process [9,15]. However, the ability of pure metaheuristics to solve hard combinatorial optimization problems has been studied as well [11]. Anyway, TSP is now recognized as a testbed for new metaheuristic algorithms [16]. However, in order to provide an informative comparison between candidate algorithms, it is important to know the *hardness* (complexity) of test TSP instances (in particular random ones [8]). Simplistic measures such as the number of cities or the minimum (maximum,

P. Krömer et al. (eds.), *Proceedings of the Fourth Euro-China Conference on Intelligent Data Analysis and Applications*, Advances in Intelligent Systems and Computing 682, DOI 10.1007/978-3-319-68527-4_41

average) route length do not provide enough information to capture the overall hardness of the optimization problem represented by a particular TSP instance.

In this work, we use a density-based clustering method, the DBSCAN algorithm [10], to assess the hardness of different TSP instances. The clustering process is applied to 2-dimensional TSP instances and the relationship between clustering properties and the ability of a pure metaheuristic method, Ant Colony Optimization (ACO) [5], to solve the problem is studied.

The rest of this paper is structured in the following way: Sect. 2 summarizes the TSP and provides an overview of methods for TSP complexity assessment. Section 3 introduces the fundamentals of cluster analysis and the principles of the DBSCAN algorithm. The proposed approach to TSP instance complexity evaluation by DBSCAN and its experimental analysis are detailed in Sect. 4. Finally, major conclusions are drawn in Sect. 5.

2 Traveling Salesman Problem

Traveling salesman problem is a hard combinatorial optimization problem [1,13, 16]. Informally, the TSP consists in finding the shortest (least expensive) route between n cities. In mathematical terms, the TSP looks in a weighted graph, $G = (V, E, w)$, with a set of vertices V, number of vertices $n = |V|$, set of edges E, and a set of edge weights $w = \{w_e \in \mathbb{R} \mid \forall e \in E\}$, for a Hamiltonian cycle, $\mathcal{L}$, with a minimum sum of weights on the edges of the cycle [13].

A particular instance of the TSP is often represented by a cost matrix, $C^{n \times n} = (c_{ij})$, where $c_{ij} = w_e$ for all edges $e = (ij)$. If G is not complete, missing edges are in C often represented by an arbitrary *large weight*. The TSP can be also formulated as a permutation problem [13]. For a set of all permutations of a set of n objects, S_n, find $\pi = (\pi(1), \pi(2), \ldots, \pi(n))$ such that the cost of the permutation (i.e. objective function),

$$f_{\text{obj}}(\pi) = c_{\pi(n)\pi(1)} + \sum_{i=1}^{n-1} c_{\pi(i)\pi(i+1)}, \tag{1}$$

is minimized.

2.1 TSP Complexity

The problem of TSP and in general combinatorial optimization problem instance complexity assessment has been addressed by several studies [8,12].

Hernando et al. [8] investigated the complexity of TSP instances with the help of 2-exchange neighbor system (i.e. a mapping that assigns to each candidate solution 2 other solutions that form its neighborhood). The basins of attraction and local optima of search spaces corresponding to different random TSP instances created with respect to the 2-exchange neighbor system were studied and two new TSP complexity measures were proposed. The first one was the ratio of the size of the basin of attraction of the global optimum to the size of the search space and the second one was the ratio of the number of different local optima that appear in the instance to the size of the search space. The properties of the measures were studied and their usefulness for TSP instances with different sizes was discussed in [8].

Another approach was presented by Mihalak et al. in [12]. The authors focused on the metric TSP (i.e. TSP where distances between cities form a metric and satisfy the triangle inequality). The authors centered their work around the notion of stability, i.e. the property of combinatorial optimization problems to retain the same solution if the input parameters are multiplied by a constant factor. The work proved that any 1.8-stable metric TSP instance can be solved by a greedy approach in polynomial time and defined a class of 2-stable metric TSP instances for which algorithms based on simple local search fail.

It can be seen that TSP complexity assessment can be conducted using different approaches. However, the current work focuses on measures that somehow rely on the solution of the problem. That makes them impractical for assessment of complexity of TSP instances for which the solution is not known. In this work, a clustering-based approach that relies exclusively on the structure of the TSP instance is proposed and studied.

3 Cluster Analysis

Cluster analysis (clustering) is a fundamental analytical task that involves separation of a set of objects into meaningful groups. A hard clustering of a data set $D = \{x_1, x_2, x_3, \ldots, x_n\}$ is a set $\mathcal{C} = \{C_1, C_2, \ldots, C_k\}$ composed of k clusters C_i subject to $C_i \subset D$, $C_i \neq \emptyset$ for each $C_i \in \mathcal{C}$, $\bigcup_{i=1}^{k} C_i = S$, and $C_i \cap C_j = \emptyset$ for each $C_i, C_j \in \mathcal{C}, i \neq j$.

There are different clustering methods that can be used for different types of data and for different kinds of data analysis and processing (e.g. for object classification [3] vs. outlier detection). Most often used clustering algorithms include hierarchical clustering [6], centroid (medoid)-based clustering, and density-based clustering [10]. Unsupervised clustering of large data sets is a complicated NP-hard task. However, it is very attractive due to many applications in various fields of data science, machine learning, and e.g. data mining [3].

3.1 DBSCAN

The DBSCAN algorithm [3,10] is a widely-used spatial density-based clustering method. Density-based clustering is useful due to its ability to discover clusters with arbitrary shapes but it suffers from high computational costs of cluster formation and clustering evaluation. Informally, a density based cluster C_i is a set of points in the problem space that are *density connected*, i.e. for each pair of points in C_i there is a chain of points with distance between two consecutive points smaller than a constant ϵ. Second parameter of the algorithm is the minimum number of points required to form a cluster, *minPts*. DBSCAN is outlined in Algorithm 1 [3].

Algorithm 1. The DBSCAN algorithm

1 Cluster index $i = 0$
2 **while** *Not All points have been labeled* **do**
3 Randomly select an unvisited *seed point*, c_i
4 Form the ϵ-neighborhood of c_i, $\mathcal{N}(\epsilon, c_i)$. That is, find the set of all points that are *density connected* to c_i
5 **if** $|\mathcal{N}(\epsilon, c_i)| > minPts$ **then**
6 $C_i = \mathcal{N}(\epsilon, c_i)$
7 Expand C_i. That is, iteratively add to C_i all unvisited points that are density connected to any point from C_i
8 $i = i + 1$
9 **else**
10 Mark c_i as an outlier (noise)
11 **end**
12 **end**

4 Computational Experiments

In this work, the DBSCAN algorithm is used to analyze TSP instances and the relationship between their clustering properties and the ability of Ant Colony Optimization [5] algorithm to solve the problem is studied. A nature-inspired metaheuristic algorithm without local search or greedy steps was used in order to see whether any clustering properties correspond to the efficiency of a generic stochastic TSP solver that makes no assumptions about problem structure.

In order to illustrate and study the proposed approach, a toy dataset of random TSP instances was created. 10 instances of the TSP with 60 cities were created so that the first 5 of them, named TSP01 - TSP05, contained cities structured into 2–4 clusters and the last 5 instances (TSP06 - TSP10) contained cities without any apparent structure. An example of test TSP instances is shown in Fig. 1. A greedy algorithm was used to find a baseline TSP solutions [1]. Although it does not guarantee optimum solutions, it provides solid TSP solutions at reasonable time and works well with small TSP instances. In this work, it was used as a baseline to assess the error of TSP solutions obtained by the metaheuristic method, ACO. An illustration of greedy solutions of test TSP instances is provided in Fig. 2.

The DBSCAN algorithm was applied to all test TSP instances with neighborhood size, ϵ, set to the average distance between every pair of cities and minimum cluster size, $minPts$, set to 3. The parameters were set so that the size of each density-connected structure (cluster) is at least three and a potential solver has to choise from multiple options for connecting the cities. Cities are considered density-connected when their distance is below the average distance between every pair of cities in the problem instance. Three measures, describing the clustering obtained by DBSCAN were observed for each test TSP instance:

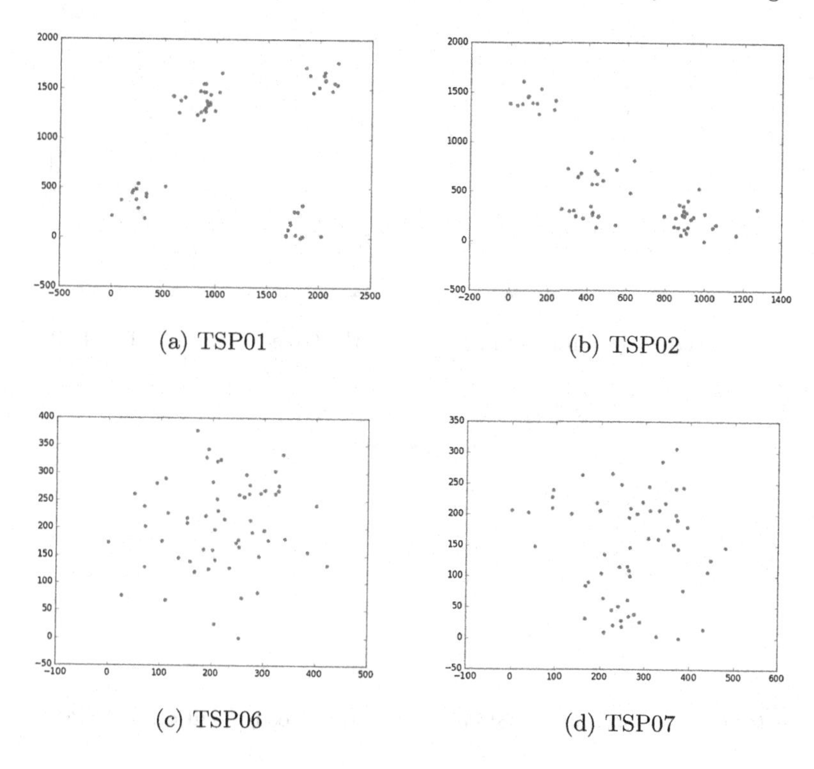

Fig. 1. Examples of test TSP instances.

- the number of clusters, *NoClust*,
- average cluster size, *AClustSize*, and
- the ratio of noise to problem size, *Noise2Size*.

Each test problem instance was solved by a stochastic nature-inspired method, Ant Colony Optimization. The ACO variant used in the experiments was the elitist ant system [4]. The algorithm utilized a-priori information $\eta_{ij} = c_{ij}$, and pheromone amplification rate, α, and a-priori information amplification rate, β, set to 1. The amount of deposited pheromone was proportional to the quality of solutions and the evaporation rate, ρ, was set to 0.1. The number of artificial ants, m, was fixed to 60 and the algorithm was executed for 100,000 iterations. The selected ACO variant and its parameters are based on best practices, authors' past experience, and extensive experimental trial-and-error runs. The ACO algorithm was for every test TSP instance executed 30 times independently.

Clustering results and the properties of test TSP instances' solutions obtained by ACO are summarized in Table 1. The table shows the three observed clustering measures as well as the error of ACO solutions when compared to the solution obtained by the greedy algorithm.

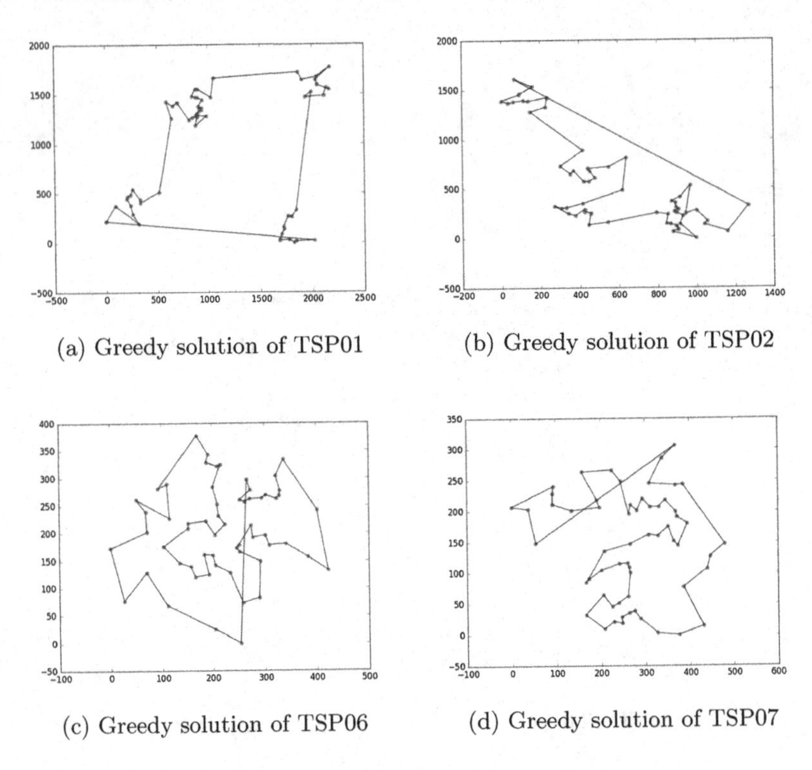

(a) Greedy solution of TSP01

(b) Greedy solution of TSP02

(c) Greedy solution of TSP06

(d) Greedy solution of TSP07

Fig. 2. Examples of greedy solutions of test TSP instances.

Several things can be immediately seen in Table 1. First, it can be noticed that the results of the ACO algorithm are indeed different for the two groups of test TSP instances, as also illustrated in Fig. 3. The first set of test TSP instances (TSP01 – TSP05) is apparently harder to solve and the local clusters represent a challenge for ACO (i.e. a global optimization method without local search). The second set of TSP instances (TSP06 – TSP10) is for the generic metaheuristic method apparently easier to solve. Second, the trends in the observed clustering measures correspond with the sets of test TSP instances as well. The distribution of $NoClust$, $AClustSize$, and $Noise2Size$ values for both groups of test TSP instances are visualized in Fig. 4. The correspondence between average solution error and clustering measures of test TSP instances was further validated by Spearman's rank correlation test [14]. The test showed that all three measures correlate with the average error of TSP solution found by ACO. The results of the statistical analysis are shown in Table 2 and confirm that all measure strongly correlate, either positively or negatively, with the average error of TSP solution found by ACO, with p-value lower than 0.05. In terms of the statistical analysis, the average cluster size, $AClustSize$, yields very strong ($\rho = 0.9245$) correlation with the average error of TSP solution found by ACO and has also the lowest p-value (0.000130).

Table 1. Clustering measures and ACO results.

TSP instance	Clustering measures			Error of ACO solutions [%]		
	NoClust	AClustSize	No2Size	Minimum	Average	Maximum
TSP01	6	8.333	0.1667	24.39	35.98	54.48
TSP02	5	7.600	0.3667	18.18	32.39	44.97
TSP03	6	8.167	0.1833	28.83	39.75	53.06
TSP04	5	8.400	0.3000	26.30	43.36	61.38
TSP05	6	7.833	0.2167	17.21	31.02	40.66
TSP06	3	3.000	0.8500	4.003	12.14	25.78
TSP07	1	3.000	0.9500	8.085	14.13	25.99
TSP08	1	3.000	0.9500	11.58	18.00	33.09
TSP09	1	3.000	0.9500	1.867	13.47	32.54
TSP10	2	3.000	0.9000	16.04	25.23	35.55

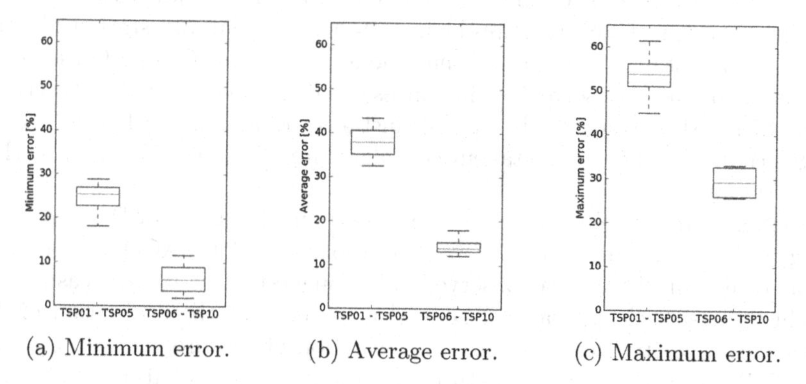

(a) Minimum error. (b) Average error. (c) Maximum error.

Fig. 3. Boxplot representation of the minimum, average, and maximum error of TSP solutions found by ACO.

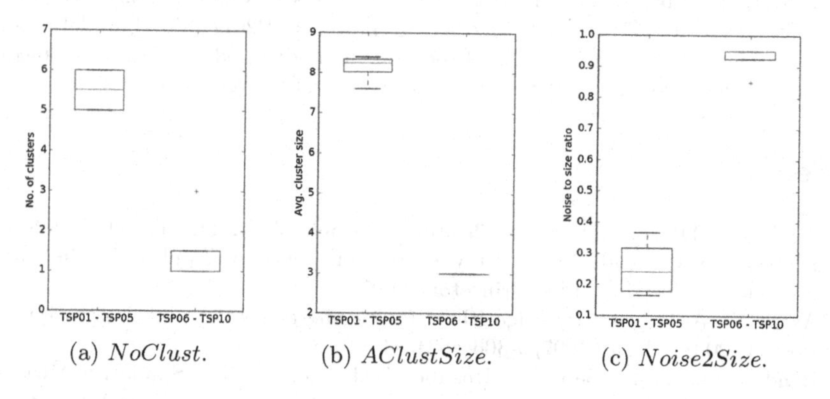

(a) *NoClust*. (b) *AClustSize*. (c) *Noise2Size*.

Fig. 4. Boxplot representation of clustering measures.

Table 2. Spearman's rank correlation between clustering measures and the average error of TSP solution found by ACO.

Measure	ρ	p-value
$NoClust$	0.7230	0.018138
$AClustSize$	0.9245	0.000130
$Noise2Size$	−0.7669	0.009639

5 Conclusions

A new approach to TSP instance complexity assessment was studied in this work. The DBSCAN algorithm was applied to a set of TSP instances with different structure and several measures based on the result of the clustering were computed. The TSP instances were solved by a nature-inspired metaheuristic method, Ant Colony Optimization. No local search was employed in order to see if and how the error of the solutions found by the metaheuristic approach correspond to the structure of problem instances. Visual and statistical analysis of experimental results showed that the ability of Ant Colony Optimization without local search to solve TSP instances is indeed affected by the structure of the problem. More structured test TSP instances represented a harder problem (bigger challenge) for the metaheuristic algorithm then test instances without local structures.

The relation between three measures based on the results of DBSCAN clustering with the average error of TSP solutions found by ACO was analyzed and a strong correlation was observed. That suggests that these measures and the clustering-based approach in general are useful for the estimation of TSP instance complexity. However, further evaluation of the proposed approach on other TSP instances is a must. Future work on this topic will also include experimental evaluation of other clustering methods.

Acknowledgement. This work was supported by the Czech Science Foundation under the grant no. GJ16-25694Y and by the projects SP2017/100 "Parallel processing of Big Data IV" and SP2017/85 "Processing and advanced analysis of bio-medical data II," of the Student Grant System, VŠB-Technical University of Ostrava.

References

1. Applegate, D.L., Bixby, R.E., Chvatal, V., Cook, W.J.: The Traveling Salesman Problem: A Computational Study (Princeton Series in Applied Mathematics). Princeton University Press, Princeton (2007)
2. Arora, S.: Approximation Algorithms for Geometric TSP, pp. 207–221. Springer, Boston (2007). doi:10.1007/0-306-48213-4_5
3. Bandyopadhyay, S., Saha, S.: Unsupervised Classification: Similarity Measures, Classical and Metaheuristic Approaches, and Applications. SpringerLink: Bücher, Heidelberg (2012). https://books.google.cz/books?id=Vb21R9_rMNoC

4. Dorigo, M., Stützle, T.: Ant Colony Optimization. MIT Press, Cambridge (2004)
5. Engelbrecht, A.: Computational Intelligence: An Introduction, 2nd edn. Wiley, New York (2007)
6. Everitt, B., Landau, S., Leese, M., Stahl, D.: Cluster Analysis. Wiley Series in Probability and Statistics, Wiley, New York (2011). https://books.google.cz/books?id=w3bE1kqd-48C
7. Fischetti, M., Lodi, A., Toth, P.: Exact Methods for the Asymmetric Traveling Salesman Problem, pp. 169–205. Springer, Boston (2007). doi:10.1007/0-306-48213-4_4
8. Hernando, L., Pascual, J.A., Mendiburu, A., Lozano, J.A.: A study on the complexity of TSP instances under the 2-exchange neighbor system. In: 2011 IEEE Symposium on Foundations of Computational Intelligence (FOCI), pp. 15–21, April 2011
9. Kefi, S., Rokbani, N., Krömer, P., Alimi, A.M.: Ant supervised by PSO and 2-opt algorithm, AS-PSO-2Opt, applied to traveling salesman problem. In: 2016 IEEE International Conference on Systems, Man, and Cybernetics, SMC 2016, Budapest, Hungary, 9–12 October 2016, pp. 4866–4871. IEEE (2016). doi:10.1109/SMC.2016.7844999
10. Kriegel, H.P., Krüger, P., Sander, J., Zimek, A.: Density-based clustering. Wiley Interdiscip. Rev. Data Min. Knowl. Discov. **1**(3), 231–240 (2011). doi:10.1002/widm.30
11. Krömer, P., Platos, J., Snásel, V.: Traditional and self-adaptive differential evolution for the p-median problem. In: 2nd IEEE International Conference on Cybernetics, CYBCONF 2015, Gdynia, Poland, 24–26 June 2015, pp. 299–304. IEEE (2015). doi:10.1109/CYBConf.2015.7175950
12. Mihalák, M., Schöngens, M., Šrámek, R., Widmayer, P.: On the complexity of the metric TSP under stability considerations. In: Proceedings of 37th International Conference on Current Trends in Theory and Practice of Computer Science, SOFSEM 2011, pp. 382–393. Springer, Heidelberg (2011). http://dl.acm.org/citation.cfm?id=1946370.1946402
13. Punnen, A.P.: The traveling salesman problem: applications, formulations and variations. In: Gutin, G., Punnen, A. (eds.) The Traveling Salesman Problem and Its Variations Combinatorial Optimization, vol. 12, pp. 1–28. Springer, Boston (2007). doi:10.1007/0-306-48213-4_1
14. Salkind, N.: Encyclopedia of Measurement and Statistics. SAGE Publications, Upper Saddle River (2006). https://books.google.cz/books?id=HJ91CgAAQBAJ
15. Stützle, T., Grün, A., Linke, S., Rüttger, M.: A comparison of nature inspired heuristics on the traveling salesman problem, pp. 661–670. Springer, Heidelberg (2000). doi:10.1007/3-540-45356-3_65
16. Stützle, T., Hoos, H.H.: Analysing the Run-Time Behaviour of Iterated Local Search for the Travelling Salesman Problem, pp. 589–611. Springer, Boston (2002). doi:10.1007/978-1-4615-1507-4_26

Real-Time Path Planning Based on Dynamic Traffic Information

Rong Hu[1(✉)], Ye Xia[1], and Fang-jun Kuang[2]

[1] Fujian University of Technology, No. 3 Xueyuan Road, University Town, Fuzhou 350118, China
hurong@fjut.edu.cn

[2] School of Information Engineering, Wenzhou Business College, Wenzhou City 325035, Zhejiang Province, China

Abstract. Traditional path planning method only consider the shortest path, as a result the planning paths may not be the best path if there is a congestion occurred in the planning path. This paper proposed a path planning based on dynamic traffic information to apply in vehicle navigation system. The dynamic information which includes the status of the road section in real-time can be applied in real time navigation. The optimal dynamic routing according to dynamic traffic information can avoid traffic jams automatically and save cost for user. The experiments showed the feasibility and effectiveness of real-time path planning based on dynamic traffic information.

Keywords: Dynamic traffic information · Path planning · Navigation system · Dynamic routing · Vehicle navigation

1 Introduction

The traditional navigation equipment can only realize static navigation [1], because during the process of the path planning, only the static traffic information is considered. So the routing would not be the real optimal path if a jam occurred in the path. This kind of navigation system always consider only the shortest path as the best routing, as a result the recommended routing is always the most busy route line, thus it can not provide the convenience for user even it may exacerbate the congestion [2]. The shortest planning path focuses on finding the optimum path between a current position and a destination. The optimum path in the traditional navigation system pursues a minimum travel distance or minimum travel time. The crucial point of this method is developing increasingly efficient algorithms. That is to say it is very important to study and find a more effective approach to find out an optimal route in the road network [3]. In order to solve this problem, the Dijkstra algorithm [4] and the A-star algorithm [5] are proposed. However, they are not effective in dynamic transport

P. Krömer et al. (eds.), *Proceedings of the Fourth Euro-China Conference on Intelligent Data Analysis and Applications*, Advances in Intelligent Systems and Computing 682, DOI 10.1007/978-3-319-68527-4_42

system due to the condition of road changing fast. The quality of a recommendation path depends on the path planning strategy applied in a vehicle navigation system. Conventional distance-based route planning has not been suitable for the modern traffic environment due to the complex and time-dependent traffic conditions. This paper proposed a real-time dynamic path planning approach based on dynamic traffic information. The proposed approach take into consider the dynamic information for path discovery. The rest of this paper is organized as follows. Section 2 gives related works. In Sect. 3 describes the proposed approach of the road network model design. Section 4 presents the evaluation of the proposed approach, and a conclusion is given in the last section.

2 Related Works

There are A* heuristic search, visibility graph method, generalized Voronoi diagram, and artificial potential field (APF) [6] for path-planning algorithms. The APF method formulates a relationship between the motion of the autonomous vehicle and the sum of the applied forces [7]. A road network can be represented as a directed graph which consists of vertices and edges. Path planning is to find the path that costs lest from the origin to the destination vertex. This area has attract much study in area of computer science and transportation [8]. The most well-known algorithm is Dijkstra algorithm for solving this shortest path problem. The performance of Dijkstra are compared with Bellman-Ford algorithms in [9]. It showed that the performance of Dijkstra is better than that of other algorithms. However these methods do not consider the condition of road real time. This condition of network includes traffic controls, commuting congestions, output events and so on. The common methods for path tracking of dynamic autonomous vehicles include sliding-mode control [10], fuzzy logic [11,12], and robust control [13]. However, many of these control applications worked under an assumption that the saturation limits of the actuators would never be reached. In practice, vehicles are on road with various constraints. It should be considered nonlinear characteristics of vehicles and their interactions with the road. To tackle this problem, the model predictive control (MPC) is presented [14,15]. Because it has the capability to systematically handle input constraints and admissible states, MPC with multiconstraints is adopted to track the planned trajectory for collision avoidance.

3 The Road Network Model Design

A road network is the basic of a traffic route path planning. The quality of road network model influence the accuracy of the planning path directly. The proposed network model contains a network topology and corresponding dynamic traffic information. The topology of road network main descripts the connection of the road nodes while the traffic information puts on the weight of nodes and lines. Traffic information includes static information and dynamic information. Static information is the changeless information such as the road types, number of lanes, section lengths, passing directions, speed limits, etc. Dynamic information, on the contrary, is about traffic conditions which include traffic flows, controls, accidents, roadworks, etc. The major objective of planning a traffic path is minimizing the traveling time. It is highly related with the traffic conditions, so it is very necessary to take the dynamic traffic information to be considered while planning a traffic path in real time. In this paper, the road network model is defined. First, the set of intersection road point is defined as:

$$G = \{c_i \,|i = 1, 2, ...N_c\} \tag{1}$$

where c_i is road intersection, i is sequence of the network, N_c is the number of road intersection. And the road intersection is defined as:

$$c = (L, E, O, P) \tag{2}$$

where $L = (x, y)$ represents the coordinates of intersection road, E is input entrance of the intersection, O is the export of the intersection and P is the connectedness of the intersection. P is defined from E and O. E and O is defined respectively as:

$$E = (e_m \,|m = 1, 2, ...N_e) \tag{3}$$

$$O = (O_n \,|n - 1, 2, ...N_o) \tag{4}$$

where N_e represents the number of entrances, and O_n is the number of exports. So t is defined as Fig. 1.

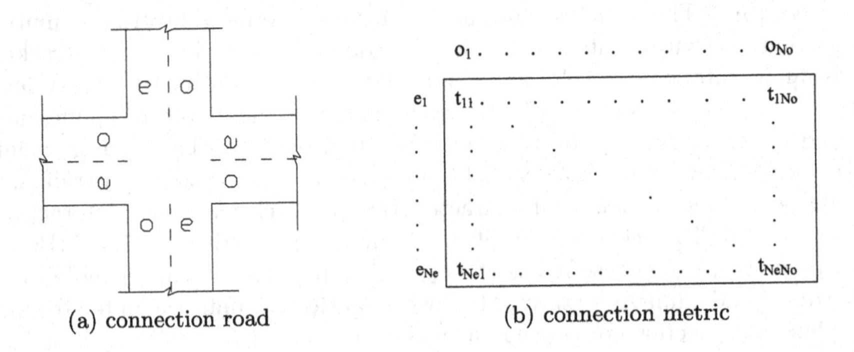

(a) connection road (b) connection metric

Fig. 1. The connection condition on road

$$t_{ij} = \begin{cases} 1 & \text{if } e_i \text{ } connect \text{ } with \text{ } o_j \\ o & others \end{cases} \tag{5}$$

Dynamic traffic information includes controls, speed limits, and accidents; roadworks, etc. the Dynamic transportation system model is designed as follow:

$$I = \{R, U, C, t\} \tag{6}$$

where R is road information about speeds, U represents accidents, C is the status of road, here $R = \{ID, l, v\}$, ID is the label of road; V is limit speed of the road and l is the length of the road. $U = \{l, d\}$, U is the location of accident, d is description.

$$C = \begin{cases} 0 & normal \\ 1 & slow \\ 2 & congestion \\ \infty & no\ thorough\ fare \end{cases} \tag{7}$$

In dynamic navigation, dynamic traffic information is applied in path planning. In order to search the optimal path between source node and destination node, the path cost need to be decided. In this paper, the path cost is described as:

$$Z = \sum_{i=1}^{N-1} z_{i,j} \tag{8}$$

where, N is the number of intersection, $z_{i,j}$ is the cost of adjacent two nodes between c_i and c_j.

$$z_{ij} = w_1 l/v + w_2 C + U + w_3 z_j \tag{9}$$

where W is the weight, $w_1 + w_2 + w_3 = 1$, z_j is the cost of intersection that is time cost for a vehicle pass the intersection. So from the above description, the most optimal path is obtained by:

$$Z_0 = \min Z \tag{10}$$

In regard to the practical applications, the proposed approach can be used in an on-line vehicle navigation system. The system in the vehicle should have a wireless connection to the traffic center sever or an on-line web system to acquire the real-time traffic information. According to the request of a source-to-destination path planning, the system utilizes the on-line traffic information to construct the road network model and dynamic transportation system and then calculate the optimal planning path.

4 Performance Evaluations

This section descripts the results of the proposed approach network modeling based on dynamical traffic information. We used the road network E-map of Fuzhou City as shown in Fig. 2. The dynamical traffic conditions of congestion status were obtained from Baidu traffic map system by the Baidu Open SDK. The performance results of path discovery for a vehicle navigation based on the real-time dynamical traffic information compared with traditional path planning based on static information. When there no congestion occurs on the road, the path planning results are same as shown in Fig. 3(a), they both select the same path because it is the optimal path which cost lest time and distance. Suppose at some time, there is a jam or congestion occurred on some point 1 as shown in Fig. 3(a). We select another path as shown in Fig. 3(b). The method based

Fig. 2. Traffic road network of Fuzhou City in China

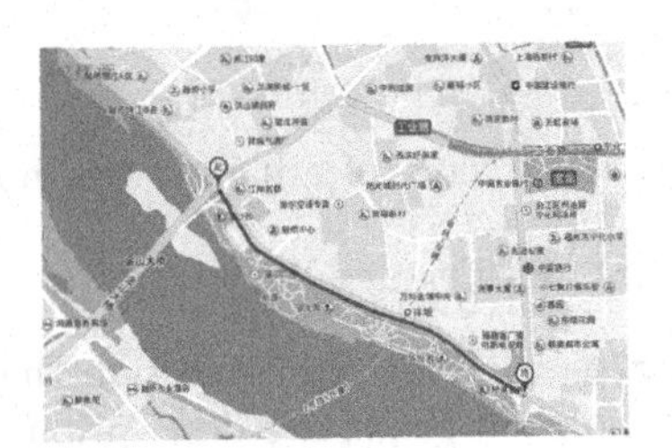

(a) path planning based on static information

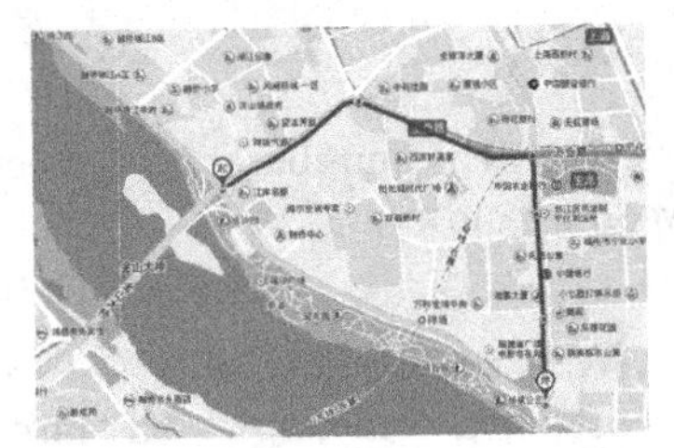

(b) Path planning based on dynamic information

Fig. 3. The connection condition on road

on dynamic traffic information succeed avoid the jam point, and it meets the requirement in practice.

This indicates that the proposed path planning method based-on dynamical traffic system can significantly reduce the time cost of users and efficiently avoid the exacerbate of jam or congestion, and it meet the vehicle navigation requirements.

5 Conclusions

The traditional path discovery for vehicle navigation only considers the distance between the source and destinations and neglects the important dynamic transportation information such as the road speed limits, congestions, controls and etc. It does not meet modern road traffic condition requirements. This paper proposed a road network modeling algorithm that concern both dynamical traffic condition and static road network information. During the process of path planning it can efficiently avoid the jam or congestion point for users and significantly reduce the time cost of users. The performance evaluations show that the proposed approach can obtain a satisfied result than the tradition static path planning method.

Acknowledgment. This work was supported in part by Fujian Provincial Department of Science and Technology, Granted No. 2017J01729 and Fujian University of Science and Technology, Granted No. GY-Z13103.

References

1. Wang, H., Jin, Z., Wei, W., et al.: GPS guidance system development based on dynamic transportation information. JBH **1**, 33–35 (2009)
2. Yang, D., Fan, S., Li, T., et al.: In-vehicle navigation system based on dynamic traffic information. In: International Conference on Information Science and Engineering, pp. 6696–6699. IEEE (2011)
3. Pan, J.-S., Wang, C., Sung, T.W.: Road network modeling with layered abstraction for path discovery in vehicle navigation systems. J. Inf. Hiding Multimed. Sig. Process. **7**(4), 791–801 (2016)
4. Dijkstra, E.W.: A note on two problems in connexion with graphs. Numer. Math. **1**(1), 269–271 (1959)
5. Hart, P.E., Nilsson, N.J., Raphael, B.: A formal basis for the heuristic determination of minimum cost paths. IEEE Trans. Syst. Sci. Cybern. **4**(2), 100–107 (1968)
6. Kunchev, V., Jain, L., Ivancevic, V., Finn, A.: Path planning and obstacle avoidance for autonomous mobile robots: a review. In: Lecture Notes in Computer Science, vol. 4252, pp. 537–544. Springer, Berlin (2006)
7. Pamosoaji, A.K., Hong, K.S.: A path-planning algorithm using vector potential functions in triangular regions. IEEE Trans. Syst. Man Cybern. Syst. **43**(4), 832–842 (2013)
8. Zhan, F.B., Noon, C.E.: Shortest path algorithms: an evaluation using real road networks. Transp. Sci. **32**(1), 65–73 (1998)
9. Wagner, D., Willhalm, T.: Speed-up techniques for shortest-path computations. In: Lecture Notes in Computer Science, vol. 4393, pp. 23–36 (2007)
10. Rovira-Más, F., Zhang, Q.: Fuzzy logic control of an electrohydraulic valve for auto-steering off-road vehicles. Proc. Inst. Mech. Eng. Part D: J. Automob. Eng. **222**(6), 917–934 (2008)
11. Tabatabaei Oreh, S.H., Kazemi, R., Azadi, S.: A sliding-mode controller for directional control of articulated heavy vehicles. Proc. Inst. Mech. Eng. Part D: J. Automob. Eng. **228**(3), 245–262 (2014)
12. Xia, Y., Rong, H.: Fuzzy neural network based energy efficiencies control in the heating energy supply system responding to the changes of user demands. J. Netw. Intell. **2**(2), 186–194 (2017)
13. Nam, K., Oh, S., Fujimoto, H., Hori, Y.: Robust yaw stability control for electric vehicles based on active front steering control through a steer-bywire system. Int. J. Automot. Technol. **13**(7), 1169–1176 (2012)
14. Raffo, G.V., Gomes, G.K., Normey-Rico, J.E., Kelber, C.R., Becker, L.B.: A predictive controller for autonomous vehicle path tracking. IEEE Trans. Intell. Transp. **10**(1), 92–102 (2009)
15. Meng, S.H., Hu, S.B., Chiu, H.C., Chang, C.H., Chu, S.-C.: A broadband ASE light source-based FTTX RoF-WDM optical network system. J. Netw. Intell. **2**(1), 162–170 (2017)

Author Index

P. Krömer et al. (eds.), *Proceedings of the Fourth Euro-China Conference on Intelligent Data Analysis and Applications*, Advances in Intelligent Systems and Computing 682, DOI 10.1007/978-3-319-68527-4

Printed in the United States
By Bookmasters